Ohio State University Mathematical Research Institute Publications 8

Editors: Gregory R. Baker, Walter D. Neumann, Karl Rubin

Ohio State University Mathematical Research Institute Publications

1 Topology '90, *B. Apanasov, W. D. Neumann, A. W. Reid, L. Siebenmann* (Eds.)

2 The Arithmetic of Function Fields, *D. Goss, D. R. Hayes, M. I. Rosen* (Eds.)

3 Geometric Group Theory, *R. Charney, M. Davis, M. Shapiro* (Eds.)

4 Groups, Difference Sets, and the Monster, *K. T. Arasu, J. F. Dillon, K. Harada, S. Sehgal, R. Solomon* (Eds.)

5 Convergence in Ergodic Theory and Probability, *V. Bergelson, P. March, J. Rosenblatt* (Eds.)

6 Representation Theory of Finite Groups, *R. Solomon* (Ed.)

7 The Monster and Lie Algebras, *J. Ferrar, K. Harada* (Eds.)

Groups and Computation III

Proceedings of the International Conference
at The Ohio State University
June 15–19, 1999

Editors

William M. Kantor
Ákos Seress

Walter de Gruyter · Berlin · New York 2001

Editors
William M. Kantor
Department of Mathematics, University of Oregon, Eugene, OR 97403-1222, USA
Ákos Seress
Department of Mathematics, The Ohio State University, Columbus, OH 43210-1174, USA

Mathematics Subject Classification 2000: 20-06; 20-04, 20-B40, 20-D06, 20-D08, 20-F05, 20-G40, 20-P05, 68-W20

Printed on acid-free paper which falls within the guidelines of the ANSI to ensure permanence and durability.

Library of Congress — Cataloging-in-Publication Data

Groups and computation III : proceedings of the international conference at the Ohio State University, June 15–19, 1999 / editors, William M. Kantor, Ákos Seress.
p. cm. — (Ohio State University Mathematical Research Institute publications, ISSN 0942-0363 ; 8)
ISBN 3-11-016721-2 (alk. paper)
1. Group theory — Data processing — Congresses. 2. Algebra — Data processing — Congresses.
I. Title: Groups and computation 3. II. Kantor, W. M. (William M.), 1944– . III. Seress, Ákos, 1958– . IV. Series
QA174.7.D36 G765 2001
512'.2—dc21 00-065758

Die Deutsche Bibliothek — Cataloging-in-Publication Data

Groups and computation III : proceedings of the international conference at the Ohio State University, June 15 – 19, 1999 / ed. William M. Kantor ; Ákos Seress. — Berlin ; New York : de Gruyter, 2001
(Ohio State University Mathematical Research Institute publications ; 8)
ISBN 3-11-016721-2

Printed in Germany.
Typeset using the authors' TEX files: I. Zimmermann, Freiburg.
Printing: WB-Druck GmbH & Co., Rieden/Allgäu. Binding: Lüderitz & Bauer-GmbH, Berlin.
Cover design: Thomas Bonnie, Hamburg.

Preface

The conference "Groups and Computation" took place at The Ohio State University in Columbus, Ohio, on June 15–19, 1999. This conference was the successor of two workshops on "Groups and Computation" held at DIMACS in 1991 and 1995; this is reflected by the "III" in the title of the present volume.

The main objective of the conference was to provide a forum for merging theory and practice of computations with groups: a continuation of the ongoing dialogue between researchers studying the complexity of group computations and those engaged in the development of practical algorithms. Based on comments of the participants, both during and after the conference, we can safely state that this goal was achieved.

There were talks by experts as well as by younger researchers. These proceedings contain accounts of most of the presentations of the various speakers. Nearly half of the participants were from abroad, reflecting the fact that most of the principal researchers in this field were able to attend the conference.

In addition to the lectures, the scientific program contained an evening devoted to software demonstrations, another evening for a Magma workshop, and an afternoon for discussion of the future of the matrix group project. Computation with matrix groups has attracted a great deal of significant research in the last few years, and many of the talks were related to various aspects of that subject. Numerous other areas of Computational Group Theory were well-represented: permutation group algorithms, finitely presented groups, polycyclic groups, and applications in science and engineering.

We are indebted to Marilyn Radcliff for her invaluable help with the complicated conference logistics. The meeting was generously supported by the Mathematical Research Institute of The Ohio State University, the National Science Foundation, and the National Security Agency.

William M. Kantor
Ákos Seress

Table of contents

Probabilistic recognition of orthogonal and symplectic groups

Christine Altseimer and Alexandre V. Borovik*

Abstract. We propose a fast one-sided Monte Carlo algorithm to distinguish, to any given degree of certainty, the symplectic group $C_n(q) = \mathrm{PSp}_{2n}(q)$ from the orthogonal group $B_n(q) = \Omega_{2n+1}(q)$ where $q > 3$ is odd and n and q are given. The algorithm does not use an order oracle and works in polynomial, of $n \log q$, time.

1991 Mathematics Subject Classification: primary 51F15; secondary 20G40.

1. Introduction

The aim of this paper is to propose a fast Monte Carlo algorithm to distinguish, to any given degree of certainty, the symplectic group $C_n(q) = \mathrm{PSp}_{2n}(q)$ from the orthogonal group $B_n(q) = \Omega_{2n+1}(q)$ where $q > 3$ is odd and n and q are given. Namely we prove the following theorem.

Theorem 1.1. *There is a polynomial time one-sided Monte Carlo algorithm which, when given a black box group G isomorphic to a simple classical group of type B_n or C_n defined over some field of given size $q > 3$ and odd characteristic, can decide whether G is isomorphic to C_n or not. If G is C_n then, with probability $> 1/2$, the algorithm produces a witness against isomorphism $G \simeq B_n(q)$.*

This results answers a question which L. Babai asked the second author.

The method of the present paper can be easily extended to the case $q = 3$ at the expense of adding a few technical details. We decided not to do this because the cases of small fields are covered in the fundamental work by Kantor and Seress **[14]**.

A earlier version of our algorithm has appeared in **[2]**. It was one-sided only in the case of odd n.

Black box groups were introduced by Babai and Szemerédi **[5]** as the ideal setting for some group computational problems and their importance has grown ever since. For the following compare Babai **[4]** and Beals **[6]**.

*Research supported by the German Academic Exchange Service (DAAD, Doktorandenstipendium HSPII/AUFE). The work on the paper had started when the first author was visiting UMIST.

A *black-box group* G is a group whose elements are encoded as 0-1 strings of uniform length N, and the group operations are performed by an oracle (the "black box"). Note that we have the upper bound 2^N for $|G|$. Given strings representing $g, h \in G$, the black box can compute strings representing gh and g^{-1}, and decide whether $g \doteq h$. We denote by μ an upper bound for the time requirement of these group operations, and by ξ an upper bound for the time requirement per element of constructing independent, (nearly) uniformly distributed elements of G. In the following we assume for the sake of argument that the black box gives us independent uniformly distributed elements of G. Compare Kantor and Seress **[14]** for similar assumptions and Babai **[4]** for a critical discussion.

A *Monte Carlo algorithm* is a randomised algorithm, where the result is correct with a probability at least $1 - \epsilon$ with $0 < \epsilon < 1/2$. After repeating a Monte Carlo algorithm several times we can obtain the answer with an arbitrarily small probability of error ϵ. A Monte Carlo algorithm with two possible answers 'yes' and 'no' is *one-sided* if the answer 'yes' is always correct.

1.1. Why is the case of $B_n(q)$ versus $C_n(q)$ so special? Our algorithm deals with one of the more difficult cases in the recognition of black box groups with order oracle. The case of $B_n(q)$ versus $C_n(q)$ cannot be handled by a direct study of statistics of orders of elements since these statistics are for all practical purposes the same in both groups. Instead, our algorithm is based on a more detailed study of involutions in these groups. The justification for our approach can be formulated in short as follows:

- The simple group of Lie type $B_n(q)$, q odd, cannot be distinguished from $C_n(q)$ by statistics of orders of elements in a random sample of length bounded by a polynomial in $n \log q$. Indeed, these statistics are controlled by the Dynkin diagram which is the same in these groups. A more detailed explanation is given in Section 1.4.

- However, the distribution of involutions in conjugacy classes and the structure of centralisers of involutions are controlled by the extended Dynkin diagrams (Iwahori **[13]**) which are different for B_n and C_n (Figure 1). Our Monte Carlo algorithm detects the difference in the behaviour of involutions.

Figure 1: *Extended Dynkin diagram for B_n (left) and C_n (right), $n \geqslant 3$.*

1.2. Description of the algorithm: the easy case, n is odd. Recall that $\Omega_{2n+1}(q)$ denotes the kernel of the spinor norm

$$SO_{2n+1}(q) \longrightarrow \{\pm 1\};$$

it is a subgroup of index 2 in $SO_{2n+1}(q)$ and coincides with the commutator of $SO_{2n+1}(q)$ if $n \geqslant 2$ and $q \geqslant 3$ (for details see, for example, Artin [3]). It is well known that $PSp_2(q) \simeq \Omega_3(q) \simeq PSL_2(q)$ and $PSp_4(q) \simeq \Omega_5(q)$. Hence it is obviously enough to give an algorithm for $n \geqslant 3$.

The easy case of the algorithm deals with the situation when n is odd, and the hard case is concerned with even n.

If x is an element of even order, we use notation $i(x)$ for the involution in the cyclic group $\langle x \rangle$.

The structure of the algorithm in the easy case is very simple.

INPUT: A black box group G isomorphic to $PSp_{2n}(q)$ or $\Omega_{2n+1}(q)$, $n \geqslant 3$ and $q > 3$ given, n is odd.

OUTPUT: If G is $PSp_{2n}(q)$ then, with probability $> 1/2$, the algorithm produces a witness against isomorphism $G \simeq \Omega_{2n+1}(q)$.

(1) Find an element x of *good order* (the definition is given below). This can be done with probability $1 - \epsilon$ in time $O((\mu n^3 \log q + \xi n) \log(1/\epsilon))$. An element of good order is contained in one of two specific maximal tori of even order.

(2) Compute the involution $i = i(x)$ in the cyclic group generated by x, this will take time $O(\mu n^2 \log q)$.

(3) *This step is vacuous in the easy case.*

(4) Compute $y := i^g i^h$ for random elements $g, h \in G$. If $o(y)$ divides $q(q+1)$ or $q(q-1)$ then go to Step 5. Otherwise return '*G is* $PSp_{2n}(q)$', with the pair (i^g, i^h) being a witness against the isomorphism $G \simeq \Omega_{2n+1}(q)$.

(5) Repeat Step 4 at least $20n$ times. If $o(y)$ always divides $q(q+1)$ or $q(q-1)$ then return '*G is not* $PSp_{2n}(q)$'. The error probability is $< 1/2$ in this case.

The number of repetitions in Step 5 is a very crude estimate, which was sufficient for our theoretical purposes. In practice, if G is of type C_n and $q > 3$, then one does get an element $i^g \cdot i^h$ whose order does not divide $q(q+1)$ or $q(q-1)$ at the very first iteration.

1.3. Description of the algorithm: the hard case, n is even

INPUT: A black box group G isomorphic to $PSp_{2n}(q)$ or $\Omega_{2n+1}(q)$, $n \geqslant 3$ and q given, n is even (and hence $q^n \equiv 1 \bmod 4$).

OUTPUT: If G is $\mathrm{PSp}_{2n}(q)$ then, with probability $> 1/2$, the algorithm constructs a witness against the isomorphism $G \simeq \Omega_{2n+1}(q)$.

(1) Find an element x of *good order* as in the easy case.

(2) Compute the involution $i = \mathrm{i}(x)$ in the cyclic group generated by x. It will be shown that i belongs to one of the two types t_n and $t_{n/2}$ (in notation of **[12]**, Section 4.5), and, correspondingly, x belongs to a torus of one of the types $\frac{1}{2}T_{q^n-1}$ and $\frac{1}{2}\left(T_{q^{n/2}+1} \times T_{q^{n/2}-1}\right)$. See Table 1 for details (which are mostly taken from **[12]**).

Involution i	t_n	$t_{n/2}$
Torus $\langle x \rangle$	$\frac{1}{2}T_{q^n-1}$	$\frac{1}{2}\left(T_{q^{n/2}+1} \times T_{q^{n/2}-1}\right)$
$F^*(C_G(i))$ in $G = \mathrm{PSp}_{2n}(q)$	$\frac{1}{2}\mathrm{GL}_n(q)$	$\mathrm{Sp}_n(q) \circ \mathrm{Sp}_n(q)$
$F^*(C_G(i))$ in $G = \Omega_{2n+1}(q)$ $q \equiv 1 \bmod 4$	$\Omega^+_{2n}(q)$	$\Omega^+_n(q) \times \Omega_{n+1}(q)$
$F^*(C_G(i))$ in $G = \mathrm{PSp}_{2n}(q)$ $F^*(C_G(i))$ in $G = \Omega_{2n+1}(q)$ $q \equiv -1 \bmod n$	$\frac{1}{2}\mathrm{GU}_n(q^2)$ $\Omega^-_{2n}(q)$	$\mathrm{Sp}_n(q) \circ \mathrm{Sp}_n(q)$ $\Omega^-_n(q) \times \Omega_{n+1}(q)$

Table 1: Involutions of type t_n and $t_{n/2}$. Here $F^*(C_G(i))$ denotes the generalised Fitting subgroup of $C_G(i)$. In all cases $C_G(i)/F^*(C_G(i))$ is an elementary abelian group of order $\leqslant 4$.

(3) Our next task is to make sure that i is an involution of type t_n. For this purpose, construct uniformly distributed random elements of $C_G(i)$ as explained in Section 3. The possibilities for $C_G(i)$ are given in Table 1. Notice that the groups $C_G(t_n)$ have much bigger orders of semisimple elements than $C_G(t_{n/2})$. Namely, with probability $> O(1/n)$ we can find in $C_G(t_n)$ a 'big' element, that is, an element of order divisible by a prime primitive divisor of $q^{n-1} \pm 1$ or $q^n + 1$. However 'big' elements are not present in $C_G(t_{n/2})$. As soon as a 'big' element in $C_G(i)$ is found, we conclude that i is of type t_n and move to Step 4. If we cannot find a 'big' element, we return to Step 1 and start our search for an involution of type t_n from the beginning.

(4–5) These steps are the same as in the easy case.

The first author has written a short GAP programme, which used an earlier version of this algorithm **[2]**. It did not include Step (3) and was one-sided only when n was odd. Tests have shown that it is very fast and reliable. In addition, further tests with GAP have shown that the algorithm for computation of centralisers of involutions (Section 3) is also fast and reliable.

1.4. Regular semisimple elements dominate the order statistics. The nature of the difficulty in distinguishing between $B_n(q)$ and $C_n(q)$ on the basis of order statistics can be easily explained in the special case of very large q.

Let k be an algebraically closed field of odd characteristic p and let H be either $\mathrm{PSp}_{2n}(k)$ or $\mathrm{PSO}_{2n+1}(k)$. Thus H is the adjoint Chevalley group of type C_n or B_n defined over the field k. Let F be a standard Frobenius map on k such that the number of elements of the base field is q. Let H^F be the set of fixed points of F on H. Then $G = [H^F, H^F]$ where $H^F/G \simeq \mathbb{Z}_2$. The group H^F is the group of points of H over $\mathbb{F}_q$.

It is well known that regular semisimple elements constitute a Zarisski open subset R of H, so that the complement $C = H \setminus R$ is Zarisski closed. By the Lang–Weyl theorem the number of points over $\mathbb{F}_q$ in C behaves asymptotically as $O(q^{\dim C})$. Since the number of points in H^F asymptotically behaves as $O(q^{\dim H})$ and $\dim C < \dim H$, we see that probability for an element in H^F to be a regular semisimple element is bigger than $1 - O(\frac{1}{q})$. Therefore for large q a random sample (of size polynomial in $n \log q$) of elements from G will contain only regular semisimple elements.

If x is a regular semisimple element in H^F, then, by definition, $T = C_H(x)$ is a maximal F-invariant torus in H. Let T_{reg} denote the set of regular elements in T. Then if T and S are two distinct tori in H, $T_{\mathrm{reg}} \cap S_{\mathrm{reg}} = \emptyset$. It is easy to deduce from this observation that the set $R \cap H^F$ of regular semisimple elements in H^F is partitioned into disjoint sets of the form $H^F \cap T_{\mathrm{reg}}$ for maximal F-invariant tori in H, every such set being conjugate in H^F to $|H^F : N_{H^F}(T \cap H^F))|$ sets of the same nature. Thus the statistics of orders of regular semisimple elements in H^F is entirely determined by the distribution of maximal F-invariant tori in H in H^F-conjugate classes, by isomorphism types of maximal tori $T \cap H^F$ in H^F and size of their normalisers $N_{H^F}(T \cap H^F)$. As it is shown in Proposition II.1.2 of Springer and Steinberg **[16]** (see also Carter **[11]**), H^F-conjugacy classes of F-invariant maximal tori in H are in one-to-one correspondence with cocycles in $H^1(\mathrm{Gal}(k/\mathbb{F}_q), W)$ for W the Weyl group of H. Since in our situation the action of $\mathrm{Gal}(k/\mathbb{F}_q)$ on W is trivial, $H^1(\mathrm{Gal}(k/\mathbb{F}_q), W)$ can be identified with the set of all conjugacy classes in W. Let T_w denotes a representative of a class of maximal F-invariant tori which corresponds to $w \in W$. Then $|T_w \cap H^F| = |\chi_w(q)|$ where $\chi_w(t)$ is the characteristic polynomial of w in the reflection representation of W. As it will be shown in Section 2.1, the action of w in the reflection representation also determines the structure of $T_w^F = T_w \cap H^F$ as a group. Furthermore $N_{H^F}(T_w \cap H^F)/(T_w \cap H^F) \simeq C_W(w)$.

But the algebraic groups B_n and C_n have the same Weyl group; summarising the above discussion, we see that the statistics of orders of regular semisimple elements is exactly the same for the finite algebraic groups H^F of types $B_n(q)$ and $C_n(q)$. To reach the same conclusion about the group $G = [H^F, H^F]$ one needs to check that the isomorphism type of the torus $T_w \cap G$ is the same in both cases $H = B_n$ and $H = C_n$. This will be seen from a more detailed description of tori in G (Section 2.1).

2. Tori and semisimple elements in groups B_n and C_n

Definitions and notation used in the paper are standard and borrowed from Carter **[10]** and Gorenstein, Lyons and Solomon **[12]**. We also systematically use notation and results from Carter **[9, 11]** and Springer and Steinberg **[16]**. We retain the notation of Section 1.4.

2.1. Maximal tori in H. The Weyl group W of H is isomorphic to the wreath product $\mathrm{Sym}_2 \wr \mathrm{Sym}_n \simeq \mathbb{Z}_2^n \rtimes \mathrm{Sym}_n$ and each element determines a product of disjoint cycles $(j_1 \dots j_m)$ or $(j_1 \dots j_m - j_1 \dots - j_m)$ where $j_i \in \{1, \dots, n, -1, \dots, -n\}$. The first kind of cycle is said to be positive of length m, the second kind negative of length m. The signed cycle type of w is the set of lengths of the disjoint cycle together with their signs. Two elements of W are conjugate if and only if they have the same cycle type. The characteristic polynomial $|\chi_w(q)|$ of an element $w \in W$ (in the natural reflection representation of W on the $\mathbb{Q}$-vector space spanned by the characters of the maximal torus T in H) is the product of the following polynomials:

(i) $(q^m - 1)$ for every positive cycle of length m, and

(ii) $q^m + 1$ for every negative cycle of length m,

see Carter **[9]** for details.

We choose the maximal torus T to be the diagonal subgroup of H. Denote by Φ the set of all roots of H and by Δ a fundamental root system. Let furthermore T_w be a maximal torus of H^F which is twisted by $w \in W$. The order of T_w is $|\chi_w(q)|$ by Springer and Steinberg **[16]** and thus is the product of numbers of the form $q^k \pm 1$ for $k \leqslant n$. As $|H^F/G| = 2$, a maximal torus T_0 of G has order $|T_w|/2$ for some $w \in W$.

We shall now see that the isomorphism type of T_w is also determined by the cycle structure of w.

Lemma 2.1. *Let $w = w_1 w_2 \in W$, where w_1 and w_2 operate on disjoint sets and set $H_m := C_T(w_m)$ and $C_m := C_G(H_m)'$ for $m = 1, 2$. Then $(C_1 C_2)^F$ contains a torus T_w twisted by w. Furthermore T_w can be chosen to be $T_1 T_2 S$ where T_m is a torus in C_m^F twisted by w_m for $m = 1, 2$ and $S = T \cap H_1 \cap H_2$.*

Proof. C_1 and C_2 are connected semisimple groups which are generated by the root subgroups $\{ X_\alpha \mid \alpha \in S_1 \}$ and $\{ X_\beta \mid \beta \in S_2 \}$ respectively where

$$S_m := \{ \pi \in \Pi \mid \exists \rho \in \Pi : \pi = \rho - w_m^{-1}(\rho) \}$$

by Proposition I.4.1 of Springer and Steinberg **[16]**. It is easy to see, that $\alpha + \beta \notin \Pi$ for all $\alpha \in S_1$ and $\beta \in S_2$. Hence C_1 and C_2 commute by Theorem 4.1.2 of Carter

[10]. Since furthermore $S_1 \cap S_2 = \emptyset$, $C_1 \cap C_2$ is finite. As C_m^F contains a torus T_m twisted by w_m the claim follows. □

We see from the lemma that if $w \in W$ and $w = w_1 \dots w_r$ is the decomposition of w into the product of disjoint positive or negative cycles of lengths l_i (including trivial cycles of length 1), then $T_w = T_1 \dots T_r$ for tori T_i of order $q^{l_i} \pm 1$. Notice further that T_i is a maximal torus in the group C_i^F for $C_i = C_G(C_T(w_i))$. It immediately follows from the classification of the centralisers of tori in reductive groups (Borel and Tits [7]) that C_i is a semisimple group corresponding to a subdiagram of the Dynkin diagram of G. It follows from here that C_i is a symplectic or orthogonal group of dimension $2l_i$ or $2l_i + 1$ arising from the natural embeddings $\mathrm{Sp}_{2l_i}(k) \leqslant \mathrm{Sp}_{2n}(k)$ and $\mathrm{SO}_{2l_i+1}(k) \leqslant \mathrm{SO}_{2n+1}(k)$, correspondingly. The torus T_i now is a maximal torus in C_i^F twisted by a cycle of maximal length in the Weyl group of C_i. It is easy to identify T_i with a cyclic subgroup of order $q^{l_i} \pm 1$ in C_i^F arising from the natural embeddings $\mathrm{SL}_2(q^{l_i}) = \mathrm{Sp}_2(q^{l_i}) \leqslant \mathrm{Sp}_{2l_i}(q)$ and $\mathrm{SO}_2^{\pm}(q^{l_i}) \leqslant \mathrm{SO}_{2l_i+1}(q)$.

We shall denote the torus corresponding to a positive (correspondingly negative) cycle of length l by T_{q^l-1} (correspondingly T_{q^l+1}).

Finally we have to move from the description of maximal tori in H^F to maximal tori in $G = \mathrm{PSp}_{2n}(q)$ or $\Omega_{2n+1}(q)$. Notice that in the symplectic case we can do the above analysis in the simply connected group $\mathrm{Sp}_{2n}(k)$ instead of the adjoint group $\mathrm{PSp}_{2n}(q)$ and see that every maximal torus in $\mathrm{Sp}_{2n}(\mathbb{F}_q)$ is a direct product of cyclic tori $T_{q^{l_1}\pm 1} \times \dots \times T_{q^{l_r}\pm 1}$ with $l_1 + \dots + l_r = n$. To get a maximal torus in $G = \mathrm{PSp}_{2n}(q)$, we have to take its factor

$$\frac{1}{2}\left(T_{q^{l_1}\pm 1} \times \dots \times T_{q^{l_r}\pm 1}\right)$$

by the group of scalar matrices $\{\pm I\}$ of order 2. An analogous transition from $\mathrm{SO}_{2n+1}(q)$ to $\Omega_{2n+1}(q)$ means taking the kernel of the spinor norm

$$T_{q^{l_1}\pm 1} \times \dots \times T_{q^{l_r}\pm 1} \longrightarrow \{\pm 1\}$$

which, as one can easily see, produces the torus in $\Omega_{2n+1}(q)$ isomorphic to the corresponding torus $\frac{1}{2}\left(T_{q^{l_1}\pm 1} \times \dots \times T_{q^{l_r}\pm 1}\right)$ in $\mathrm{PSp}_{2n}(q)$.

2.2. Tori of order $\frac{1}{2}(q^n - 1)$. The crucial Steps 1 and 2 of our algorithm are aimed at finding involutions belonging to tori $\frac{1}{2}T_{q^n\pm 1}$ twisted by an element of the cycle type $\pm n$. We have to distinguish the following cases:

(i) $q^n \equiv -1 \bmod 4$, i.e. n odd and $q \equiv -1 \bmod 4$.

(ii) $q^n \equiv 1 \bmod 4$, i.e. n even or $q \equiv 1 \bmod 4$.

In the first case the maximal tori of order $\frac{1}{2}(q^n + 1)$ contain involutions, in the second case the maximal tori of order $\frac{1}{2}(q^n - 1)$ contain involutions. Note that if n is even, the tori of order $\frac{1}{2}(q^n - 1)$ form at least two conjugacy classes:

$$\frac{1}{2}T_{q^n-1},\ \frac{1}{2}\left(T_{q^{n/2}+1} \times T_{q^{n/2}-1}\right),\ \frac{1}{2}\left(T_{q^{n/2}+1} \times T_{q^{n/4}+1} \times T_{q^{n/4}-1}\right), \ldots$$

We call the tori $\frac{1}{2}T_{q^n-1}$ and $\frac{1}{2}\left(T_{q^{n/2}+1} \times T_{q^{n/2}-1}\right)$ in the first two conjugacy classes *maximal non-split tori of even order* and of *Type* 1 and 2, correspondingly.

To distinguish other tori of order $\frac{1}{2}(q^n - 1)$ from the maximal non-split tori of even order, we introduce the concept of *2-height*. Let X be a group and $x \in X$ such that $o(x) = 2^k m$ where m is odd. Then k is called the *2-height of* x. The 2-height of X is the maximum of heights of elements in X.

Lemma 2.2. *The tori $\frac{1}{2}T_{q^n-1}$ and $\frac{1}{2}\left(T_{q^{n/2}+1} \times T_{q^{n/2}-1}\right)$ have the same 2-height. It is bigger than the 2-height of other tori of order $\frac{1}{2}(q^n - 1)$.*

Proof. Since the maximal tori in $\mathrm{PSp}_{2n}(q)$ are isomorphic to tori in $\Omega_{2n+1}(q)$ which are twisted by the same element of the Weyl group, it is enough to give a proof only in the symplectic case.

The Sylow 2-subgroup in $T_{q^{n/2}+1} \times T_{q^{n/2}-1}$ is isomorphic to $\mathbb{Z}_2 \times \mathbb{Z}_{2^s}$ for some s. The group $\frac{1}{2}\left(T_{q^{n/2}+1} \times T_{q^{n/2}-1}\right)$ is obtained from $T_{q^{n/2}+1} \times T_{q^{n/2}-1}$ by taking the factor group over the subgroup of order 2 generated by the diagonal involution in the direct product. Therefore the Sylow 2-subgroup in $\frac{1}{2}\left(T_{q^{n/2}+1} \times T_{q^{n/2}-1}\right)$ is cyclic. The tori $\frac{1}{2}T_{q^n-1}$ and $\frac{1}{2}\left(T_{q^{n/2}+1} \times T_{q^{n/2}-1}\right)$ have the same order and cyclic Sylow 2-subgroups. Hence they have the same 2-height.

But the tori $\frac{1}{2}\left(T_{q^{n/2}+1} \times T_{q^{n/4}+1} \times T_{q^{n/4}-1}\right), \ldots$, etc., contain an elementary abelian subgroup of order $\geqslant 2^3$. Hence the factor by a subgroup of order 2 cannot have a cyclic Sylow 2-subgroup, and the 2-height of these tori is smaller than that of $\frac{1}{2}T_{q^n-1}$ and $\frac{1}{2}\left(T_{q^{n/2}+1} \times T_{q^{n/2}-1}\right)$. □

Lemma 2.3. *If T_1 and T_2 are tori of type $\frac{1}{2}T_{q^n\pm1}$ and $\frac{1}{2}\left(T_{q^{n/2}+1} \times T_{q^{n/2}-1}\right)$ then*

$$|G : N_G(T_1)| = 2n \text{ and } |G : N_G(T_2)| = n^2.$$

Proof. The tori T_1 and T_2 are constructed by twisting a maximal split torus in H with the elements w_1 and w_2 of cycle types $(\pm n)$ and $(n/2, -n/2)$, correspondingly. Then by Section II.1.10 in **[16]**, $N_G(T_i)/T_i \simeq C_W(w_i)$ and the result follows by an easy computation in $W \simeq \mathrm{Sym}_2 \wr \mathrm{Sym}_n$. □

2.3. Good elements. In the following we are looking for elements of maximal 2-height in a maximal non-split torus of even order. To find elements in a maximal non-split torus the notion of *primitive prime divisor elements* of G plays a crucial role as in Neumann and Praeger **[15]** and Kantor and Seress **[14]**. Let p be a prime and $m \geqslant 2$. A prime dividing $p^m - 1$ but not $p^i - 1$ for $1 \leqslant i < m$ is called a *primitive prime divisor of* $p^m - 1$. In our case primitive prime divisors always exist by Zsigmondy **[19]**. Furthermore an integer j is called a $\mathrm{ppd}(p, m)$ if $j \mid p^m - 1$ and j is divisible by a primitive prime divisor of $p^m - 1$. An element of G of order $\mathrm{ppd}(p, en)$ is called *primitive prime divisor element*. Let $q = p^e$. Let P be a maximal non-split torus of even order and let $|P| = 2^k m$ where k and m are natural numbers and m is odd. We say that an element $g \in P$ is of *maximal 2-height*, if the 2-height of g is k. Since P is cyclic, elements of maximal 2-height exists. Furthermore half of the elements of P have maximal 2-height as the cyclic group generated by a 2-element x of maximal 2-height contains 2^{k-1} elements of maximal 2-height, namely x^s for s an odd integer. We call g *good* if it satisfies the following conditions:

- g has $\mathrm{ppd}(p, 2en)$ order when $q^n \equiv -1 \bmod 4$ and $\mathrm{ppd}(p, en)$ order when $q^n \equiv 1 \bmod 4$.
- $o(g) \nmid q^{n/2} + 1$ if n is even and $q \equiv -1 \bmod 4$.
- g has maximal 2-height.

The following important observation immediately follows from Lemma 2.2.

Lemma 2.4. *Good elements are regular. The unique torus containing a good element is a maximal non-split torus of even order.*

Lemma 2.5. *The number of good elements of maximal non-split tori of even order in G is at least $\frac{1}{4n}|G|$.*

Proof. Since good elements are regular the result is proven as the corresponding estimate in Neumann and Praeger **[15]**. □

Lemma 2.6. *If n is even then a good element belongs to a torus of type $\frac{1}{2}T_{q^n-1}$ with probability $\rho = \frac{1}{1+(2/n)}$.*

Proof. Since tori T_1 and T_2 of type $\frac{1}{2}T_{q^n-1}$ and $\frac{1}{2}(T_{q^{n/2}+1} \times T_{q^{n/2}-1})$ are isomorphic, they contain the equal number of good elements each. We compute using Lemma 2.3:

$$\begin{aligned}
\rho &= \frac{\#\{\, T_1^G \,\}}{\#\{\, T_1^G \,\} + \#\{\, T_2^G \,\}} \\
&= \frac{|G : N_G(T_1)|}{|G : N_G(T_1)| + |G : N_G(T_2)|} \\
&= \frac{\frac{1}{|N_G(T_1):T_1|}}{\frac{1}{|N_G(T_1):T_1|} + \frac{1}{|N_G(T_2):T_2|}} \\
&= \frac{\frac{1}{2n}}{\frac{1}{2n} + \frac{1}{n^2}} = \frac{1}{1 + (2/n)}.
\end{aligned}$$

□

2.4. Regular twisted elements. A regular element in a torus of type $\frac{1}{2}T_{q^n-1}$ or $\frac{1}{2}T_{q^n+1}$ will be called a *regular twisted* element. Notice that good elements are regular twisted elements.

Lemma 2.7. *The number of regular twisted elements in G conjugate to an element in a torus T of the given type $\frac{1}{2}T_{q^n-1}$ or $\frac{1}{2}T_{q^n+1}$ is at least $\frac{1}{4n}|G|$.*

Proof. An argument with ppd elements shows that at least half of the elements in T do not belong to any other torus and hence are regular. By Lemma 2.3 $|N_G(T)/T| = 2n$. Therefore there are $\frac{|G|}{2n|T|}$ tori conjugate to T, each of which contains at least $\frac{1}{2}|T|$ regular element. Since a regular element belongs to exactly one torus, their number is at least

$$\frac{|G|}{2n|T|} \cdot \frac{1}{2}|T| = \frac{1}{4n}|G|.$$

2.5. Involutions of type t_n in $\mathrm{PSp}_{2n}(q)$. For the following compare the description of conjugacy classes of involutions in $\mathrm{PSp}_{2n}(k)$ and $\mathrm{SO}_{2n+1}(k)$ as given in Theorems 11.52 and 11.53 of Taylor **[18]** and Section 4.5 of **[12]**.

The group $\mathrm{PSp}_{2n}(q)$ has $[\frac{n}{2}]+1$ conjugacy classes of involutions. Let $s \in \mathrm{Sp}_{2n}(q)$ represent an element of order 2 in $\mathrm{PSp}_{2n}(q)$. Then either $s^2 = -I$ or $s^2 = I$. All involutions which are represented by an element s such that $s^2 = -I$ form one conjugacy class of involutions of type t_n. Furthermore for every even number $2 \leqslant m \leqslant n$ there is one conjugacy class of involutions of type $t_{m/2}$. If s is an involution of type $t_{m/2}$, then the eigenvalues of s on the underlying natural module are -1 with multiplicity m and 1 with multiplicity $2n - m$.

The following lemma is part of the more general theory of reflexive tori (that is, tori inverted by some involution) as developed by Altseimer **[1**, Chapter 5**]**.

Lemma 2.8. *Let $G := \mathrm{PSp}_{2n}(q)$ and $i \in G$ an involution of type t_n. If H is a torus of G then there exists an involution $j \in i^G$ such that H is inverted by j.*

Proof. We use a slight variation of the classical Galois cohomology computation (see, for example, **[16**, Chapter I**]**).

Let $e_1, e_1^*, \ldots, e_n, e_n^*$ be a hyperbolic basis in the underlying vector space V for $\mathrm{Sp}_{2n}(q)$. The symplectic scalar product $(\,.\,,\,.\,)$ is given by $(e_i, e_i^*) = 1$, $(e_i^*, e_i) = -1$, $i = 1, \ldots, n$, while all other products equal zero. It is easy to see (for example, by comparing the centralisers) that any involution of type t_n is conjugate to the involution represented by the matrix

$$x := \begin{pmatrix} 0 & -1 & & & & \\ 1 & 0 & & & & \\ & & \ddots & & & \\ & & & \ddots & & \\ & & & & 0 & -1 \\ & & & & 1 & 0 \end{pmatrix}.$$

When viewed as an element in $\mathrm{Sp}_{2n}(q)$, x is of order 4. Abusing notation, we shall use for elements and subgroups of $G = \mathrm{Sp}_{2n}(q)$ the same notation as for $G = \mathrm{PSp}_{2n}(q)$. It will suffice to prove that every torus H in $\mathrm{Sp}_{2n}(q)$ is inverted by some conjugate y of x, that is, y normalises H and $h^y = h^{-1}$ for all $h \in H$. Let T be the maximal split torus

$$T := \left\{ \begin{pmatrix} \lambda & 0 & & & & \\ 0 & \lambda^{-1} & & & & \\ & & \ddots & & & \\ & & & \ddots & & \\ & & & & \lambda & 0 \\ & & & & 0 & \lambda^{-1} \end{pmatrix}, \quad \lambda \neq 0 \right\}.$$

Let k be the algebraic closure of $\mathbb{F}_q$; we shall use notation $\widetilde{G}$, $\widetilde{T}$, etc, for the groups of points over k of the corresponding (algebraic) subgroups in G. Denote $W = N_G(T)/T$ and $\widetilde{W} = N_{\widetilde{G}}(\widetilde{T})/\widetilde{T}$. We know that W covers $\widetilde{W}$ in the sense that $N_G(T)\widetilde{T} = N_{\widetilde{G}}\widetilde{T}$. Thus $xT \in W$ corresponds to the permutation $(1\ -1)\dots(n\ -n)$ in $\widetilde{W} = \mathrm{Sym}_2 \wr \mathrm{Sym}_n$ hence is central in W and $\widetilde{W}$. Since x inverts $\widetilde{T}$ and $\widetilde{T}$ is a 2-divisible group, every element xt in xT is conjugate to x in $\widetilde{G}$: $(\sqrt{t})^{-1}x\sqrt{t} = x\sqrt{t}\sqrt{t} = xt$.

Let F be a standard Frobenius map on k such that the number of elements of the base field is q. is trivial on $\widetilde{W}$.

For the purpose of the proof we can assume that H is a maximal torus of G, so that the torus $\widetilde{H}$ is maximal in $\widetilde{G}$ and F-invariant. Let $g \in \widetilde{G}$ be such that $\widetilde{T}^g = \widetilde{H}$. Applying the Frobenius map F to this identity, we have $\widetilde{T}^{g^F} = \widetilde{H}$ and $\widetilde{T} = \widetilde{T}^{g^F g^{-1}}$. Denote $n_w = g^F g^{-1} \in N_{\widetilde{G}}(\widetilde{T})$, so that $w = n_w \widetilde{T} \in \widetilde{W}$.

The isomorphism

$$\begin{aligned} n_w \circ F : \widetilde{T} &\longrightarrow \widetilde{T} \\ t &\mapsto (t^F)^{n_w} \end{aligned}$$

is rational and fixes only finitely many points. Indeed, if $Z = C_{\widetilde{T}}(n_w \circ F)$ is infinite then $t^F = t^{n_w^{-1}}$ for $t \in Z$, while the automorphism F is of infinite order on Z and n_w^{-1} is of finite order.

Now we start our computation; our task is to find an F-invariant element $(xt)^g \in N_{\widetilde{G}}(\widetilde{H})$. This element will invert the torus $\widetilde{H} = \widetilde{T}^g$ and will also belong to $N_G(H)$. Since $n_w \circ F$ is a rational endomorphism of $\widetilde{T}$ with finite number of fixed point, the map

$$t \mapsto t \cdot (n_w \circ F(t)^{-1})$$

is surjective by the Steinberg–Lang Lemma **[17**, §10**]**. We can take $t_0 \in \widetilde{T}$ with $x^{n_w} = xt_0$ and find $t \in \widetilde{T}$ such that

$$t_0 = t \cdot (n_w \circ F(t)^{-1}).$$

The element t satisfies, after a simple rearrangement,

$$t_0^g \cdot (t^F)^{g^F} = t^g.$$

We claim that $(xt)^g$ is the desired element. Indeed, since $g^F = n_w g$, we have

$$\begin{aligned} \left((xt)^g\right)^F &= (xt^F)^{g^F} \\ &= x^{g^F} \cdot (t^F)^{g^F} \\ &= x^{n_w g} \cdot (t^F)^{g^F} \\ &= x^g \cdot t_0^g \cdot (t^F)^{g^F} \\ &= x^g \cdot t^g \\ &= (xt)^g. \end{aligned}$$

This proves the lemma. □

Lemma 2.9. *Let $G := \mathrm{PSp}_{2n}(q)$ and $i \in G$ be an involution of type t_n. Then $i^g i^h$ is a regular twisted element of odd order with probability $p \geqslant \frac{1}{20n}$ for random $g, h \in G$.*

Proof. One of the tori $\frac{1}{2}T_{q^n \pm 1}$ has odd order. Let us first assume $q^n \equiv -1 \bmod 4$, then $|T| = \frac{1}{2}(q^n - 1)$ has odd order. Consider the map

$$\begin{aligned} \varphi : i^G \times i^G &\longrightarrow G \\ (g, h) &\mapsto i^g \cdot i^h. \end{aligned}$$

We are interested in the fibers of this map. Let $x \in G$ be a regular twisted element from T. Then x is contained in $\varphi(i^G \times i^G)$ by Lemma 2.8. Furthermore

$$|\varphi^{-1}(x)| = |Ti^g| = |T|,$$

where the generalised centraliser $C_G^*(x)$ is $T\langle i^g \rangle$ and involutions inverting x constitute the coset Ti^g. By Lemma 2.7 at least $\frac{1}{4n}|G|$ elements of G are regular twisted elements conjugate to elements in T. Let X be the collection of all regular twisted elements conjugate to elements in T, then

$$|\varphi(X)^{-1}| \geqslant \frac{q^n - 1}{8n} \cdot |G|.$$

On the other hand,

$$|i^G \times i^G| = |G|^2 |C_G(i)|^{-2}$$

and the share of pairs of involutions which are sent to X is

$$\frac{|C_G(i)|^2 (q^n - 1)}{4n \cdot |G|}.$$

Since $q \equiv -1 \bmod 4$, the centraliser of i is $\frac{1}{2}\mathrm{GU}_n(q^2)$ Thus

$$|C_G(i)| = \frac{1}{2}(q+1) q^{\frac{n(n-1)}{2}} (q^2 - 1)(q^3 + 1) \dots (q^n + 1)$$

by **[10]**. Hence

$$\frac{|C_G(i)|^2 (q^n - 1)}{4n \cdot |G|} = \frac{1}{8n} \frac{q^n + 1}{q^n} \cdot \frac{(q+1)(q^2 - 1) \dots (q^n + 1)}{(q-1)(q^2 + 1) \dots (q^n - 1)}.$$

If we denote the last fraction on right hand side as

$$x(n, q) := \prod_{i=1}^{n} \frac{q^i - (-1)^i}{q^i + (-1)^i}$$

then

$$x(n, q) = \exp \left(\sum_{i=0}^{n-1} (-1)^i f_i(q) \right),$$

where

$$f_i(q) := \ln(q^{i+1} + 1) - \ln(q^{i+1} - 1).$$

Since $f_i(q)$ is a strictly decreasing sequence for all $q > 1$, the sum

$$\sum_{i=0}^{n-1} (-1)^i f_i(q)$$

remains positive as n grows and $x(n, q) \geqslant 1$.

Hence the share of pairs of involutions which are sent to good elements is

$$\begin{aligned} \frac{|C_G(i)|^2 (q^n - 1)}{4n \cdot |G|} &= \frac{1}{8n} \frac{q^n - 1}{q^n} x(n, q) \\ &\geqslant \frac{1}{8n} \frac{3^3 - 1}{3^3} \geqslant \frac{1}{9n}; \end{aligned}$$

we use here our assumption that $q \geqslant 3$ and $n \geqslant 3$.

If $q^n \equiv 1 \bmod 4$, similar arguments apply and, after application of a bit of elementary calculus, produce the combined result $p \geqslant \frac{1}{20n}$. □

As an immediate corollary we have the following lemma.

Lemma 2.10. *If i is an involution of type t_n in $G = \mathrm{PSp}_{2n}(q)$ then the share of elements $g \in G$ such that the element ii^g is of odd order is at least $\frac{1}{20n}$.*

Proof. Apply the previous lemma, using the obvious observation that $i^g \cdot i^h = (ii^{hg^{-1}})^g$. □

2.6. Involutions of type t_n in $\Omega_{2n+1}(q)$. The group $\Omega_{2n+1}(q)$ has n conjugacy classes of involutions. There is one conjugacy class of involutions of type $t_{m/2}$ for each even number $2 \leqslant m \leqslant 2n$. An involution s is of type t_m if the eigenvalues of s on the underlying natural module are -1 with multiplicity m and 1 with multiplicity $2n + 1 - m$.

Lemma 2.11. *Let $G := \Omega_{2n+1}(q)$ and $i \in G$ an involution of type t_n. Then the element $i^g \cdot i^h$ acts trivially on a subspace of codimension 2 in the natural module for G. In particular, $o(i^g i^h) \mid q(q+1)$ or $o(i^g i^h) \mid q(q-1)$ for all $g, h \in G$.*

Proof. Let V be the natural module for G. Since i is of type t_n the dimension of the eigenspace V_- of i for the eigenvalue -1 is $2n$, while the eigenspace V_+ for the eigenvalue $+1$ is 1-dimensional. Therefore $i^g i^h$ belongs to $C = C_{\mathrm{GL}(V)}(V_-g \cap V_-h)$ and $W = V_-g \cap V_-h$ has codimension $\leqslant 2$ in V. Notice that $C/O_p(G)$ is a subgroup of $\mathrm{GL}_2(q)$ (here $p = \mathrm{char}\, \mathbb{F}_q$). Notice also that the element $i^g i^h$ leaves invariant the subspace $V_+g \oplus V_+h$. The result now easily follows. □

Lemma 2.12. *Let $G := \Omega_{2n+1}(q)$ and $i \in G$ an involution of type t_n. Then for at least*

$$\frac{1}{4(1 + 2/q + o(1/q))}$$

elements $g \in G$, the element ii^g has odd order.

Proof. We use notation of the previous lemma with $W = V_- \cap V_-g)$ and $U = V_+ + V_+g$. Let $q \equiv 1 \bmod 4$. Then a direct computation with the help of information about finite orthogonal geometries contained in Section III.6 of Artin [3] shows that, with probability

$$\rho > \frac{1}{4(1 + 2/q + o(1/q))},$$

the subspace U is nondegenerate and has index 0. Hence the group $\langle i, i^g \rangle$ is isomorphically imbedded into $O(U) \simeq \mathbb{Z}_{q+1} \rtimes \mathbb{Z}_2$ and half of the elements ii^g have odd order, yielding the desired estimate. The case $q \equiv -1 \bmod 4$ can be treated analogously. □

2.7. Involutions produced by good elements

Lemma 2.13. *Let i be the involution in a maximal non-split torus $T < G$ of even order.*

(i) *If $T = \frac{1}{2}T_{q^n \pm 1}$ then i is of type t_n.*

(ii) *If $T = \frac{1}{2}\left(T_{q^{n/2}+1} \times T_{q^{n/2}-1}\right)$ then i is of type $t_{\frac{n}{2}}$.*

Proof. The easiest way to prove the claim is to look which centralisers of involutions of type t_k can contain maximal non-split tori of even order. Conjugacy classes of maximal tori and their orders are given by the tables in Section 4.5 of Gorenstein, Lyons and Solomon **[12]** together with Carter **[9]** and Springer and Steinberg **[16]**. Thus a torus $\frac{1}{2}T_{q^n+1}$ can be only contained in centralisers of involutions of type t_n for both groups. On the other hand, we have two types of maximal non-split tori of order $\frac{q^n-1}{2}$ and they arise in different ways. It can be contained in centralisers of involutions of type t_n or $t_{\frac{n}{2}}$ if n is even. Assume that n is even and that i is of type $t_{\frac{n}{2}}$. Then a torus in $C_G(i)$ of order $\frac{1}{2}(q-1)$ arises as central product of a torus of order $q^{n/2}-1$ and $q^{n/2}+1$. That means that it is a torus of type $\frac{1}{2}\left(T_{q^{n/2}+1} \times T_{q^{n/2}-1}\right)$ (compare Lemma 2.1). This gives us the claim. □

3. Construction of centralisers of involutions

A crucial ingredient of our algorithm is the construction of the centraliser of a certain involution in $G = \mathrm{PSp}_{2n}(q)$ or $\Omega_{2n+1}(q)$. We use a very general construction which originates in the group-theoretical folklore; see also Bray **[8]** for application to computations in sporadic groups.

Let X be an arbitrary black box finite group and assume that $x^m = 1$ for all elements $x \in X$. Write $m = 2^t \cdot r$ with r odd. Let x be a random element in X. Notice that

- if x is of odd order, then $x^r = 1$ and $y = x^{(r+1)/2}$ is a square root of x:
$$y^2 = x^{(r+1)/2} \cdot x^{(r+1)/2} = x^{r+1} = x;$$
- if x is of even order then x^r is a 2-element and the consecutive squaring of x^r produces the involution $\mathrm{i}(x)$ from the cyclic group $\langle x \rangle$.

We remark that we can test whether some $x \in X$ has odd order without knowing the exact order of x, by testing whether $x^r = 1$. Furthermore, the elements $y = \sqrt{x}$ and $\mathrm{i}(x)$ can be found by $O(\log m)$ multiplications.

Let now i be an involution in X. Construct a random element x of X and consider $z = ii^x$.

- If z is of odd order and $y = \sqrt{z}$ then
$$i^y = y^{-1}iy = iyy = iz = i \cdot ii^x = i^x$$
and $yx^{-1} \in C_X(i)$. We write $yx^{-1} = \zeta_1(x)$.

- if z is of even order then $\mathrm{i}(z)$ lies in the center of the dihedral group $\langle i, i^x \rangle$ and thus $\mathrm{i}(z) \in C_X(i)$. We write $\mathrm{i}(z) = \zeta_0(x)$.

Thus we have a map $\zeta = \zeta_1 \sqcup \zeta_0$ defined by

$$\begin{aligned} \zeta : X &\longrightarrow C_X(i) \\ x &\mapsto \begin{cases} \zeta_1(x) = (ii^x)^{(r+1)/2} \cdot x^{-1} & \text{if } o(ii^x) \text{ is odd} \\ \zeta_0(x) = \mathrm{i}(ii^x) & \text{if } o(ii^x) \text{ is even.} \end{cases} \end{aligned}$$

If $c \in C_X(i)$ then

$$\begin{aligned} \zeta_0(xc) &= \mathrm{i}(i \cdot i^{xc}) = \mathrm{i}(i^c \cdot i^{xc}) = \mathrm{i}((i \cdot i^x)^c) = \mathrm{i}(ii^x)^c \\ &= \zeta_0(x)^c, \end{aligned}$$

$$\begin{aligned} \zeta_1(cx) &= (ii^{cx})^{(r+1)/2} \cdot x^{-1}c^{-1} = (ii^x)^{(r+1)/2} \cdot x^{-1}c^{-1} \\ &= \zeta_1(x) \cdot c^{-1}. \end{aligned}$$

Hence if the elements $x \in X$ are uniformly distributed and independent in X then

- the distribution of elements $\zeta_1(x)$ in $C_X(i)$ is invariant under right multiplication by elements $c \in C_X(i)$, that is, if $A \subset C_X(i)$ and $c \in C_X(i)$ is an arbitrary element then the probabilities $P(\zeta_1(x) \in A)$ and $P(\zeta_1(x) \in Ac)$ coincide.

- The distribution of involutions $\zeta_0(x)$ is invariant under the action of $C_X(i)$ on itself by conjugation, that is,
$$P(\zeta_0(x) \in A) = P(\zeta_0(x) \in A^c).$$

Therefore if the elements $x \in X$ are uniformly distributed and independent in X then the elements $\zeta_1(x)$ are uniformly distributed and independent in $C_X(i)$. It might happen that the share of elements $x \in X$ for which the function $\zeta_1(x)$ is defined is too small to use ζ_1 as an efficient way to generate elements in $C_X(i)$. However, in the special situation considered in this paper we have sufficiently frequent elements $i \cdot i^x$ of odd order (Lemma 2.10 and 2.12), and thus ζ_1 is good enough for our purposes.

The function ζ_0 is interesting on its own and can be used for effective generation of a normal subgroup in $C_X(i)$; details will be published elsewhere.

Lemma 3.1. *Let i be an involution of type t_n in $G = \mathrm{PSp}_{2n}(q)$ or $\Omega_{2n+1}(q)$. Then we can find, in time $O(\mu n^3 \log q)$, in the symplectic case and in time $O(\mu n^2 \log q)$ in the orthogonal case, an element $g \in G$ such that ii^g is of odd order. Elements $\zeta_1(g)$ constructed in that way are independent and uniformly distributed in $C_G(i)$.*

The time estimate in Lemma 3.1 in case $G = \mathrm{PSp}_{2n}(q)$ is very crude; we believe that the actual timing is $O(\mu n^2 \log q)$ and is determined mostly by time required for checking that an element ii^g is of odd order.

4. Detailed description of the algorithm

4.1. Step 1: search for a good element x. The algorithm spends most of its timing testing if elements are good. The following lemma corresponds to Lemma 2.6 of Kantor and Seress **[14]**.

Lemma 4.1. *Let $m_1 := \prod_{i=2}^{2n-1} q^i - 1$ and $m_2 := \prod_{i=2}^{n-1} q^i - 1$. An element $g \in G$ is good*

- *in case $q^n \equiv -1 \bmod 4$, if $o(g) \mid q^{2n} - 1$ but $o(g) \nmid m_1$ and $o(g) \nmid \frac{1}{4}(q^n + 1)$,*
- *in case $q^n \equiv 1 \bmod 4$ for n odd, if $o(g) \mid q^n - 1$ but $o(g) \nmid m_2$ and $o(g) \nmid \frac{1}{4}(q^n - 1)$,*
- *in case $q^n \equiv 1 \bmod 4$ for n even, if $o(g) \mid q^n - 1$ but $o(g) \nmid m_2$ and $o(g) \nmid \frac{1}{4}(q^n - 1)$ and $o(g) \nmid q^{n/2} + 1$.*

Whether an element is good or not can be tested using $O(n^2 \log q)$ multiplications.

Proof. By definition of good element, an element is good if it satisfies the conditions in the lemma. Furthermore, we can decide if $o(g)$ divides k for any natural number k by computing g^k and comparing g^k with 1. This requires $O(\log k)$ multiplications if we use the squaring method. □

It easily follows from Lemma 2.5 that a sample of N random elements of G contains a good element with probability

$$1 - \left(1 - \frac{1}{4n}\right)^N > 1 - e^{-\frac{N}{4n}}.$$

Therefore if we are given fixed margin of error ϵ, $0 < \epsilon < 1$, and take a sample of size $O(n \log(1/\epsilon))$, then we can be satisfied that it contains a good element with probability $1 - \epsilon$. Production of a random element and testing it for goodness require time $O(\mu n^2 \log q + \xi)$. Hence we can find a good element x with probability $1 - \epsilon$ in time $O((\mu n^3 \log q + \xi n) \log(1/\epsilon))$.

4.2. Step 2: involution $i = \mathrm{i}(x)$. As soon as we have a good element x and know that it is an element of maximal 2-height in a cyclic group of order $\frac{1}{2}(q^n \pm 1)$, the involution $x = \mathrm{i}(x)$ can be obtained by raising x to the power $\frac{1}{4}(q^n \pm 1)$. This requires time $O(\mu n \log q)$. Lemma 2.13 asserts that i is one of the types t_n or $t_{n/2}$.

4.3. Step 3: involution of type t_n. Notice that if n is odd then the involution i is of type t_n.

However, if n is even then i can happen to be of type $t_{n/2}$ (with a rather small probability $1 - \frac{1}{1+(2/n)}$, see Lemma 2.6).

We can check whether i is of type t_n or not by constructing uniformly distributed random elements $\zeta_1(x)$ of $C_G(i)$ as described in Section 3. If i is of type t_n then this method produces (in polynomial of $n \log q$ time) random and uniformly distributed elements of $C_G(i)$. Let $L = F*(C_G(i))$. Depending on whether L is one of the groups listed in Table 1: $\frac{1}{2}\mathrm{GL}_n(q)$, $\Omega^+_{2n}(q)$, $\frac{1}{2}\mathrm{GU}_n(q^2)$, or $\Omega^-_{2n}(q)$, the group $C_G(i)$ contains 'big' elements, that is, elements of order divisible by a primitive prime divisor of $q^{n-1}-1, q^{n-1}+1, q^{n-1}-1, q^n+1$. It can be observed directly from the formulae for the orders of the corresponding groups, say, in **[9]**, pp. 2–8, that 'big' elements are not present in $C_G(t_{n/2})$. It is easy to see that 'big' elements are regular elements in maximal tori of orders $\frac{1}{2}(q \pm 1)(q^{n-1} \pm 1)$ or $\frac{1}{2}(q^n + 1)$. Using the same technique as in Sections 2.1–2.3, we can see that the probability of a random element to be 'big' is $O(1/n)$ and hence one of them has to pop up from the black box for $C_G(i)$ after $O(n)$ iterations. As soon as we have found a 'big' element, we know that i is of type t_n and can go to Step 4.

If a 'big' element cannot be found after $O(n)$ tests of elements from $C_G(i)$, then i is likely to be of type $t_{n/2}$. In that case return to Step 1 and start search for an involution of type t_n from the beginning.

4.4. Steps 4 and 5: Behaviour of an involution of type t_n. The involution i of type t_n behaves very differently in $\mathrm{PSp}_{2n}(q)$ and $\Omega_{2n+1}(q)$. Namely, in $\mathrm{PSp}_{2n}(q)$ an element $i^g i^h$ for random $g, h \in G$ is a regular twisted element with probability $\geqslant \frac{1}{20n}$ (Lemma 2.9), while in $\Omega_{2n+1}(q)$ we always have $o(i^g i^h) \mid q(q \pm 1)$ (Lemma 2.11). Obviously, a regular twisted element cannot have order dividing $q(q \pm 1)$. As soon as we have that $o(i^g i^h)$ does not divide $q(q \pm 1)$, the pair (i^g, i^h) is a witness against the isomorphism $G \simeq \Omega_{2n+1}(q)$ and in that case $G \simeq \mathrm{PSp}_{2n}(q)$.

On the other hand, if in $20n$ tests we had the same result $o(i^g i^h) \mid q(q \pm 1)$, then $G \simeq \Omega_{2n+1}(q)$ with probability $\geqslant 1 - e^{-1}$. Timing estimate here is very crude and can be considerably improved. In practice, if $G \simeq \mathrm{PSp}_{2n}(q)$ then $o(i^g i^h) \nmid q(q \pm 1)$ at the very first attempt.

Acknowledgement. The authors would like to thank Bill Kantor and Ákos Seress for many useful comments.

References

[1] C. Altseimer, Contributions to the Classification of Tame K*-groups of Odd Type and Other Applications of 2-local Theory, Ph.D. thesis, Freiburg 1998.

[2] C. Altseimer and A. Borovik, A non-deterministic algorithm for recognition of orthogonal and symplectic groups, Manchester Centre for Pure Mathematics, preprint 1999/1, 11 pp.

[3] E. Artin, Geometric Algebra, Interscience Publishers, New York, 1957.

[4] L. Babai, Randomization in group algorithms: conceptual questions, in: Groups and Computations II (L. Finkelstein and W. Kantor, eds.), DIMACS Ser. Discrete Math. Theoret. Comput. Sci. 28, Amer. Math. Soc., Providence, RI, 1997, 1–17.

[5] L. Babai and E. Szemerédi, On the complexity of matrix group problems, in: Proc. 25th IEEE Sympos. Foundations Comput. Sci., 1984, 229–240.

[6] R. Beals, Towards polynomial time algorithms for matrix groups, in: Groups and Computations II (L. Finkelstein and W. Kantor, eds.), DIMACS Ser. Discrete Math. Theoret. Comput. Sci. 28, Amer. Math. Soc., Providence, RI, 1997, 31–54.

[7] A. Borel and J, Tits, Groupes réductifs, Inst Hautes Études Sci. Publ. Math. 27 (1965), 55–151.

[8] J. N. Bray, An improved method for generating the centralizer of an involution, Arch. Math. 74 (2000), 241–245.

[9] R. W. Carter, Conjugacy Classes in the Weyl Group, in: Seminar on Algebraic Groups and Related Finite Groups (A. Borel et al., eds.) Lecture Notes in Math. 131, Springer-Verlag, Berlin, 1970, G1–G22.

[10] R. W. Carter, Simple Groups of Lie Type, John Wiley and Sons, London a.o., 1972.

[11] R. Carter, Finite Groups of Lie Type: Conjugacy Classes and Complex Characters, Wiley, 1985.

[12] D. Gorenstein, R. Lyons and R. Solomon, The classification of the finite simple groups 3, Math. Surveys Monographs 40, Amer. Math. Soc., Providence, RI, 1998.

[13] N. Iwahori, Centralizers of Involutions in finite Chevalley Groups, in: Seminar on Algebraic Groups and Related Finite Groups (A. Borel et al., eds.), Lecture Notes in Math. 131, Springer, Berlin, 1970, F1–F29.

[14] W. M. Kantor and Á. Seress, Black box classical groups, Mem. Amer. Math. Soc. 149, Nr. 708, Amer. Math. Soc., Providence, RI, 2001.

[15] P. Neumann and C. Praeger, A recognition algorithm for special linear groups, Proc. London Math. Soc. (3) 65 (1992), 555–603.

[16] T. A. Springer and R. Steinberg, Conjugacy Classes, in: Seminar on Algebraic Groups and Related Finite Groups (A. Borel et al., eds.), Lecture Notes in Math. 131, Springer-Verlag, Berlin, 1970, E1–E100.

[17] R. Steinberg, Endomorphisms of Linear Algebraic Groups, Mem. Amer. Math. Soc. 80, Amer. Math. Soc., Providence, RI, 1968.

[18] D. E. Taylor, The Geometry of the Classical Groups, Heldermann, Berlin, 1992.

[19] K. Zsigmondy, Zur Theorie der Potenzreste, Monatsh. Math. Phys. (3) (1892), 265–284.

Christine Altseimer
Allianz Lebensversicherungs-AG, MPE/Research
Reinsburgstr. 19, 70178 Stuttgart, Germany
christine.altseimer@allianz.de

Alexandre V. Borovik
Department of Mathematics, UMIST
PO Box 88, Manchester M60 1QD, U.K.
Alexandre.Borovik@umist.ac.uk

Janko's simple groups J_2 and J_3 are irreducible subgroups of $SL_{85}(5)$ with equal centralizers of an involution

Roberto M. Avanzi, Mathias Kratzer and Gerhard O. Michler

Abstract. Let $\mathcal{F} = \langle S, F, X \rangle$ be the free group generated by the letters S, F and X. In this article we construct two 85-dimensional representations $\varrho_i : \mathcal{F} \to SL_{85}(5)$ of $\mathcal{F}$ over the prime field GF(5), $i = 2, 3$, with the following properties:

(1) $\varrho_2|_{\langle S,F \rangle} = \varrho_3|_{\langle S,F \rangle}$, and $H = \langle \varrho_2(S), \varrho_2(F) \rangle \cong 2^{1+4}{:}PSL_2(5)$.

(2) $M_i = \langle \varrho_i(S^F)^3, \varrho_i(S)^2, \varrho_i(X) \rangle \cong 2^{2+4}{:}(S_3 \times 3)$ for $i = 2, 3$.

(3) However, $G_2 = \mathcal{F}/\ker(\varrho_2) = \langle H, M_2 \rangle \not\cong \langle H, M_3 \rangle = \mathcal{F}/\ker(\varrho_3) = G_3$.

In fact, we prove that G_2 and G_3 are isomorphic to Janko's second and third sporadic simple groups J_2 and J_3, respectively. This is done by computing permutation representations of the matrix groups $G_2, G_3 \le SL_{85}(5)$ with stabilizer H. The matrices $\varrho_2(x), \varrho_3(x) \in SL_{85}(5)$ are also calculated from H by means of a uniform construction method.

1991 Mathematics Subject Classification: primary 20D08; secondary 20H30.

1. Introduction

By the Brauer–Fowler theorem there are only finitely many simple groups G having a 2-central involution $z \neq 1$ such that its centralizer $C_G(z)$ is isomorphic to a given group H. Unfortunately its proof does not yield an algorithm to construct all these groups from H.

In **[15]** the third author has described a deterministic algorithm giving a partial answer to that problem without using the classification of the finite simple groups. Starting from a given finite group H this algorithm constructs all the finite simple groups G having a 2-central involution t and the following properties:

1. There exists an isomorphism $\tau : C_G(z) \to H$.
2. There exists an elementary abelian normal subgroup A of a fixed Sylow 2-subgroup S of H of maximal order $|A| \ge 4$ such that

$$G = \langle C_G(t), N = N_G(\tau^{-1}(A)) \rangle.$$

Groups and Computation III
Ohio State Univ. Math. Res. Inst. Publ. 8

3. For some prime $p > 0$ not dividing $|H||N|$ the group has an irreducible p-modular representation M with multiplicity-free restriction $M_{|H}$.

In **[15]** it is also shown that all sporadic simple groups but the O'Nan group ON satisfy all three conditions for at least one prime $p > 0$ not dividing $|H||N|$. In all these 25 cases it turns out that each sporadic simple group G with a given centralizer $H = C_G(z)$ of a 2-central involution z is uniquely determined by its local data: H, $N = N_G(A)$ for some maximal abelian normal subgroup A of a Sylow 2-subgroup S of H, the prime $p > 0$, and some semisimple multiplicity-free p-modular representation M of H.

In this article we give an example of a non semisimple 5-modular 85-dimensional representation φ of $H = 2^{1+4} : A_5$ which extends irreducibly to two non-isomorphic simple groups G_i, $i = 2, 3$, each having a 2-central involution z_i, respectively, such that their centralizers $C_{G_i}(z_i) \cong H$ for $i = 2, 3$. The main results of this article are summarized in the following theorem.

Theorem 1.1. *The finite group $H = 2^{1+4} : A_5$ has a faithful 5-modular representation*

$$\varphi : H \to \mathrm{SL}_{85}(5)$$

which is afforded by the two matrices s, $f \in \mathrm{SL}_{85}(5)$ having orders 6, 5, respectively; they are given in the appendix.

Using the methods described in Remarks 5.3 and 5.4 two new matrices t_2, $t_3 \in \mathrm{SL}_{85}(5)$ of order 3 are constructed from H; they are given in the appendix. Let $z = s^3 f^3 s^3 f^2$, $e_1 = s^3$, $e_2 = (e_1)^f$, and $t = s^2$.

Let $G_2 = \langle s, f, t_2\rangle$ and $G_3 = \langle s, f, t_3\rangle$. Then the following assertions hold:

(a) $C_{G_2}(z) = H = \langle s, f\rangle = C_{G_3}(z)$, $z^2 = 1$.

(b) *$A = \langle z, e_1\rangle$ is a maximal elementary abelian normal subgroup of a Sylow 2-subgroup S of H such that*

$$M_2 = N_{G_2}(A) = \langle (s^3)^f, t, t_2\rangle \cong \langle (s^3)^f, t, t_3\rangle = N_{G_3}(A) = M_3,$$

and $M_2 \cap H = M_3 \cap H = D = N_H(A)$.

(c) *G_2 is a finite simple group of order 604800 with 2 conjugacy classes of involutions.*

(d) *G_3 is a finite simple group of order 50232960 with 1 conjugacy class of involutions.*

In particular, the representation $\varphi : H \to \mathrm{SL}_{85}(5)$ has two extensions

$$\varphi_i : G_i \to \mathrm{SL}_{85}(5),\ i = 2, 3,$$

to two non-isomorphic simple groups $G_2 \cong \mathrm{J}_2$ and $G_3 \cong \mathrm{J}_3$, where J_2 and J_3 denote Janko's sporadic simple groups of **[12]**.

The assertions concerning the uniqueness of J_2 and J_3 are quoted from Hall–Wales **[8]** and Higman–McKay **[9]**, respectively. Other existence and/or uniqueness proofs for J_2/J_3 have been given in **[1]**, **[2]**, **[7]**, **[14]**, **[16]**, **[17]**, **[18]**.

In order to determine the structure of G_2 we construct the permutation representation $(\mathbb{1}_H)^{G_2}$ in Section 3. This permutation representation enables us to provide a complete set of representatives y_j for the conjugacy classes of G_2 as words in s, f and t_2, the lengths of these classes and the respective power maps, confer Proposition 3.1. Using standard tools of MAGMA **[3]** and the results of **[12]** and **[8]** it follows that G_2 is a simple and isomorphic to Janko's second sporadic group.

Following the same strategy in Section 4, we construct the permutation representation $(\mathbb{1}_H)^{G_3}$ from the extension φ_3 of φ. The conjugacy classes $(y_j)^{G_3}$ and the order of G_3 are given in Proposition 4.1. There it is also shown that G_3 is simple and isomorphic to Janko's third sporadic group (see **[12]**, **[9]**). The remaining assertion (b) of Theorem 1.1 is proved in Proposition 5.6.

Concerning our notation and terminology we refer to the books of Butler **[4]**, Feit **[6]** and Isaacs **[11]**.

2. The structure of the common subgroup $H = \langle s, f \rangle$

Lemma 2.1. *Let $s, f \in SL_{85}(5)$ be the two blocked diagonal matrices given in the appendix of orders $o(s) = 6$ and $o(f) = 5$, respectively. Let $H = \langle s, f \rangle$ be the subgroup of $SL_{85}(5)$ generated by s and f. Let $t = s^2$, $e_i = (s^3)^{f^{i-1}}$ for $1 \le i \le 5$, $v_1 = e_1e_3$, $v_2 = v_1e_2$, $v_3 = v_1e_4$, and $z = v_1^2$. Then:*

(a) *$A_5 = \langle t, f \rangle \le H$ is isomorphic to the alternating group of degree 5.*
(b) *$v_1^4 = v_2^4 = v_3^4 = 1$, and $v_2^2 = v_3^2 = z$.*
(c) *$(v_1)^{e_1} = v_1^{-1}$, and $D = \langle v_1, e_1 \rangle \le H$ is a dihedral group of order 8.*
(d) *$(v_2)^{v_3} = v_2^{-1}$, and $Q = \langle v_2, v_3 \rangle \le H$ is a quaternion group of order 8.*
(e) *$(v_1)^t = v_1$, $(v_2)^t = v_2v_3$, $(v_2v_3)^t = v_3$, and $(v_3)^t = v_2$.*
(f) *$[e_1, v_2] = [v_1, v_2] = [e_1, v_3] = [v_1, v_3] = 1$, $e_5 = e_1e_2e_3e_4z$.*

Moreover, let E be the subgroup of H generated by e_1, e_2, e_3, e_4 and e_5. Then:

(g) *E is a central product $D * Q$ with amalgamated subgroup $D \cap Q = \langle z \rangle$.*
(h) *E is an extraspecial normal 2-subgroup of H of order $|E| = 2^5$.*
(i) *$H = EA_5$, and $E \cap A_5 = 1$.*
(j) *$[z, s] = [z, f] = 1$, and the center of H is $Z(H) = \langle z \rangle$.*

Proof. (a) Using MAGMA we verify that $f^5 = t^3 = (tf)^2 = 1$. Thus $A_5 = \langle t, f \rangle$ is isomorphic to the alternating group of degree 5 by Huppert **[10]**, p. 140.

All the equations stated in (b), (c), (d), (e) and (f) can be checked by means of MAGMA. Therefore (c) and (d) hold by Huppert **[10]**, p. 91. Furthermore, the relations given in (f) imply that the dihedral subgroup D and the quaternion subgroup Q commute elementwise, and $D \cap Q = \langle z \rangle$ by (b). Thus (g) holds.

(h) By Huppert [10], p. 355, $D * Q$ is an extraspecial subgroup of order 32 in E. As $e_3 = e_1 v_1 \in D$, $v_1, v_2 \in D * Q$, it follows that $e_2 = v_1^{-1} v_2 \in D * Q$, and $e_4 = v_1^{-1} v_3 \in D * Q$. Using MAGMA one gets that $e_5 = z \cdot e_1 e_2 e_3 e_4 \in D * Q$. Hence $E = \langle e_1, e_2, e_3, e_4, e_5 \rangle = D * Q$. As $D \leq C_H(t)$ and $Q^t = Q$ by (e) it follows that E is normal in $H = \langle s, f \rangle$, because $E^f = E$.

(i) $s^{-1} = e_1 t = t e_1 \in EA_5$, and $f \in A_5$. Hence $H = EA_5$, and $E \cap A_5 = 1$.

(j) By means of MAGMA we get $[z, f] = [z, s] = 1$, and $z \in Z(H)$. Hence $Z(H) = \langle z \rangle$ by Janko's character table of H given in [12], p. 40. □

With the notation of Lemma 2.1 the structure of a Sylow 2-subgroup of H is described in the following subsidiary result.

Lemma 2.2. *In $H = \langle f, s \rangle \leq \mathrm{SL}_{85}(5)$ let $t = s^2$, $r_1 = (tf)^f$, $r_2 = r_1^t$, $r_3 = e_2 e_3 r_1$, $r_4 = e_2 e_4 r_2$ and $c = t^{r_1}$. Then:*

(a) *The element $z = (e_1 e_3)^2$ is a central involution of the Sylow 2-subgroup $S = \langle e_1, e_2, e_3, e_4, r_1, r_2 \rangle$ of H with $|S| = 2^7$.*

(b) *$A = \langle z, e_1 \rangle$ is the maximal elementary abelian normal subgroup of S.*

(c) *$R = \langle z, e_1, r_1, r_2, r_3, r_4 \rangle$ is a normal subgroup of $S = \langle R, e_2 \rangle$ with $|S : R| = 2$ and $R' = Z(R) = A$.*

(d) *$D = N_H(A) = \langle R, c, e_2 \rangle$*

(e) *$C = C_H(A) = \langle R, c \rangle \cong R : \langle c \rangle$*

(f) *$D/C \cong \langle e_2 \rangle$*

Proof. All three assertions have been checked by means of MAGMA. □

3. Determining the structure of $G_2 = \langle H, t_2 \rangle$

Let $t_2 \in \mathrm{SL}_{85}(5)$ be the first dense matrix of order $o(t_2) = 3$ given in the appendix. Let $H = \langle s, f \rangle$ be the subgroup of $\mathrm{SL}_{85}(5)$ constructed in Lemma 2.1. It is the purpose of this section to show that $G_2 = \langle H, t_2 \rangle \leq \mathrm{SL}_{85}(5)$ is a finite simple group of order 604800 which is isomorphic to Janko's sporadic simple group J_2.

Proposition 3.1. *Let $H = \langle s, f \rangle$, $G_2 = \langle H, t_2 \rangle \leq \mathrm{SL}_{85}(5)$, where the matrices s, f and t_2 are given in the appendix. Then the following assertions hold:*

(a) *$|G_2| = 2^7 \cdot 3^3 \cdot 5^2 \cdot 7 = 604800$.*

(b) *G_2 has 21 conjugacy classes $(y_j)^{G_2}$, $1 \leq j \leq 21$, with representatives, orders, sizes and power maps as follows:*

Representative	Order	Size	2P	3P	5P	7P
$y_1 = t_2^3$	1	1	1	1	1	1
$y_2 = s^3$	2_1	315	1	2_1	2_1	2_1
$y_3 = s^4 f$	2_2	2520	1	2_2	2_2	2_2
$y_4 = s^2$	3_1	560	3_1	1	3_1	3_1
$y_5 = t_2$	3_2	16800	3_2	1	3_2	3_2
$y_6 = (sf)^2$	4	6300	2_1	4	4	4
$y_7 = (sft_2)^2$	5_1	2016	5_2	5_2	1	5_2
$y_8 = (f^2 t_2)^2$	5_2	2016	5_1	5_1	1	5_1
$y_9 = f$	5_3	12096	5_4	5_4	1	5_4
$y_{10} = f^2$	5_4	12096	5_3	5_3	1	5_3
$y_{11} = s$	6_1	25200	3_1	2_1	6_1	6_1
$y_{12} = s^2 t_2$	6_2	50400	3_2	2_2	6_2	6_2
$y_{13} = st_2^2 f$	7	86400	7	7	7	1
$y_{14} = sf$	8	75600	4	8	8	8
$y_{15} = sft_2$	10_1	30240	5_1	10_2	2_2	10_2
$y_{16} = f^2 t_2$	10_2	30240	5_2	10_1	2_2	10_1
$y_{17} = s^3 f^2$	10_3	60480	5_3	10_4	2_1	10_4
$y_{18} = sfsf^2$	10_4	60480	5_4	10_3	2_1	10_3
$y_{19} = ft_2$	12	50400	6_1	4	12	12
$y_{20} = st_2 ft_2 f$	15_1	40320	15_2	5_1	3_1	15_2
$y_{21} = f^2 t_2 ft_2$	15_2	40320	15_1	5_2	3_1	15_1

(c) *G_2 is a simple group which is isomorphic to the Janko group* J_2.

Proof. The permutation representation $(\mathbb{1}_H)^{G_2}$ of $G_2 = \langle H, t_2 \rangle \leq SL_{85}(2)$ has been computed by the MAGMA-function `CosetAction`. It follows that G_2 operates transitively and faithfully on $\Omega = \{1, 2, \ldots, 315\}$, and that $H = \langle s, f \rangle$ is the stabilizer $\mathrm{stab}_{G_2}(1)$. Hence $|G_2| = |H||\Omega|$ which proves a).

(b) The representatives of the conjugacy classes are constructed as representatives of the order classes of G_2 by means of an efficient tracing algorithm on the wordtree of the three generators s, f and t_2, which has been developed and implemented as a MAGMA function by the second author. An explicit description of this algorithm will be published elsewhere.

Using the permutation representation $(\mathbb{1}_H)^{G_2}$

MAGMA is able both to distinguish representatives of different conjugacy classes within the same order class and to compute the size of each conjugacy class once it is identified. The sum of these sizes meets the order of G_2 iff all the conjugacy classes of G_2 have been found. Then it is straight forward to determine the power maps.

(c) Applying the MAGMA-function `NormalClosure` to all cyclic subgroups generated by the 21 representatives of the conjugacy classes of G_2 it follows that G_2 is a simple group. By Lemmas 2.1 and 2.2 $z = (s^3 f^3 s^3 f^2)^2$ is a central involution with $H = \langle s, f \rangle \leq C_{G_2}(z)$. Since $|z^{G_2}| = |315|$ by (b), it follows that $C_{G_2}(z) = H \cong 2^{1+4} : A_5$, see Lemma 2.1. Furthermore, G_2 has exactly two conjugacy class of involutions. Therefore G_2 is isomorphic to Janko's second sporadic group J_2 by the main results of Janko **[12]** and Hall/Wales **[8]**. □

4. Determining the structure of $G_3 = \langle H, t_3 \rangle$

Let $t_3 \in \mathrm{SL}_{85}(5)$ be the second dense matrix of order $o(t_3) = 3$ given in the appendix. Let $H = \langle s, f \rangle$ be the subgroup of $\mathrm{SL}_{85}(5)$ constructed in Lemma 2.1. It is the purpose of this section to show that $G_3 = \langle H, t_3 \rangle \le \mathrm{SL}_{85}(5)$ is a finite simple group of order 50232960 which is isomorphic to Janko's sporadic simple group J_3.

Proposition 4.1. *Let $H = \langle s, f \rangle$, $G_3 = \langle H, t_3 \rangle \le \mathrm{SL}_{85}(5)$, where the matrices s, f and t_3 are given in the appendix. Then the following assertions hold:*

(a) $|G_3| = 2^7 \cdot 3^5 \cdot 5 \cdot 17 \cdot 19 = 50232960.$

(b) *G_3 has 21 conjugacy classes $(y_j)^{G_3}$ with representatives, orders, sizes and power maps as follows:*

Representative	Order	Size	2P	3P	5P	17P	19P
$y_1 = t_3^3$	1	1	1	1	1	1	1
$y_2 = s^3$	2	26163	1	2	2	2	2
$y_3 = s^2$	3_1	46512	3_1	1	3_1	3_1	3_1
$y_4 = t_3$	3_2	206720	3_2	1	3_2	3_2	3_2
$y_5 = (sf)^2$	4	523260	2	4	4	4	4
$y_6 = f$	5_1	1674432	5_2	5_2	1	5_2	5_1
$y_7 = f^2$	5_2	1674432	5_1	5_1	1	5_1	5_2
$y_8 = s$	6	2093040	3_1	2	6	6	6
$y_9 = sf$	8	6279120	4	8	8	8	8
$y_{10} = s^2(ft_3)^2$	9_1	1860480	9_3	3_2	9_2	9_1	9_1
$y_{11} = s(ft_3)^2t_3$	9_2	1860480	9_1	3_2	9_3	9_2	9_2
$y_{12} = s(ft_3)^3$	9_3	1860480	9_2	3_2	9_1	9_3	9_3
$y_{13} = sft_3$	10_1	5023296	5_2	10_2	2	10_2	10_1
$y_{14} = f^2t_3$	10_2	5023296	5_1	10_1	2	10_1	10_2
$y_{15} = sf^2$	12	4186080	6	4	12	12	12
$y_{16} = s(t_3f)^2$	15_1	3348864	15_2	5_1	3_1	15_2	15_1
$y_{17} = f(ft_3)^2$	15_2	3348864	15_1	5_2	3_1	15_1	15_2
$y_{18} = st_3^2f$	17_1	2954880	17_1	17_2	17_2	1	17_1
$y_{19} = s(ft_3)^2$	17_2	2954880	17_2	17_1	17_1	1	17_2
$y_{20} = ft_3$	19_1	2643840	19_2	19_2	19_1	19_1	1
$y_{21} = f^2t_3^2$	19_2	2643840	19_1	19_1	19_2	19_2	1

(c) *G_3 is a simple group which is isomorphic to the Janko group* J_3.

Proof. (a) The permutation representation $(\mathbb{1}_H)^{G_3}$ has been computed by the MAGMA-function `OrbitAction` starting from the 5-dimensional H-invariant subspace of the natural G_3-module over GF(5). It follows that G_3 operates transitively and faithfully on $\Omega = \{1, 2, \dots 26163\}$, and that $H = \langle s, f \rangle$ is the full stabilizer of point 1 in G_3. Hence, $|G_3| = |H||\Omega|$.

(b) The conjugacy class representatives of G_3 have been computed as described in the proof of Proposition 3.1.

(c) Applying the MAGMA-function `NormalClosure` to all cyclic subgroups generated by the 21 representatives of the conjugacy classes of G_2 it follows that G_2

is a simple group. By Lemmas 2.1 and 2.2 $z = (s^3f^3s^3f^2)^2$ is a central involution with $H = \langle s, f\rangle \le C_{G_3}(z)$. Since $|z^{G_3}| = |26163|$ by (a), it follows that $C_{G_3}(z) = H \cong 2^{1+4} : A_5$, see Lemma 2.1. Furthermore, G_3 has only one conjugacy class of involutions. Therefore G_3 is isomorphic to Janko's third sporadic group J_3 by the main results of Janko **[12]** and G. Higman/McKay **[9]**. □

5. How to construct the matrices t_2 and t_3 from $H = \langle s, f\rangle$

By Lemma 2.1 we know that $H = \langle s, f\rangle \cong 2^{1+4} : A_5$. Janko **[12]** has shown that such a group is isomorphic to a 2-central involution centralizer in any of his simple sporadic groups J_2 and J_3. Therefore we apply the third author's algorithm 4.6 of **[15]** which constructs from a given group H of even order all the finite simple groups G having a 2-central involution $t \neq 1$ with the following properties:

(1) There is an isomorphism $\tau : C_G(t) \to H$

(2) There is some elementary abelian normal subgroup A of a fixed Sylow 2-subgroup S of H of maximal order $|A| \ge 4$ such that

$$G = \langle C_G(t), N = N_G(t^{-1}(A))\rangle$$

(3) For some prime $p > 0$ not dividing $|H||N|$ the group G has an irreducible p-modular representation M with multiplicity-free restriction $M_{|H}$.

Since the characteristic 5 divides the order of $H \cong 2^{1+4} : 5$ we compute a presentation of $H = \langle s, f\rangle \le SL_{85}(5)$ in terms of generators and relations, where the matrices s and f are given in the appendix.

Let $t = s^2$, and $e_i = (s^3)^{f^{i-1}}$ for $1 \le i \le 5$. Then Lemma 2.1 (f) and Lemma 2.2 assert that

$$z = e_4e_3e_2e_1e_5$$

is the unique central involution of $H = \langle s, f\rangle$. As $s = te_1$ it follows from Lemma 2.1 that H has the following presentation which is due to the second author **[13]**.

Proposition 5.1.

(a) *The group* $H \cong 2^{1+4} : A_5$ *can be generated by the five involutions* z, e_1, e_2, e_3, e_4, *and two elements* t *and* f *of orders* 3 *and* 5, *respectively, such that the following relations hold:*

(a1) $[z, e_i] = 1$ *for* $1 \le i \le 4$
(a2) $[z, t] = 1 = [z, f]$
(a3) $[e_1, t] = 1 = [e_3, t]$
(a4) $(e_2)^t = e_1e_2e_3e_4$
(a5) $(e_4)^t = e_2$
(a6) $(e_i)^f = e_{i+1}$ *for* $i = 1, 2, 3$
(a7) $(e_4)^f = e_1e_2e_3e_4z$
(a8) $(tf)^2 = 1$

(b) *Let* $r_1 = (tf)^f$, $r_2 = (r_1)^t$, $r_3 = e_2e_3r_1$, $r_4 = e_2e_4r_2$ *and* $c = t^{r_1}$. *Then* $A = \langle z, e_1 \rangle$ *is a maximal elementary abelian normal subgroup of the Sylow* 2*-subgroup* $S = \langle e_1, e_2, e_3, e_4, r_1, r_2 \rangle$ *of* H *with central involution* $z = (e_1e_3)^2$.

(c) $D = N_H(A) = \langle R, c, e_2 \rangle$, *where* $R = \langle z, e_1, r_1, r_2, r_3, r_4 \rangle$ *is a special normal subgroup of* S *with center* $Z(R) = A$, *and* $|R| = 2^6$.

(d) $C = C_H(A) = R : \langle c \rangle$, $R \lhd C$, $R \cap \langle c \rangle = 1$.

Proof. (a) has been proved in **[13]**.

(b)–(d) follow from (a), Lemmas 2.1 and 2.2. □

Using Theorem 2.3 of **[15]** the second author **[13]** has also given a presentation of an extension

$$1 \to C \to M \to \Phi \to 1$$

where $C = C_H(A)$, $A = \langle z, e_1 \rangle$ of Proposition 5.1, and $\Phi = \mathrm{GL}_2(2) \cong S_3$. His result is restated as

Proposition 5.2. *Let* $H \cong 2^{1+4} : A_5$ *be generated by the involutions* z, e_1, e_2, e_3, e_4 *and the elements* t, f *of orders* 3 *and* 5*, respectively, such that these generators satisfy all the relations of Proposition 5.1 a). Let* $r_1 = (tf)^f$, $r_2 = (r_1)^t$, $r_3 = e_2e_3r_1$, $r_4 = e_2e_4r_2$ *and* $c = t^{r_1}$.

Let $D = N_H(A)$ *and* $C = C_H(A)$, *where* $A = \langle z, e_1 \rangle$ *is a maximal elementary abelian normal subgroup of a Sylow* 2*-subgroup* S *of* H. *Then:*

(a) *Let* $\eta : D \to \mathrm{GL}_2(2)$ *be defined by the conjugation action of* D *on* A *with* $\Delta = \eta(D) \cong \langle e_2 \rangle$. *Then* η *has an extension* $\hat{\eta} : M \to \mathrm{GL}_2(2)$ *with kernel* $\ker(\hat{\eta}) = C \lhd D$ *such that* $\Phi = \hat{\eta}(M) \cong \langle x, e_2 \mid x^3 = 1, e_2^2 = 1, x^{e_2} = x^2 \rangle \cong S_3$, *and the following diagram commutes*

$$\begin{array}{ccccccccc}
1 & \longrightarrow & C & \longrightarrow & M & \xrightarrow{\hat{\eta}} & \Phi & \longrightarrow & 1 \\
 & & \uparrow{\scriptstyle \mathrm{id}} & & \uparrow{\scriptstyle \mathrm{id}} & & \uparrow{\scriptstyle \mathrm{id}} & & \\
1 & \longrightarrow & C & \longrightarrow & D & \xrightarrow{\eta} & \Delta & \longrightarrow & 1
\end{array}$$

The extension group M *of* C *is uniquely determined by* Φ *up to isomorphism.*

(b) $M = \langle z, e_1, r_1, r_2, r_3, r_4, c, e_2, x \rangle$, *and these generators of* M *satisfy the following relations:*

(b1) $[z, e_1] = 1$
(b2) $r_1^2 = r_2^2 = r_3^2 = r_4^2 = e_1^2 = e_2^2 = z^2 = 1$
(b3) $c^3 = 1$
(b4) $x^3 = 1$
(b5) $[r_1, r_2] = [r_3, r_4] = 1$
(b6) $[r_1, r_3] = [r_2, r_4] = z$
(b7) $[r_1, r_4] = e_1$
(b8) $[r_2, r_3] = ze_1$
(b9) $z^{e_2} = z$, $e_1^{e_2} = ze_1$, $r_1^{e_2} = r_3$, $r_2^{e_2} = r_4$

(b10) $z^c = z$, $e_1^c = e_1$, $r_1^c = r_2$, $r_2^c = r_1r_2$, $r_3^c = r_4$, $r_4^c = r_3r_4$
(b11) $[e_2, c] = 1$
(b12) $z^x = e_1$, $e_1^x = ze_1$
(b13) $r_1^x = r_2$, $r_2^x = r_1r_2$, $r_3^x = r_3r_4$, $r_4^x = r_3$
(b14) $x^{e_2} = x^2$
(b15) $[c, x] = 1$

Furthermore, $M = \langle D, x\rangle = C \cdot \Phi$, $C \cap \Phi = 1$ *and* $C \lhd M$.

Remark 5.3. Using the methods of algorithm 4.6 of **[15]** it can be shown that the free product $H *_D M$ with amalgamated subgroup $D = N_H(A) = C_M(z)$ of the groups H and M constructed in Propositions 5.1 and 5.2, respectively, has an irreducible 14-dimensional representation

$$\tau : H *_D M \to \mathrm{GL}_{14}(5)$$

over $F = \mathrm{GF}(5)$. Since $H = \langle s, f\rangle$ with $s = te_1$, and $M = \langle D, x\rangle$ by Proposition 5.2, it follows that

$$E = \tau(H *_D M) = \langle \tau(s), \tau(f), \tau(x)\rangle.$$

Let $V = F^{14}$ be the corresponding irreducible FE-module. Forming then the tensor product $V \otimes V$ an application of the meataxe yields a 21-dimensional irreducible constituent V_{21}. Applying now the meataxe to $V_{21} \otimes V_{21}$ we obtain an 85-dimensional irreducible FE-module U. With respect to a fixed basis B of U one obtains the 3 matrices s_1, f_1 and x_1 of the 3 generators $\tau(s)$, $\tau(f)$ and $\tau(x)$ of E, respectively.

Let C be the canonical basis of $W = F^{85}$. Then $U_{|H_1} \cong W_{|H}$. An application of MAGMA now provides a transformation matrix m such that $m^{-1}s_1m = s$ and $m^{-1}f_1m = f$. The matrix $t_2 = m^{-1}x_1m$ is the one given in the appendix.

Remark 5.4. Let $H = \langle z, e_1, e_2, e_3, e_4, t, f\rangle \cong 2^{1+4} . A_5$ with all the generators given in Proposition 5.1, and

$$M = \langle z, e_1, r_1, r_2, r_3, r_4, c, e_2, x\rangle \cong 2^{2+4} : (3 \times S_3)$$

with all the generators given in Proposition 5.2. Furthermore,

$$D = H \cap M = \langle z, e_1, r_1, r_2, r_3, r_4, c, e_2\rangle.$$

Computing the character tables of D, H, M, and the fusion patterns of the conjugacy classes of D in H and also in M, the second author has determined two faithful characters $\nu \in \mathrm{Irr}_{\mathbb{C}}(H)$ and $\omega \in \mathrm{Irr}_{\mathbb{C}}(M)$ of degree $\nu(1) = 18 = \omega(1)$ with equal restrictions to D. After this application of step (3) of algorithm 4.6 of **[15]** he then has determined faithful permutation representations of H and M, from which he was able to construct two semisimple 5-modular representations

$$\varphi_\nu : H \to \mathrm{GL}_{18}(25)$$

$$\varphi_\omega : H \to \mathrm{GL}_{18}(25)$$

such that

$$\varphi_{\nu|D} = \varphi_{\omega|D} \text{ and}$$

$$C_{\mathrm{GL}_{18}(25)}(D) = C_{\mathrm{GL}_{18}(25)}(H) \cdot C_{\mathrm{GL}_{18}(25)}(M),$$

where D, H and M are identified with their images $\varphi_\nu(D) = \varphi_\omega(D)$, $\varphi_\nu(H)$ and $\varphi_\omega(M)$ in $\mathrm{GL}_{18}(25)$. Furthermore, φ_ν is the semisimple representation corresponding to the 5-modular reduction of $\nu \in \mathrm{Irr}_{\mathbb{C}}(H)$. It is a direct sum of an 8-dimensional and a ten dimensional irreducible FH-module, where z acts trivially on the 8-dimensional representation, and $F = \mathrm{GF}(25)$. Similarly φ_ω splits into a 6-dimensional and a 12-dimensional irreducible FM-module.

Let $E = \langle H, M \rangle \leq \mathrm{GL}_{18}(25)$. Using the meataxe algorithm it is easy to see that $V = F^{18}$ is an irreducible FE-module.

Let $s = te_1$. Then $E = \langle \tau(s), \tau(f), \tau(x) \rangle \leq \mathrm{GL}_{18}(25)$, where

$$\tau(s) = \begin{pmatrix}
0 & 1 & 1 & 0 & 0 & 3 & 0 & 2 & 0 & 0 & 0 & 0 & 0 & 0 & 0 & 0 & 0 & 0 \\
0 & 4 & 0 & 0 & 0 & 1 & 0 & 1 & 0 & 0 & 0 & 0 & 0 & 0 & 0 & 0 & 0 & 0 \\
1 & 1 & 0 & 0 & 0 & 4 & 0 & 2 & 0 & 0 & 0 & 0 & 0 & 0 & 0 & 0 & 0 & 0 \\
0 & 3 & 0 & 0 & 0 & 2 & 4 & 4 & 0 & 0 & 0 & 0 & 0 & 0 & 0 & 0 & 0 & 0 \\
0 & 3 & 0 & 0 & 0 & 3 & 0 & 4 & 0 & 0 & 0 & 0 & 0 & 0 & 0 & 0 & 0 & 0 \\
0 & 4 & 0 & 0 & 4 & 1 & 0 & 3 & 0 & 0 & 0 & 0 & 0 & 0 & 0 & 0 & 0 & 0 \\
0 & 2 & 0 & 4 & 0 & 4 & 0 & 1 & 0 & 0 & 0 & 0 & 0 & 0 & 0 & 0 & 0 & 0 \\
0 & 4 & 0 & 0 & 0 & 1 & 0 & 0 & 0 & 0 & 0 & 0 & 0 & 0 & 0 & 0 & 0 & 0 \\
0 & 0 & 0 & 0 & 0 & 0 & 0 & 0 & 0 & 1 & 0 & 0 & 4 & 0 & 1 & 4 & 0 & 0 \\
0 & 0 & 0 & 0 & 0 & 0 & 0 & 0 & 1 & 0 & 0 & 0 & 1 & 0 & 4 & 0 & 0 & 0 \\
0 & 0 & 0 & 0 & 0 & 0 & 0 & 0 & 0 & 0 & 0 & 0 & 0 & 0 & 4 & 1 & 0 & 0 \\
0 & 0 & 0 & 0 & 0 & 0 & 0 & 0 & 0 & 0 & 0 & 0 & 0 & 0 & 4 & 0 & 0 & 1 \\
0 & 0 & 0 & 0 & 0 & 0 & 0 & 0 & 0 & 0 & 0 & 0 & 0 & 1 & 4 & 0 & 0 & 0 \\
0 & 0 & 0 & 0 & 0 & 0 & 0 & 0 & 0 & 0 & 0 & 0 & 4 & 0 & 0 & 4 & 0 & 0 \\
0 & 0 & 0 & 0 & 0 & 0 & 0 & 0 & 0 & 0 & 1 & 0 & 4 & 0 & 0 & 0 & 0 & 0 \\
0 & 0 & 0 & 0 & 0 & 0 & 0 & 0 & 0 & 0 & 0 & 0 & 0 & 0 & 0 & 4 & 1 & 0 \\
0 & 0 & 0 & 0 & 0 & 0 & 0 & 0 & 0 & 0 & 0 & 0 & 0 & 0 & 0 & 4 & 0 & 0 \\
0 & 0 & 0 & 0 & 0 & 0 & 0 & 0 & 0 & 0 & 0 & 1 & 4 & 0 & 0 & 0 & 0 & 0
\end{pmatrix}$$

$$\tau(f) = \begin{pmatrix}
1 & 0 & 0 & 0 & 3 & 0 & 3 & 4 & 0 & 0 & 0 & 0 & 0 & 0 & 0 & 0 & 0 & 0 \\
0 & 0 & 4 & 1 & 1 & 0 & 1 & 2 & 0 & 0 & 0 & 0 & 0 & 0 & 0 & 0 & 0 & 0 \\
0 & 1 & 4 & 0 & 4 & 0 & 4 & 2 & 0 & 0 & 0 & 0 & 0 & 0 & 0 & 0 & 0 & 0 \\
0 & 0 & 4 & 0 & 2 & 1 & 3 & 2 & 0 & 0 & 0 & 0 & 0 & 0 & 0 & 0 & 0 & 0 \\
0 & 0 & 0 & 0 & 3 & 0 & 4 & 0 & 0 & 0 & 0 & 0 & 0 & 0 & 0 & 0 & 0 & 0 \\
0 & 0 & 4 & 0 & 1 & 0 & 0 & 4 & 0 & 0 & 0 & 0 & 0 & 0 & 0 & 0 & 0 & 0 \\
0 & 0 & 0 & 0 & 4 & 0 & 4 & 0 & 0 & 0 & 0 & 0 & 0 & 0 & 0 & 0 & 0 & 0 \\
0 & 0 & 0 & 0 & 1 & 0 & 1 & 1 & 0 & 0 & 0 & 0 & 0 & 0 & 0 & 0 & 0 & 0 \\
0 & 0 & 0 & 0 & 0 & 0 & 0 & 0 & 1 & 1 & 0 & 0 & 0 & 4 & 1 & 0 & 0 & 4 \\
0 & 0 & 0 & 0 & 0 & 0 & 0 & 0 & 0 & 4 & 1 & 0 & 0 & 1 & 4 & 0 & 0 & 0 \\
0 & 0 & 0 & 0 & 0 & 0 & 0 & 0 & 0 & 4 & 0 & 1 & 0 & 0 & 4 & 0 & 0 & 1 \\
0 & 0 & 0 & 0 & 0 & 0 & 0 & 0 & 0 & 4 & 0 & 0 & 1 & 0 & 0 & 0 & 0 & 0 \\
0 & 0 & 0 & 0 & 0 & 0 & 0 & 0 & 0 & 4 & 0 & 0 & 0 & 0 & 4 & 0 & 0 & 0 \\
0 & 0 & 0 & 0 & 0 & 0 & 0 & 0 & 0 & 0 & 0 & 0 & 0 & 4 & 1 & 1 & 0 & 4 \\
0 & 0 & 0 & 0 & 0 & 0 & 0 & 0 & 0 & 0 & 0 & 0 & 0 & 4 & 1 & 0 & 0 & 0 \\
0 & 0 & 0 & 0 & 0 & 0 & 0 & 0 & 0 & 0 & 0 & 0 & 0 & 0 & 0 & 0 & 0 & 4 \\
0 & 0 & 0 & 0 & 0 & 0 & 0 & 0 & 0 & 0 & 0 & 0 & 0 & 0 & 1 & 0 & 0 & 4 \\
0 & 0 & 0 & 0 & 0 & 0 & 0 & 0 & 0 & 0 & 0 & 0 & 0 & 4 & 0 & 0 & 1 & 0
\end{pmatrix}$$

$$\tau(x) = \left(\begin{array}{cccccccccccccccccc}
1 & 0 & 4 & 2 & 0 & 0 & 2 & 0 & [4] & [16] & 0 & [16] & 0 & 0 & 0 & [22] & [16] & [4] \\
0 & 4 & 0 & 0 & 4 & 0 & 0 & 3 & 0 & 0 & [10] & [4] & 0 & [22] & 0 & [22] & [10] & [16] \\
1 & 0 & 4 & 2 & 0 & 0 & 2 & 0 & [16] & [4] & 0 & 0 & 0 & 0 & 0 & [4] & [4] & 0 \\
0 & 0 & 0 & 4 & 0 & 4 & 4 & 0 & 0 & 0 & 0 & [16] & [10] & 0 & [22] & [16] & [10] & [4] \\
0 & 1 & 0 & 2 & 1 & 0 & 2 & 2 & 0 & 0 & [10] & [16] & 0 & [22] & 0 & [4] & [10] & [4] \\
0 & 0 & 0 & 0 & 0 & 0 & 0 & 0 & 0 & 0 & 0 & [4] & 0 & 0 & 0 & [4] & 0 & [16] \\
0 & 0 & 0 & 1 & 0 & 1 & 1 & 0 & 0 & 0 & 0 & [10] & [10] & 0 & [22] & [10] & [10] & [22] \\
0 & 0 & 0 & 4 & 0 & 0 & 4 & 0 & 0 & 0 & 0 & [22] & 0 & 0 & 0 & [22] & 0 & [10] \\
[2] & 0 & [2] & [14] & 0 & 0 & [2] & [2] & [10] & [10] & 0 & 0 & 0 & 0 & 0 & [10] & [16] & 0 \\
[14] & 0 & [14] & [2] & 0 & 0 & [14] & [14] & [10] & [10] & 0 & 0 & 0 & 0 & 0 & [22] & [10] & 0 \\
0 & [8] & 0 & [20] & [20] & 0 & [8] & 0 & 0 & 0 & [2] & [14] & 0 & [2] & 0 & [22] & [16] & [14] \\
0 & 0 & 0 & [20] & 0 & [8] & [8] & [14] & 0 & 0 & 0 & [14] & [2] & 0 & [2] & 0 & [10] & [14] \\
0 & 0 & 0 & [20] & 0 & 0 & [8] & [20] & 0 & 0 & 0 & [14] & 0 & 0 & 0 & [22] & [10] & [14] \\
0 & [20] & 0 & [8] & [8] & 0 & [20] & 0 & 0 & 0 & [2] & [14] & 0 & [2] & 0 & [10] & [4] & [14] \\
0 & 0 & 0 & [8] & 0 & 0 & [20] & [8] & 0 & 0 & 0 & [14] & 0 & 0 & 0 & [10] & [22] & [14] \\
0 & 0 & 0 & 0 & 0 & 0 & 0 & 0 & 0 & 0 & 0 & 0 & 0 & 0 & 0 & 0 & [4] & 0 \\
0 & 0 & 0 & 0 & 0 & 0 & 0 & 0 & 0 & 0 & 0 & 0 & 0 & 0 & 0 & [16] & [4] & 0 \\
0 & 0 & 0 & [8] & 0 & [20] & [20] & [2] & 0 & 0 & 0 & [14] & [2] & 0 & [2] & 0 & [22] & [14]
\end{array}\right)$$

with $[n] := \zeta^n$, ζ a generator of the multiplicative group of GF(25).

The 85-dimensional irreducible FG-module V_{85} of $G = E/Z(E)$ is constructed as follows. The FE-module $V \otimes_F V = W_{171} \oplus W_{153}$, where W_n denotes an n-dimensional irreducible FE-module. Now applying the meataxe algorithm $W_{153} \otimes W_{153}$ we obtain a 1530-dimensional irreducible FE-module W_{1530} as composition factor. Another application of the meataxe to $V \otimes V_{1530}$ yields an 85-dimensional irreducible composition factor V_{85}.

Using then the same routines as in Remark 5.3 the matrix t_3 is computed by means of the operation of $\tau(x)$ on V_{85} with respect to a suitable basis of V_{85}.

Remark 5.5. The matrices $s, f \in SL_{85}(5)$ given in the appendix can easily be computed from the presentation of $H \cong 2^{1+4} : A_5$ given in Proposition 5.1 and the character table of H. All one has to know is that s and f describe a projective FH-module of dimension 85 over $G\hat{O}F = GF(5)$ which is a direct sum $S \oplus P$ of an indecomposable projective FH-module P of dimension 40, and a multiplicity-free semisimple FH-module S with projective irreducible composition factors of degrees 5, 10_1, 10_2 and 20.

Proposition 5.6. *Let* $H = \langle f, s\rangle \leq SL_{85}$, $t = s^2$, $e_1 = s^3$, $z = s^3 f^3 s^3 f^2$ *and* $e_2 = (s^f)^3$. *Let* t_2 *and* t_3 *be the dense matrices given in the appendix. Then*

$$M_2 = \langle e_2, t, t_2\rangle \cong 2^{2+4} : (S_3 \times 3) \cong \langle e_2, t, t_3\rangle = M_3,$$

and $M_2 \cap H = M_3 \cap H = D = N_H(A)$, *where* $A = \langle z, e_1\rangle$ *is the maximal elementary abelian normal subgroup of a Sylow 2-subgroup* S *of* H.

Proof. By Lemma 2.2 we know that $A = \langle z, e_1\rangle$ is a maximal elementary abelian normal subgroup of the Sylow 2-subgroup $S = \langle e_1, e_2, e_3, e_4, r_1, r_2\rangle$ of H. Let $C = C_H(A)$, and $D = \langle R, c, e_1\rangle = N_H(A)$, where R and c are defined in Lemma 2.2. Then $C = \langle R, c\rangle$, $D = C \cdot \langle e_2\rangle$, $C \cap \langle e_2\rangle = 1$, and $C \lhd D$.

By Proposition 5.2 there is an extension

$$1 \to C \to M \to \Phi \to 1$$

of C such that $M = C \cdot \Phi$, $C \lhd M$, $C \cap \Phi = 1$, and $M = \langle D, x \rangle$, where

$$\Phi = \langle x, e_2 \mid x^{e_2} = x^2 \rangle \cong S_3 .$$

In particular, $M = \langle z, e_1, r_1, r_2, r_3, r_4, c, e_2, x \rangle$, and these 9 generators of M satisfy all the relations b1) until b15) of Proposition 5.2. As $R \cong 2^{2+4}$, $C = R : \langle c \rangle$, and $\Phi \times \langle c \rangle = \langle x, e_2 \rangle \times \langle c \rangle$, it follows that

$$M \cong 2^{2+4} : (S_3 \times 3).$$

Choosing $x = t_2$ it follows that the 9 matrices $\{z, e_1, r_1, r_2, r_3, r_4, c, e_2, t_2\}$ satisfy all the relations b1) until b15) of Proposition 5.2. Furthermore, using MAGMA it is easy to see that $M_2 = \langle e_2, t, t_2 \rangle = \langle z, e_1, r_1, r_2, r_3, r_4, c, e_2, t_2 \rangle$ and $M_2 \cap H = \langle z, e_1, r_1, r_2, r_3, r_4, c \rangle = D$. Hence $M_2 \cong M$.

Choosing $x = t_3$ it follows similarly that $M_3 \cong M$. □

Acknowledgements. The authors are grateful to W. Kantor and Á. Seress and the referees for their critical remarks and raising the question whether the matrices t_2, $t_3 \in \mathrm{SL}_{85}(5)$ given in the appendix can be calculated from $H = \langle s, f \rangle \cong 2^{1+4} : A_5$ in a uniform manner. The positive answer is given in section 5 which was not part of the original version.

We also would like to thank W. Lempken (Essen) for some group theoretical information.

Appendix: The Matrices $s, f, t_2, t_3 \in SL_{85}(5)$

```
4431................................................................................
1.44................................................................................
..323................................................................................
..13.................................................................................
..224................................................................................
.......4.1.4.2.......................................................................
......122.234.4......................................................................
......122..1.4.......................................................................
......113.44341......................................................................
.....4.4.4.44........................................................................
.......31..321.......................................................................
.......14...1........................................................................
.......41..412.......................................................................
.......41....4.......................................................................
.......4...11........................................................................
...............4.42..4.34............................................................
...............3.22141.34............................................................
...............2.4.232131............................................................
...............1211.43443............................................................
................314311.31............................................................
...............234.31114.............................................................
...............22221313.4............................................................
................334..4432............................................................
...............42.3.14434............................................................
...............2141221343............................................................
.........................441.31.2.2.322.14222........................................
..........................342434.14...2433134........................................
.........................4..2.1.4243412..1443........................................
.........................2.4424.4112.41..4.12........................................
.........................41424.4.4.231.421231........................................
.........................1.2213.411.23.3.3421........................................
.........................221.124.1.3414233.24........................................
.........................314231323441..213232........................................
.........................121.1.44..44.4134113........................................
.........................133332134124.2112341........................................
.........................12.14242241...341.24........................................
.........................3421124.3.14221.4.42........................................
.........................244.1121213434331223........................................
.........................41.211.213332.322111........................................
..........................3.2.344.1.1341.2132........................................
.........................21.442411..3344.13.4........................................
.........................4.13.4.2.341.1323242........................................
.........................121324.33112.3342131........................................
.........................32234113222422.3123.........................................
..........................313.1.341224.3.313.........................................
.............................................441.2.3333223233........................
.............................................3.1.324..4314134........................
.............................................1212.224111444.2........................
.............................................1.242..344111313........................
.............................................242424232322334.........................
.............................................4.12..24321123..........................
...............................................243.4.3.2432.3........................
.............................................12.243332132244.........................
.............................................21123244241111..........................
.............................................3342.3..3.43.342........................
.............................................1144313243..4231........................
..............................................241232123.44..4........................
..............................................2.1.4424.412.43........................
.............................................2311411.121121..........................
.............................................4231121.3.2313..........................
.............................................1.4.33434213.133........................
.............................................4132.33.1142.4312441224.................
..............................................314442.3412143.243..313................
.............................................11444...21331313444332.2................
.............................................44.211323.31.413.41443.4................
..............................................4.311.44.44.214124.2413................
..............................................3..33424..333.332214.3.................
..................................................................4..................
.............................................422.42244224.2414342.334................
.............................................341.3424.2121423124..2.3343.4441114223.4
.............................................34.23313.11333.2.22221132.233.2.1411.313
.............................................2334.43422..424.34.1.43434214444434.1122
.............................................3443..2.34443.24.3.2.3344.2324134434341.
.............................................43324314....233324.3.244.243.3.31443.13.
.............................................434232213.24124232.34.2.441424..1..12..4
.............................................3234111.13.2.4143313.122134441141..21314
.............................................31443.14.32232431341444.343.34314.11.2.1
.............................................4434313421.4.31414.442113.3411.21311433.
..............................................212412.422422.213.312.43342.14.14.3314.
.............................................122.222342441113.23.3...3.114143441.1131
.............................................2234322241241..2.341.32121233431.22222..
...............................................44112.432.432121322233..42414.122.2234
..............................................4314123441241.2433.211..44.1.3...43223.
.............................................4434122231321.32..32333.22213211141.2323
..............................................43..1442..1..223322.23113.3323...4.2222
```

The order 6 matrix s

```
2.123................................................................................
32.2.................................................................................
24.1.................................................................................
12...................................................................................
.1.21................................................................................
.....1442.1.313......................................................................
.....2.2.12.223......................................................................
.......1.213412......................................................................
.....1..14143.2......................................................................
.....1433.2423.......................................................................
.....34.413442.......................................................................
.....3.3.434232......................................................................
.....2.141222.2......................................................................
.....2..11..334......................................................................
.....1441.44.3.......................................................................
...............2.242.23.1............................................................
...............143231.444............................................................
...............4.4.111211............................................................
.................2332.2.1............................................................
...............2312233144............................................................
...............1.13433313............................................................
...............123244.3.3............................................................
...............1243.2................................................................
...............33.2444122............................................................
...............311431433.............................................................
...........................3224.131...432.343........................................
.........................1.44344122133331432.........................................
.........................2222..4..222134.33.2........................................
.........................4.3142233434114421.4........................................
.........................42123434443434144433........................................
.........................21.31.1.4.2.34442113........................................
.........................32443.2134..431422..........................................
.........................423.222443331334333.........................................
............................23.2.3214.33.11.4........................................
.........................123.4442443141..12.1........................................
.........................43222.44223131421123........................................
.........................2424442..34.14143144........................................
.........................11443214144231..2.22........................................
..........................14113.44321..1.32.4........................................
.........................3413211.332221114121........................................
.........................14.442232...11334222........................................
.........................1.143.43.4222221.3.2........................................
..........................421142434433.12.2..........................................
..........................4.11.23433.1.223112........................................
..........................3.2323334..3.3.2443........................................
.............................................4.24333.412.2314........................
.............................................223.4224.31.3331........................
.............................................4.4423.442331422........................
.............................................42431.313.44.313........................
.............................................3223...3212.1223........................
..............................................321.23421..423.........................
.............................................1212.143..142421........................
..............................................11121214324433.........................
.............................................4124121132444...........................
.............................................3..1.1.434424323........................
.............................................1343134311..2241........................
..............................................24.434.42.34111........................
.............................................4..42342114323.3........................
.............................................2.334233.1442424........................
.............................................31.4334221431234........................
..............................................21..431212..422........................
..............................................42313421.1.131.321.31.1................
.............................................3324.234.24434.214.2.113................
.............................................43243244212332333.2.4331................
.............................................44.22221.241133323211441................
..............................................4.411.41.11243.4.31142.................
.............................................2312.414.443.111.224134.................
..............................................41.413334322132.44.4424................
..............................................221111.144.21.221.1.2.2................
.............................................122234.431324.2211.4132312144314.232132.
.............................................22.341113441123.1123413423..3131443132.3
.............................................213332...143.24322333.4333...12.23314123
.............................................4.142133411423411.31...13331.23212.21214
.............................................14.44244....4.21244122331432443.43.41.34
..............................................41..34.341423121.42.3.2.121112424333423
.............................................42.132..2214.21....244.1..211.2.3443.231
.............................................42.224224..222212222.24.411.1.324.3.4..1
..............................................444.....3242331.1324..12.4.3..3.3422.4.
.............................................43.21314223.44231.31333412243331.31341..
.............................................424.222321134413.11.22213141344423432.24
.............................................44.442423323.44224243...331144..12.4444.
.............................................4441..14..4212312...333.441.134..221..31
.............................................1..414313434.114242433.4332412112341424.
...............................................412.4.1123.241222142244.2.3323.421.4.1
.............................................13..233.2.34.324332142.2144222..2.341..4
```

The order 5 matrix f

```
2..12331323...2..........24123.3244443.44.243........................................
.24..4.122423.3..........423.42214122321.34.4........................................
.........................341.34423244142.13.3........................................
..........................211.312144143231431........................................
.........................3.1.21423443.1.22124........................................
2311224.3.241.2..............................4223.231121232.334231324431442..42244432
.............................................2344113212.41114441433.221.3421424113313
.............................................24323.422111421112323332..4221132344.122
..............................................24.41423.3..123134332323..3..33331121.4
13131.2112.41.4..............................2.34..3132.22222243.234.3222441122222111
..............................................3413.3342.1214423432.31.1432114.314.432
.............................................41411...443..41..342141.243.1443...4234.
.............................................22.1113111.422431131332..11.342333.4213.
.............................................44133111.32221.22.14321....123314312111.
.............................................33234431.1122224.44231.3131..311111123.4
...............2244342241.2314221112422.31121.2344.2.34..3.41121.4.4.................
...............2.422.24.413221.42131132412343223133.2.441413.143.31.1................
...............24..441.2..2143231.342212144.4331214..24.2242441243222................
...............121.3.432413.12414211.33..14.1224232..41442122.2212432................
...............1.121331114311..443.324113.2.1..4233.23334..4.43441422................
...............342344.344344.31.33424..3331443221.34.2211441.22.411.3................
...............3.4444..313.2222344213424123224.4441114343.2..23144342................
...............13232.3313.34443333232.1.1341..43.334142.12.11...21142................
...............4.3.41233434414213.11.442.4.123.212424.42414.4412244.2................
...............1.323134322211..334442134.2.4.41124.2.21241221.44234.3................
..44............11.211.23.124424.31323...34434..12334312131.3121444..343224134.2.2.11
...31..........222331342241132422122.213441.1144323.23132234142..232.111431.34.442.42
...42..........34..11314...12..3.32.24432.1.122.141332123241224443.41442.332...123333
..334..........334341341212.11323.24121..3141212.33212..43114441..1221324..21..2431.3
..22...........23.12423.33213224421.4..33.4..3324.4.132..2.212.2121412.342.1433223231
..421...........2323.123.111.14213322222.123.334113.421111.2432222.3341.1.32.4.423133
..2.3...........2.3.44.32.33314..11.424333223344414..423412342.13242.133224.24.111.2.
..234..........142112123..3..2424.3.232242.41.2..2.13.3333.1.1131233...33223222324.24
..1............1.11.4314124444.23443.2.344423232341311414...1.12233.44231.32333344.34
...32...........4.3.22224.4..4.32123.22232121.122.2112214223..433414342444312442..321
..314..........122131112..2.124..42.13.113123.212233.422.141.14243.33.1133.22.4314324
...3............2341.411.12.14424413232411.41.42131122311.424324.44..1212321412421344
..123..........4222..33.11433111432...12322.232.212212.4122.4.11.432..3342.3.1.24213.
..422..........142.322232323422...1311.14.33241..3223.1131223431232.2.43343.322144143
..44...........34333.3.21..3312341.3..111443.411123.2.1.242.124.4423.423.13123.4.4342
..134..........1314.21244321214241.431.24424122.312342...1.11.1.2234113.2332113.13414
..234..........23.221443332212243324.2.241.221.131..4442.21.12242143..3342134334.1222
..41...........11.111.43133333133231232431243314.1323222433...3232434..21322.41.12311
..22...........322..333.11212113224....1231..4.234423.341443244211.34144312.14113.414
...2............23241..211442243421121441.2112.3.4.1322.411341.13.1.3.21312312421..22
.......41.44..2..........243..42.13131..44....14.3211.232...11334.24134333441221341.4
......434.432.2..............4311.1.4432142.2112.14443113141.3..44.422.1.43.132411224
......331.3.42...........11.3112133442..2.241.144414414423.34442411113424332.22223123
......31..142.2..........332122.2224.114244423.214143234.222142413.32.4.413424....213
......314..4311..........2314..2.243.34.1323432122123341.1234332423144.34432123114.43
......1.1.12143..........4232144.3.2122432..3.3.214..4421412224.31412.14232424233.34.
......131.3.231..........4.2221..11233.13413142..431113433.3.32..423423413.3244324212
......324.412.4..........422343.1.1232243223442.41.334122341.2...2...3131...324.3.124
.......22..421...........2.42....143311333..12422.2111.233444.42.112421214.2.33322142
......11..24321..........414..31.333.4.232442344411241412111243412232331323.1.3443.2.
......231.44443..........21..4.3443222.3.123.122213431..4214..4.2132224.431334.43..31
......3.3.1.224...........443.133243441.414214223.3213.1333331.143233223..314113.1.1.
......112.44323..........441.3344331413224.4.43421331.13.11233.3.23233112122124331..4
.......22.3.132..........224.1332234413.333.1.414.21..414...23421412121.112.2.1413323
......411.422.3..........33432.1.4.243.4213422.43423422111243.3.32424323.1132111.1133
......112..3.43..........44111.1411.22.42.2.2122132431122134432413..3.222214..24413.1
......432.133212141324.2141.1214134.1422334.1413..4243212114112344121.3.142.221431424
......214.233411.2312423333.213134442..212112412.3..44113343413213143232412331242.22.
......322.332341.32234.41242341423.4413..1444..2331242..12.4324322.41.1434.32333.1214
......23..3341.3441.11413.141.223.21.333.122..1431..2322.42.31243233.1424223.224.1231
......143.23.1234121.123432424.32132321.3442321411.14334314314111112412312333.2311244
......3.4.1.2.114123.3.1.413441131.4......233..13424243214.4.1.2123342324143312344242
......2.1.4122422.214.244.323.3411.11.21.124..12421132334.23442143413..........32222.
......442.2.42313.311214231.123414.31412..213114441..3.2.242.32222424.333.41.4424.1.3
......422...3...1.4..213.14112.43112.33121.142124.1213333.2.421.2324..441...141331113
.......42.11.24232.44142.3.1.11421424434431123322232312323434.13.2.1222242111.24.2.1.
.......44.2434.34241312424324.34441311.3242.234132434443221323.11132333.44434121312.3
.......21..111..32.1432.12422233143421313..31211332314333423.14.222.32334244244314.14
......3...4431.3324...1324141...22342.3213243.4.23.21323.114141.414222.23123.4113..24
......323.2.214.342.234221.324.32223141324.333.433311123..11.44341..24214424343414.21
......11..14434.11424142322..43..2334314.4422224.314121.4334.224.33132.414422..41..22
......214.2122432111.4114.441..212121111333432.244431222.4324224.4314.3324423444144.3
.......12..223.1322122113.41.2224231.4.4112.122113....41.3.2341124..321.341333322...3
......2...44.4412.432143233143341344.21..22.42..211433.1.3.2313431423223121.211224322
......312.41244313424312..1233113324124.4334341311..2242.2124.3233221411.43113321321.
......313.3...422211..21.332.3....14233.144.313...2414..322242434212241..23..2442..42
......23....44124.44322143.311221234.133212413.423411.341.31.44321.223.241.4341244134
......322.....33.43222323134322313.3223311.3421222.4313.344.41214.3243.31133.32.43131
......122.224.22.2.1.314..331.23141.3.41.34.1324422.24..11.33114.32.41.3.34.1..22444.
......14..4313244422..42.24.3...2133.11.1111.32.4.1..31.34312.3142431332113313.123112
```

The order 3 matrix t_2

```
1..31....................43241.1433331.33.431........................................
.12......................341.34423244142.13.3........................................
.........................132.13341433234.21.1........................................
..........................422.124233231412312........................................
.........................1.2.42341331.2.44243........................................
.....12.4.123.1..............................333.444....343413221.342..2.4.412221.133
.............................................21.441.3.223244444222.3411.4.1244212.242
.............................................3..443..41321231..2424.224..124223.212..
.............................................1333331322.112343.1414331.3244.23422..23
......1331.23.2..............................32.3.11.1.112.43.23333424.311.4.43413433
..............................................3232.34143..22..212443.213422.44.314211
.............................................13.3442.1.1413331.21.1132.42.31.23224211
.............................................12.2.21133233112.412.33234.441.34241.324
.............................................34131244121.332.22432.324.33212321323..4
.............................................433.132.....2..3.12431414..411..31443.21
...............3314442231.3241334443133.24434.3211.3.21..2.14434.1.1.................
...............3.34341.2442334.13424423143212332422.3.114142.412.24.4................
...............232213..34.3412324.213343411.1224341..31.3313114312333................
...............3.4441323442.43141344.22..41.4331323..14113433.3343123................
................31.2.33141244..112.231442.3.4..1322.32221..1.12114133................
.................11312422211.24.22131..2224112334.21.3344114.33.144.2................
...............411..121.12.3333211342131432331.1114441212.3..32411213................
...............34132..244.21112222323.4.4214..12.221413.43.44...34413................
................21.42322121141342.44.113.1.432.343131.13141.1143311.3................
...............4.241..4123344..221113421.3.1.14431.3.34314334.11321.2................
..22............11.211.23244.1.31.3..13223.2.4...311443234.4131.43321412224422.423414
...43..........222331342244.3.223.44.1124113412342.42...413.23334341.342.1.2.2212.234
...21..........34..11314....2.243..414331122.4.341323221311124232.23.4243.44.24..234.
..442..........3343413412113412322223224434322121114..2.2142222232434321.434.1322.421
..11...........23.12423.3.34.3.132.2343343221422.311.1.4432443.314..2..2.313.3.44333.
..213...........2323.123.313114144422.1..2341424..2313.2411331.4.12342421231142434214
..1.4...........2.3.44.322...312.12.1141.24..3..33433.124.1432..22422.44334.22.334422
..142..........142112123.3311.3123.2.212.23.24221233323331..243112.2111444331213...21
..3............1.11.431414.2444433.213...13.34444121122.1132.11424.31144134331343433.
...41...........4.3.22224213.143.4.23222.123111.2124441.24.21132..2232322.33213.14333
..432..........122131112.1...1311.2.3143333142.2.241432121234.4.4224314.42.4413121334
...4............2341.411.222.32.31131142123232134.14.24132213.213..3134.2.2.442113214
..314..........4222..33.13.431412...13.4243441.334234244.1.1..321.41.332.23.2244.4212
..211..........142.322232.123231....2443223.3.33.2131..13.21322231..122.14.3322.3222.
..22...........34333.3.21..221141.1..4.222.12.424212.4323.214..111.331441.433.34.23.3
..342..........1314.212442.4331.343.44.32..42.24.4344.42.11.14.413412...4421.24132144
..142..........23.221443342212.41243123442441342233433111114443.24.32144.421331.223.41
..23...........11.111.4311341321.212141234333.4344.233...222222222.4331411333.3222133
..11...........322..333.133444.24.313.1112233111.2142.3341.4411122332244.441131131143
...1............23241..212311..2.1222434412.11.431.212212423434113.2.13211.2122144.1.
......113.32234..........3141.3114134.1..41.23334234.432..323...22.12.13123421.3413.3
.......32..324...........233412322.1321.423232..41.444.323.3.113.1.3.31.23412.1421.2.
......4.3.4222............4..13..222.4231413331.141242242223312442434.14...14.1412433
......434.2213...........1.22114441.14..41.24143.1123.2.21412.2..321.3422114413.34222
......22..14..3..........4.1..3.1.322244434214221.31121.231412.2322342..421.21431.324
......14..1443...........2223241234.333.4224.22.3323323324.232.3443..423332212.2...4.
......244.13431..........14..341.1223.3.4.4444314..234.14212.13.2.33.111332223314312.
......13..33432..........24232..44422..243124144.143211.442.12423.1434414.333.44243.3
.......31.4231...........323..4.41.12342224323432.444.41321..44..113..3112213.3334.14
......13...4233..........3443..23422433222213334..13343333.32.2143.12..2224.1411412.2
......443.32212..........32223....3..4.3.2112..34332444.214.2333124314334.43.21.2.14.
......43..323.3..........4231134.311.442.3443.4424..423212.242.4143231.22242132..2344
......1.1.4.314..........2..1321431..13.42244333242.4.224.14431113.24.11.1434.1432.13
.......21.1.41.............232343431323214413412322.3311314344.33.331144.3.4.4.311.44
......1.4.4.333..........3334.43333124313.1.24433221...11.14...31234142.3424411.41322
......321.4333............1212221.344.324.122...3214214..2.13.2.242.323411.333.23341.
......221.3341.3414231.341433224.212424114.42144..432.314.34..123.12211..244..1132.12
......24..44...4.3243132232.443.242142323141.224211123.421...1.4.3214.444.23.2212.3.4
......144.31.314.23321.142121.2334.1242443233.4134432.424.2441133422343231433213314.2
......223.411122114.44142..21.221..22..23.32.32.41.21.331214243..22123442.21131233234
......213.13.2321434.43214.2.311.24...4331224.2422212.4.....412.2411313.41.413..43243
......1.3.2213241432.2.4.24311144413.44234133.4322.34231441224144422..444.23.22.31414
......4.2.3122333.341.311....32...342.111.2144.421...313.111.14.2.412....4....2.41.4
.......43.1313242.244341342..4412231.411244322421244.1231.1.42.1.43.44.3221414141.244
.......23..3.14..4.1..34221124321432441.2.133.1.3.2332.43422..22.1.42433.232244343141
......2.1.3.221323.11413.33.221443444423..1.1.34.2.2.11.241.13...3.1133.4343444.21331
......133.1..4.213142431343.3.44.232..12434.143.4.4321.423423231.31.1.43243423142.432
......4...4.224.23.4123.4234343412.3.34.4234.41.24433212..311.344344232111..334414244
......3.3.213.12231...42312131333122312313323.344.2122..34113312141.2322422322.42132.
......21..4..21.213.321334221...1.43.1324443211132..224244314123.122412..211443224.24
......223.22411.4413141321312323.34..34.4.3121..4.41413212314.441412113.33..1.3433.13
......234.4.42323444.1441312431.413..4.1.1141.1.1341.1...4313.31223411..44433411.1211
......413.1.44442334334424224.2.243211432.44242433432423.124.31212.313124121.24212..2
......322..223443.123412333..1..32...443324443313341433342324.2432331323231132.332122
......432.4.212242131243.4143444.2.44434.1132.4.422442..2.1..22.314344132131142.21321
......412.411.433344..34...333.242244.11.2223231443.1.4214144322112121114121342.1.122
......412.1134.31.11233411.41.333444141.22234.432423.4222.2.3111..123114.1.3.42.12434
......423..4..12.123332323..34411.3424113143.234413.12332.44344.333.44243213143244224
......4.2.431113.3.4.241.3.4224.234323.23444.43..443122.134.32.22332114..2..14.23.111
......14..3.32.11133..13...4243243.41..1.2.232..21.22.34223413124.231.334.4..122.34.2
```

The order 3 matrix t_3

References

[1] M. Aschbacher, The existence of J_3 and its embeddings in E_6, Geom. Dedicata 35 (1990), 143–154.

[2] B. Baumeister, A computer-free construction of the third group of Janko, J. Algebra 192 (1997), 780–809.

[3] W. Bosma, J. Cannon, MAGMA Handbook, Sydney, 1996.

[4] G. Butler, Fundamental algorithms for permutation groups, Lecture Notes in Comput. Sci. 559, Springer-Verlag, Heidelberg, 1991.

[5] J. H. Conway, R. T. Curtis, S. P. Norton, R. A. Parker, R. A. Wilson, Atlas of finite groups, Clarendon Press, Oxford, 1985.

[6] W. Feit, The representation theory of finite groups, North-Holland, Amsterdam, 1982.

[7] D. Frohardt, A trilinear form for the third Janko group, J. Algebra 83 (1983), 349–379.

[8] M. Hall, D. B. Wales, The simple group of order 604800, J. Algebra 9 (1968), 417–450.

[9] G. Higman, J. McKay, On Janko's simple group of order 50232960, Bull. London Math. Soc. 1 (1969), 89–94.

[10] B. Huppert, Endliche Gruppen I, Springer-Verlag, Heidelberg, 1967.

[11] I. M. Isaacs, Character theory of finite groups, Academic Press, New York, 1976.

[12] Z. Janko, Some new simple groups of finite order, Sympos. Math. 1 (1968), 25–65.

[13] M. Kratzer, Natural and uniform existence proofs for Janko's sporadic groups J_2 and J_3, Arch. Math., to appear.

[14] J. H. Lindsey II., A new lattice for the Hall-Janko group, Proc. Amer. Math. Soc. 103 (1988), 703–709.

[15] G. O. Michler, On the construction of the finite simple groups with a given centralizer of a 2-central involution, J. Algebra, to appear.

[16] J. Tits, Quaternions over $\mathbb{Q}(\sqrt{5})$, Leech's lattice and the sporadic group of Hall-Janko, J. Algebra 63 (1980), 56–75.

[17] R. Weiss, A characterization and another construction of Janko's group J_3, Trans. Amer. Math. Soc. 298 (1986), 621–633.

[18] R. Weiss, A geometric construction of Janko's group J_3, Math. Z. 179 (1982), 91–95.

Institut für Experimentelle Mathematik der Universität GH Essen
Ellernstraße 29
D-45326 Essen, Germany
mocenigo@exp-math.uni-essen.de (R. M. Avanzi)
kratzer@exp-math.uni-essen.de (M. Kratzer)
archiv@exp-math.uni-essen.de (G. O. Michler)

Recognizing simplicity of black-box groups and the frequency of p-singular elements in affine groups

László Babai and Aner Shalev***

Abstract. We consider the asymptotic complexity of manipulating matrix groups over finite fields. The question is, given a matrix group G by a list of generators, what can we say in polynomial time about the structure of G?

While considerable progress has been made recently in identifying the nonabelian composition factors of a matrix group, the fundamental question of recognizing the simplicity of a non-abelian matrix group remains open.

For the purposes of polynomial time computation, we reduce this problem to the following "affine case:"

> Let A be an elementary abelian p-group which is a non-central minimal normal subgroup of a finite group G. Assume further that the quotient G/A is a finite simple group S of Lie type of characteristic p.
>
> The algorithmic goal is to distinguish the simple group S from the non-simple group G. Both S and G are given as "black-box groups of characteristic p."

The reduction to this basic problem involves a large number of recent techniques and results which we review along the way.

We address the affine case for the groups $S = \mathrm{PSL}(2, q)$ where $q = p^f$, p prime. We show that $\mathrm{PSL}(2, p)$ can be recognized in Monte Carlo polynomial time among all black-box groups of known characteristic. The situation for $f \geq 2$ seems much harder; we exhibit challenging open cases for every $f \geq 2$ when q is large.

Along with $S = \mathrm{PSL}(2, p)$, the positive result holds for all simple groups S of Lie type of characteristic p with the property that in every nontrivial S-module in characteristic p, every element of S has a fixed point. We call these groups "unisingular." Very recently, Guralnick, Saxl, and Tiep classified these groups. They found that a large number of classes of simple groups are unisingular, including certain classes of linear and unitary groups, all symplectic groups over an odd prime field, all orthogonal groups of odd degree over an odd prime field, many orthogonal groups of even degree, and a number of classes of exceptional groups.

1. Introduction

Let G be a finite group given by a list of generators. Our goal is to design efficient algorithms to obtain structural information about G. In this paper we take "efficient" to

*Partially supported by NSF grant CCR-9732205

**Partially supported by the Israel Science Foundation

mean "Monte Carlo polynomial time." A Monte Carlo algorithm uses randomization and produces a correct output with probability $\geq 1-\epsilon$ where $\epsilon > 0$ is specified by the user. The running time of a polynomial time Monte Carlo algorithm is $O(|\log \epsilon| \cdot n^c)$ where c is a constant and n is the bit-length of the input. Throughout the paper we shall state the results under the tacit assumption that ϵ is a constant; multiplying by $|\log \epsilon|$ will adjust the timing estimates for variable ϵ.

1.1. Black-box groups. Typically G will be a quotient of a matrix group over a finite field, but the algorithms considered here work in the more general context of "black-box groups." The concept of black-box groups, introduced in **[BaS]** (1984), is critical to our work.

In a *black-box group*, a "black box" performs the group operations on codewords representing the group elements (multiplication, inverting, deciding whether a given string represents the identity.) Each codeword is a (0, 1)-string of the same length n, called the *encoding length.* Not every string encodes a group element, and several strings may encode the same element. A black-box group is "given" if in addition to the oracles for group operations, a list of generators is given. If the input is a group with k generators and the encoding length is n then the length of the input is kn and correspondingly a polynomial-time algorithm is allowed to run for $(kn)^{O(1)}$ steps.

The rudiments of polynomial-time computation in black-box groups have been worked out in **[BaCFLS]**. The results include Monte Carlo polynomial-time algorithms for normal closure, the commutator subgroup, deciding solvability, nilpotence and deciding whether or not G is a p-group.

A detailed polynomial-time theory of black-box groups was given in **[BaB]**, our standard reference. For the precise statement of the main results, the concept of "black-box groups of characteristic p" **[BaB]** is required; we give this definition in Section 4.1. Let it suffice to mention here that this concept includes all matrix groups over a field of characteristic p as well as quotients of such groups by recognizable normal subgroups. Membership testing in the normal subgroup is needed in order to decide what strings represent the identity in the quotient.

1.2. Random sampling. Most of the algorithms for matrix groups and black-box groups assume that we can choose uniformly distributed random elements from the group G. This assumption is not justified, but "nearly uniform" random sampling can be achieved (every element has $(1 \pm \epsilon)/|G|$ probability to be chosen). For the purposes of the polynomial-time theory, it is shown in **[Ba1]** that nearly uniform random sampling can be performed in Monte Carlo polynomial time, even with exponentially small ϵ. For practical purposes, variants of the "product replacement" heuristic seem to perform very well **[CLMNO]**. For concepts and caveats regarding randomized algorithms for groups we refer to **[Ba2]**.

In this paper, when using the term "pick an element at random from the (nonempty, finite) set Ω," we mean (nearly) uniform selection from Ω. For the analysis we think of the sampling as uniform, but "nearly uniform" would work just as well.

1.3. Center, outer automorphisms. A recent result from **[BaPS]** states that for every prime r and every nonabelian finite simple group S, a significant fraction of the elements of S is r-regular (of order not divisible by r) (see Theorem 4.9). This result was used in **[BaB]** to split a direct product of simple groups into its factors, a key step toward finding the nonabelian composition factors. In this paper, we give two additional applications: finding the center in a central extension of a simple group (Section 4.7) and analysing an algorithm by C. Leedham-Green to test membership in a simple normal subgroup (Section 4.4). The most important special case of the latter situation is when $S \leq G \leq \mathrm{Aut}(S)$ for some simple group S. In this case S can be constructed as the stable derivative (last member of the derived chain) of G (since, by Schreier's hypothesis, G/S is solvable); we are now also able to test elements of G for membership in S. In fact we show that any simple normal subgroup with a known set of generators is recognizable (Theorem 4.6).

1.4. Nonabelian composition factors. The main result of **[BaB]** is a polynomial-time algorithm to find a black-box representation of characteristic p for all non-abelian composition factors of a black-box group of characteristic p. Combining this with recent results on the statistical recognition of finite simple groups (**[AltB]**, **[KS2]**, **[BaKPS]**), we are now able to name the nonabelian composition factors:

Theorem 1.1 (**[BaB, AltB, KS2, BaKPS]**). *Given a black-box group G of known characteristic, the standard names of all nonabelian composition factors of G can be computed in Monte Carlo polynomial time.*

In particular, this result applies to all matrix groups over finite fields. (See Section 4.3.)

2. Recognizing simplicity: the main results

In spite of all this progress, the basic question, to *recognize the simplicity of a matrix group (or a black-box group of given characteristic)* remains open. This is the main question addressed in this paper.

Since by Theorem 1.1, every simple group can be recognized *among all simple groups*, the recognition of *simplicity* is equivalent recognizing each simple group (among all groups). We give a more precise formulation to this statement.

Theorem 2.1. *The following four computational tasks are equivalent in Monte Carlo polynomial time. In each case, a black box group G of known characteristic p is given.*

(a) *Decide whether or not G is simple ("recognizing simplicity").*

(b) *Given the standard name of a simple group S, decide whether or not $G \cong S$ ("recognizing the simple group S").*

(c) *Same as* (b) *except that S is restricted to being a simple group of Lie type of charcateristic p.*

(d) *("Affine case"). Assume that S is a simple group Lie type of characteristic p and G is known to have a normal subgroup A such that $G/A \cong S$ and either $A = 1$ or A is a non-central minimal normal subgroup of G and A an elementary abelian p-group. Decide whether or not $A = 1$.*

The equivalence of (a) and (b) follows from Theorem 1.1. It is also clear that (d) reduces to (c) and (c) reduces to (b), each being a subproblem of the preceding one.

The principal result of this paper is a reduction of (b) (general case) to (d) ("affine case"). Full details of this reduction are given in Section 4.8.

The purpose of this reduction is twofold. First, it helps in solving the problem for certain classes of simple groups.

In particular, we show, using a monumental recent work by Kantor and Seress **[KS1]**, how to recognize a simple group of Lie type (i. e., to distinguish such a group from *all* other groups, not only from other simple groups), assuming the characteristic is known and the field of definition is "tiny" (Section 4.6).

We also exhibit the first classes of Lie-type simple groups over *large fields* which can be recognized. We state the result after a definition.

Definition 2.2. Let G be a finite group. We say that G is *unisingular in characteristic p* if every nontrivial G-module M of characteristic p has the property that every element of G has a nonzero fixed point in its action on M.

In other words, G is unisingular in characteristic p if 1 is an eigenvalue of the action of every element of G on any nontrivial module of characteristic p.

We shall be particularly interested in the case when G is simple of Lie type of characteristic p. In such a case we shall simply call G "unisingular," tacitly referring to the same characteristic as the characteristic of the field of definition of G.

Theorem 2.3. *If S is a unisingular simple group of Lie type of characteristic p, then S can be recognized in Monte Carlo polynomial time among all black box groups of characteristic p.*

The definition of "black box groups of characteristic p" will be given in Section 4.1. We emphasize that Theorem 2.3 refers to recognition *among all groups*, not only among simple groups. For further discussion of the concept of recognition we refer to Section 4.2.

We complete the proof of Theorem 2.3 in Section 4.9 (Theorem 4.19).

We actually prove that the result holds more generally for those S for which every nontrivial S-module M of characteristic p has the property that a *non-negligible fraction* of the elements has a nonzero fixed point in its action on M. The running time will be inversely proportional to this "non-negligible fraction."

We prove that the groups $\mathrm{PSL}(2, p)$ are unisingular (Section 3.5), providing the first examples of simple groups over large fields which we can recognize. The power of Theorem 2.3 has very recently been greatly extended by Guralnick, Saxl and Tiep, who, answering the authors' question, classified all unisingular groups. We state their result **[GST]**.

Theorem 2.4 (Guralnick, Saxl, Tiep). *Let S be a finite simple group of Lie type of characteristic p, defined over the field $GF(q)$, where $q = p^f$. Then S is unisingular if and only if S is one of the following:*

(a) $\mathrm{PSL}_n(p)$ *with* $n | p - 1$;

(b) $\mathrm{PSU}_n(p)$ *with* $n | p + 1$;

(c) $\mathrm{PSp}_{2n}(p)$ *with p odd;*

(d) $\Omega_{2n+1}(p)$ *with p odd;*

(c) $\mathrm{P}\Omega^{\epsilon}_{2n}(p)$ *with* $\epsilon = \pm$, *p odd, and* $\epsilon = (-1)^{n(p-1)/2}$;

(f) ${}^2\mathrm{G}_2(q)$, $\mathrm{F}_4(q)$, ${}^2\mathrm{F}_4(q)$, $\mathrm{E}_8(q)$, *q arbitrary;*

(g) $\mathrm{G}_2(q)$ *with* $q = p^f$ *and* $p \geq 3$;

(h) $\mathrm{E}_6(p)$ *with* $3 | (p - 1)$;

(i) ${}^2\mathrm{E}_6(p)$ *with* $3 | (p + 1)$;

(j) $\mathrm{E}_7(p)$ *with p odd.*

As a consequence, *we can recognize all these groups* in Monte Carlo polynomial time (among black box groups of characteristic p).

The second purpose of the reduction to the affine case was that the reduction allowed us to pinpoint where the main difficulty may lie. As an example, we offer as a challenge to distinguish $\mathrm{PSL}(2, p^2)$ from certain affine extension of $\mathrm{PSL}(2, p^2)$ by an elementary abelian group of order p^8 (Section 3.6).

3. The proportion of p-singular elements in affine extensions

In this section we describe general observations regarding the affine case, providing the basis for our analysis of the recognition of unisingular groups. We also show where the same idea fails for $\mathrm{PSL}(2, p^f)$, $f \geq 2$.

3.1. The proportion of p-singular elements. An element $g \in G$ is called *p-singular* if its order is a multiple of p; otherwise g is *p-regular.*

Let $\rho_p(G)$ denote the proportion of p-singular elements in a finite group G.

A good supply of p-singular elements is very helpful in recognizing finite simple groups of Lie type of characteristic p (cf. 4.14). The trouble is, if the field of definition is large, such elements become very scarce and difficult to find.

For instance, for $\mathrm{PSL}(2, q)$, we have the following proportion.

Proposition 3.1. *$\rho_p(\mathrm{PSL}(2, q)) = 2/q$ for odd q and $\rho_p(\mathrm{PSL}(2, q)) = 1/q$ for even q.* □

Recent work by Á. Bereczky **[Ber]** on linear groups and on certain classes of exceptional groups suggested that $\rho_p(S) \leq c/q$ may hold for all finite simple groups S of Lie type of characteristic p, defined over $\mathbb{F}_q$ (with a small absolute constant c). This was confirmed by independent work by Guralnick and Lübeck **[GL]**.

Theorem 3.2 (Guralnick–Lübeck). *Let S be a finite simple group of Lie type of characteristic p defined over the field $\mathbb{F}_q$ ($q = p^f$). Then*

$$\rho'_p(S) < \frac{3}{q-1} + \frac{2}{(q-1)^2}, \tag{1}$$

where $\rho'_p(S)$ is the proportion of elements which commute with at least one element of order p.

Note that $\rho_p(S) < \rho'_p(S)$.

3.2. A naive algorithm for the affine case. It is evident that for $A \triangleleft G$ we have

$$\rho_p(G) \geq \rho_p(G/A).$$

We wish to estimate the quantity $\rho_p(G) - \rho_p(G/A)$.

If this quantity is "non-negligible" then a naive Monte Carlo algorithm will distinguish G from G/A: by sampling elements and checking for p-singularity, we statistically estimate ρ_p. The following statement gives the sample size for statistically significant estimation.

Proposition 3.3. (a) *We can statistically distinguish G from G/A by sampling*

$$\frac{2|\ln \epsilon|}{\left(\rho_p(G) - \rho_p(G/A)\right)^2}$$

elements, where ϵ is the margin of error.

(b) *If $\rho_p(G) \geq 1 - \delta$ and $\rho_p(G/A) \leq \delta$ where $\delta < 1/4$ then we can statistically distinguish G from G/A by sampling*

$$\frac{2|\ln \epsilon|}{|\ln(4\delta)|}$$

random elements. In other words, by sampling m random elements, we reduce the probability of error in our statistical inference below $(4\delta)^{m/2}$.

Proof. (a) Suppose the X_i are independent random $(0, 1)$-variables with $\Pr(X_i = 1) = \rho$ and $\Pr(X_i = 0) = 1 - \rho$. Then for $\vartheta > 0$, the probability that $\rho - (1/m)\sum_{i=1}^{m} X_i \geq \vartheta$ is less than $\exp(-2m\vartheta^2)$ and the same holds for the probability that $\rho - (1/m)\sum_{i=1}^{m} X_i \leq -\vartheta$. This follows from a Chernoff bound stated as Theorem A4 in **[AloS**, p. 235**]**.

We sample m elements from the given group; X_i will be the indicator whether or not the i-th element selected is p-singular. Let $\mu = (\rho_p(G) + \rho_p(G/A))/2$. If the number of p-singular elements selected is less than μm then we decide that the group is G/A, otherwise we declare it to be G. Setting $\vartheta = (\rho_p(G) - \rho_p(G/A))/2$, we see by Chernoff's bound that the probability that we make an erroneous decision is less than $\exp(-2m\vartheta^2)$ regardless of which of the two groups we are sampling.

This quantity is $\leq \epsilon$ for the stated value of m.

(b) This part is elementary. Suppose the X_i are independent random $(0, 1)$-variables with $\Pr(X_i = 1) \leq \delta$ and $\Pr(X_i = 0) = 1 - \Pr(X_i = 1)$. Then the probability that $(1/m)\sum_{i=1}^{m} X_i \geq 1/2$ is less than

$$\binom{m}{m/2}\delta^{m/2} < (4\delta)^{m/2}.$$

We again sample m elements from the given group; Y_i will be the indicator whether or not the i-th element selected is p-singular. If the number of p-singular elements selected is less than $m/2$ then we decide that the group is G/A, else it is G. Setting $Y_i = X_i$ if the group is G/a and $Y_i = 1 - X_i$ if the group is G we see by the above estimate that the probability that we make an erroneous decision is less than $(4\delta)^{m/2}$.

This quantity is $\leq \epsilon$ for the stated value of m. □

We shall show that if $S = \mathrm{PSL}(2, p)$ then $\rho_p(G) - \rho_p(S) > 1 - 3/p$ regardless of the extension. Therefore $\mathrm{PSL}(2, p)$ can be efficiently distinguished from its affine extensions by sampling $O(|\log \epsilon|/\log p)$ elements (Section 3.5). More precisely, by sampling m elements we can distinguish these two groups with error probability less than $(8/p)^{m/2}$ (using $\delta = 2/p$ in part (b) of Proposition 3.3, cf. (Section 3.5)).

On the other hand, we find that for $S = \mathrm{PSL}(2, p^f)$, $f \geq 2$, there exists an affine extension G with $|A| = p^{4f}$ such that $\rho_p(G) - \rho_p(S) \leq p^{1-f}$. The challenge is to distinguish G and S (as black-box groups of characteristic p) efficiently (for large p) (Section 3.6).

3.3. Split extensions suffice. The following observation saves us a lot of headache by allowing us to ignore non-split extensions.

Proposition 3.4. *Let A be an abelian p-group, S a finite group, and $1 \to A \to G \to S \to 1$ be an extension. Then the number of p-singular elements in G depends only on the S-module A_S and not on the specific extension.*

Proof. It suffices to show that the number of *p-regular* elements of G depends only on A_S. Let $\varphi : G \to S$ denote the natural epimorphism. Clearly, if $g \in G$ is p-regular then so is $\varphi(g) \in S$. It suffices to prove that for every p-regular $s \in S$, the number of p-regular elements in the coset $\varphi^{-1}(s)$ is determined by A_S.

Let $g \in \varphi^{-1}(s)$ and let $H = \langle A, g \rangle$. Now $A \lhd H$ and $|H/A|$ is the order of s, which is relatively prime to the order of A. Therefore $1 \to A \to H \to \langle s \rangle \to 1$ is a split extension, uniquely determined by the action of s on A. In particular, the distribution of orders of elements in the coset Ag is determined by the action of s on A. □

Remark 3.5. The proof shows that for every p-regular element $s \in S$, the number of p-singular elements in $\varphi^{-1}(s)$ is determined by the action of s on A.

3.4. Unisingular action of p-regular elements. Let A be a group and s an automorphism of A. By a *fixed point* of s we mean a nonidentity element of A fixed by s. If such an element exists, we call s *unisingular*. In the case that A is a vector space and s a linear transformation, s is unisingular exactly if 1 is an eigenvalue of s, i. e., $s - 1$ is singular, hence the term.

The opposite of "unisingular" is "fixed-point-free."

Next we show that the proportion of p-regular elements in $\varphi^{-1}(s)$ critically depends on whether or not s is unisingular (in its action on A).

Proposition 3.6. *Let A be an abelian p-group, S a finite group, and $1 \to A \to G \to S \to 1$ be an extension. Let $s \in S$ be p-regular.*

(a) *If s is fixed-point-free (in its action on A) then all elements of $\varphi^{-1}(s)$ are p-regular. (This case is bad for the naive algorithm.)*

(b) *If s is unisingular (not fixed-point-free) then at most a $1/p$ fraction of the elements of $\varphi^{-1}(s)$ is p-regular. (Good case.)*

Proof. As in the proof of Proposition 3.4, let $\varphi : G \to S$ denote the natural epimorphism, let $g \in \varphi^{-1}(s)$ and let $H = \langle A, g\rangle$. As before, the extension $1 \to A \to H \to \langle s\rangle \to 1$ is split, so the elements of $\varphi^{-1}(s)$ can be written as affine transformations of the form $\alpha = \alpha(s, a) : x \mapsto sx + a$ $(a, x \in A)$ where sx denotes the image of $x \in A$ under the action of s (corresponding to conjugation by any member of the coset $\varphi^{-1}(s)$). (We write the operation in A additively.) Then $\alpha^i : x \mapsto s^i x + \beta(s, i)a$ where $\beta(s, i) = s^{i-1} + s^{i-2} + \cdots + s + 1$.

Let k be the order of s. Then $\alpha^k : x \mapsto x + \beta(s, k)a$ and therefore $\alpha^{kj} : x \mapsto x + j \cdot \beta(s, k)a$. It follows that the order of α is of the form kp^ℓ $(\ell \geq 0)$.

Now in case (a) we note that $\beta(s, k) = (s^k - 1)(s - 1)^{-1} = 0$ and therefore the order of α is k regardless of the choice of a.

In case (b), $\beta(s, k) \neq 0$ since $\beta(s, k)b = kb \neq 0$ for $b \in \ker(s - 1)$, $b \neq 0$. Therefore $\ker \beta(s, k)$ is a proper subgroup of A. Consequently the proportion of those $a \in A$ for which $\beta(s, k)a = 0$ is at most $1/p$. For all other $a \in A$ we have $\beta(s, k)a \neq 0$ and therefore the order of α is divisible by kp. □

Notation 3.7. Let S be a finite group and A an abelian p-group with an S-action. We use A_S to denote A as an S-module. Let $\lambda(A_S)$ denote the proportion in S of those p-regular elements which act as unisingular (not fixed-point-free) transformations on A.

The following is evident from the foregoing.

Corollary 3.8.

$$\lambda(A_S)(1 - 1/p) \leq \rho_p(G) - \rho_p(S) \leq \lambda(A_S).$$

Remark 3.9. Corollary 3.8 tells us that the naive algorithm succeeds if $\lambda(A_S)$ is not too small.

Observation 3.10. *A Lie-type simple group S of characteristic p is unisingular if and only if*

$$\lambda(A_S) = 1 - \rho_p(S)$$

holds for every nontrivial S-module A_S of characteristic p.

Combining this observation with the Guralnick–Lübeck bound on $\rho_p(S)$ (Theorem 3.2), we obtain the following key inequality on $\lambda(A_S)$ for unisingular groups:

Corollary 3.11. *If S is a unisingular simple group of Lie type characteristic p over $\mathbb{F}_q$ then*

$$\lambda(A_S) > 1 - \frac{3}{q-1} - \frac{2}{(q-1)^2} \tag{2}$$

holds for every nontrivial S-module A_S of characteristic p.

This inequality implies that the naive algorithm performs very well on unisingular groups over large fields. We remark that for small fields, the work of Kantor and Seress quoted in Section 4.6 solves the recognition problem in a stronger sense (constructive recognition). However, no polynomial-time methods have previously been known to handle the cases when the field of definition is large.

3.5. PSL(2, p) is unisingular. In this section we prove that the naive algorithm works for the case when $S = \mathrm{PSL}(2, p)$, $p \geq 5$. In fact we prove that in this case, for any S-module A_S, *all* elements of S are unisingular. In the terminology of Definition 2.2, this means that PSL(2, p) is a *unisingular group.*

Theorem 3.12. *The groups* PSL(2, p) *(p an odd prime) are unisingular.*

Proof. Let A be an S-module where $S = \mathrm{PSL}(2, p)$. We need to prove that every element of S is unisingular (in its action on A).

Without loss of generality we may assume that A_S is irreducible. (Indeed, restriction to an irreducible quotient module can only decrease the number of unisingular elements.)

The irreducible modules of SL(2, p^f) in characteristic p have been characterized by Brauer and Nesbitt **[BrN]**. For the case $k = 1$ we have the following modules M_t, $t = 0, \ldots, p-1$:

> M_t is the space of homogeneous polynomials of degree t in the two variables $\{x, y\}$. The action of SL(2, p) is defined by the linear substitutions corresponding to the elements of SL(2, p). (This is the t-th symmetric tensor power of the natural 2-dimensional module M_1.) Note that $\dim M_t = t + 1$.

The irreducible modules of PSL(2, p) are those irreducible modules of SL(2, p) on which the center acts trivially, i. e., M_{2t}, $t = 0, \ldots, (p-1)/2$.

Let M_t' be the module obtained from M_t by extending the field of definition ($\mathbb{F}_p$) to its algebraic closure $\mathcal{F}_p$. This does not affect unisingularity.

Let s be an element of SL(2, p) with eigenvalues λ and $1/\lambda$ (in $\mathcal{F}_p$).

If $\lambda = \pm 1$ then, after a suitable change of variables, we may assume that s acts on M_1' by $x \mapsto \pm x$. Therefore the element $x^{2t} \in M_{2t}'$ is a fixed point of s, hence s is unisingular.

If $\lambda \neq \pm 1$ then $\lambda \neq 1/\lambda$ and therefore s is diagonalizable over $\mathcal{F}_p$. After a suitable change of bases, s acts on M_1' by $x \mapsto \lambda x$ and $y \mapsto (1/\lambda) y$. Therefore the element $x^t y^t \in M_{2t}'$ is a fixed point of s, hence s is unisingular. □

Corollary 3.13. *Let A be an elementary abelian p-group,* $S = \mathrm{PSL}(2, p)$, $p \geq 5$, *and let* $1 \to A \to G \to S \to 1$ *be an extension. Then* $\rho_p(G) - \rho_p(S) > 1 - (3/p)$.

Proof. Indeed, by Theorem 3.12, $\lambda(A_S) = 1 - \rho_p(S) = 1 - 2/p$, and therefore by Corollary 3.8,

$$\rho(G) - \rho(S) \geq (1 - 2/p)(1 - 1/p) > 1 - 3/p. \qquad \square$$

3.6. Case PSL(2, p^f), $f \geq 2$: challenge. For unisingular groups over large fields, $\lambda(A_S)$ is close to 1 for all nontrivial S-modules of characteristic p (Corollary 3.11). Unfortunately nothing like this holds for $S = \mathrm{PSL}(2, p^f)$, $f \geq 2$.

Proposition 3.14. *Let $S = \mathrm{PSL}(2, q)$ where $q = p^f$, $f \geq 2$. Then there exists an irreducible S-module A_S of dimension 4 over $\mathbb{F}_q$ such that $\lambda(A_S) \leq p/q$.*

Proof. We are again guided by the Brauer–Nesbitt Theorem. In the general case ($q = p^f$), the theorem says that the irreducible modules in characteristic p for $\mathrm{SL}(2, q)$ are the tensor products of the form

$$V_0 \otimes V_1^{\sigma} \otimes \cdots \otimes V_{f-1}^{\sigma^{f-1}},$$

where σ is the Frobenius automorphism of $\mathbb{F}_q$ and the V_i are modules described in the proof of Theorem 3.12.

Let now $V_0 = V_1 = M_1$ be the natural module for $\mathrm{SL}(2, q)$ (2-dimensional over $\mathbb{F}_q$). We can represent the 4-dimensional module $A = V_0 \otimes V_1^{\sigma}$ as a space of polynomials in the variables x_0, y_0, x_1, y_1, spanned by the four polynomials $z_0 z_1$ where $z_i \in \{x_i, y_i\}$.

Let now s be a p-regular element of S with eigenvalues $\lambda^{\pm 1}$. Noting that s is diagonalizable over $\mathcal{F}_p$, we see, that, after a suitable change of bases, s acts on M_1' by $x \mapsto \lambda x$ and $y \mapsto (1/\lambda)y$. It follows that $z_0 z_1 \mapsto \lambda^{\pm 1 \pm p} z_0 z_1$, therefore the four eigenvalues of the s-action on A are $\lambda^{\pm 1 \pm p}$. If the action of s on A is unisingular, one of these eigenvalues must be 1, therefore either $\lambda^{p-1} = 1$ or $\lambda^{p+1} = 1$. Standard calculations show that fewer than a p/q fraction of the elements of $\mathrm{SL}(2, q)$ satisfy this condition.

Since $-id \in \mathrm{SL}(2, q)$ acts trivially on A, we can view A as a module for $\mathrm{PSL}(2, q)$. The proportion of unisingular elements is clearly the same. $\square$

Corollary 3.15. *Let $S = \mathrm{PSL}(2, q)$ where $q = p^f$, $f \geq 2$. Then S has an extension $1 \to A \to G \to S \to 1$ by the elementary abelian group A of order $|A| = q^4 = p^{4f}$ such that the proportion of p-regular elements in G is*

$$\rho(G) < (p + 2)/q. \qquad (3)$$

Proof. Combine Corollary 3.8 with the fact that $\rho(S) \leq 2/q$. $\square$

4. Simplicity

4.1. Black-box groups of characteristic p. Following **[BaB**, Sec. 8.2], we say that H is a *black-box group of characteristic p* if H is a black-box group of some encoding length n and H is isomorphic to a section (quotient of subgroup) of $\mathrm{GL}(d, p)$ where $d = \lfloor n/\log p \rfloor$. When we say that such a group H is *given*, we tacitly assume that p is known.

The significance of this concept is that it is tailor-made to treat black-box algorithms for matrix groups, retaining the exact amount of information required for polynomial-time analysis.

Among the many advantages of working with this concept, let us highlight one.

Algorithms for matrix groups and for black-box groups often assume that we are able to determine the order of group elements. This, however, is only true if we possess a superset of the prime divisors of the order of the group. For $\mathrm{GL}(n, p)$, finding such a superset is, alas, as difficult as factoring integers. (Cf. **[BaB**, Sec. 8].)

This unjustifiable assumption seems difficult to avoid in the context of black-box groups. But for matrix groups as well as for black-box groups of a given characteristic, many significant applications survive without assuming a superset of prime divisors of the order of the group; an explicitly defined small set of "pretend-primes" suffices instead. (See Remark 4.10 for details.)

4.2. Recognition within a given class of groups. Let $\mathcal{G}$ be a class of black-box groups. We say that a procedure *recognizes a group G within the class $\mathcal{G}$* if given a black-box group $H \in \mathcal{G}$, the procedure will accept H if $H \cong G$ and reject if $H \ncong G$.

The following result was proved recently:

Theorem 4.1. *Every finite simple group of a given characteristic can be recognized among finite simple groups of the same characteristic (given as black-box groups of that characteristic) by a polynomial-time Monte Carlo algorithm.*

This result appears in **[BaKPS]** and is based on work by Altseimer, Borovik **[AltB]** and Kantor and Seress **[KS2]**. Its content is that given two nonisomorphic finite simple groups of the same (known) characteristic, a polynomial time statistical procedure can tell them apart.

We emphasize that for this procedure we need to know that *both* groups are simple.

The key question we wish to address is whether the assumption of simplicity of the members of the class $\mathcal{G}$ can be dropped. (We continue to assume that G itself is simple.)

4.3. Naming the composition factors. Finding the orders of composition factors of a cyclic group is as hard as factoring integers, which is not expected to be doable in polynomial time.

We are faring much better regarding nonabelian composition factors.

First we state the main result of **[BaB]**.

Theorem 4.2 ([**BaB**]). *Given a black-box group G of characteristic r, one can construct, in Monte-Carlo polynomial time, black-box-group representations of characteristic r of all nonabelian composition factors of G. Moreover, if a composition factor L is a sporadic group, alternating group, or a simple group of Lie type of characteristic other than r, then a permutation representation of L can be constructed in polynomial time.*

(This is a combination of Theorems 1.2 and 8.6 of **[BaB]**.)

Remark 4.3. Given a permutation representation of L, we can switch to a standard representation via an explicit isomorphism in deterministic polynomial time (Kantor **[Ka]**). Thus we learn the standard name of L in polynomial time (and virtually everything else we may wish to know about L).

Combined with Theorem 4.1, this result yields the following powerful corollary.

Corollary 4.4. *Given a black-box group G of given characteristic, one can compute, in Monte Carlo polynomial time, the standard names of all nonabelian composition factors of G.* □

In particular, this result applies to matrix groups over finite fields.

Corollary 4.5. *Given a matrix group over a finite field, one can compute, in Monte Carlo polynomial time, the standard names of all nonabelian composition factors of G.* □

4.4. Membership testing in simple normal subgroups. The following result provides a useful tool in many contexts. It settles the "Outer automorphism problem" (recognizing S within G where $S \leq G \leq \mathrm{Aut}(S)$), stated as Problem 10.3 in **[BaB]**.

Theorem 4.6. *Let S be a nonabelian simple group. Assume that S is a normal subgroup of a black-box group G of given characteristic. Given G and S, we can test membership in S of any $g \in G$ in Monte Carlo polynomial time.*

The following simple algorithm was communicated to us by Charles Leedham-Green. Our contribution is its analysis.

Algorithm (C. Leedham-Green)

> pick a reasonable number of random elements $s_i \in S$
> compute the g. c. d. of the orders of the elements gs_i
> **if** the g. c. d. is 1 **then** output "member"
> **else** output "probably not member"

Remark 4.7. We shall see that $O(\sqrt{n}\log n)$ is a "reasonable number," where n is the encoding length of the black-box group H.

Remark 4.8. This algorithm assumes that we can compute the orders of elements. See Remark 4.10 on how to avoid this assumption.

The analysis of the algorithm is based on the following result.

Theorem 4.9 ([BaPS]). *Let S be a finite simple group and r a prime number. Then at least a c/d fraction of the elements of S is r-regular where $c > 0$ is an absolute constant and $d = d(S)$ is defined as follows. For the alternating group A_t we set $d(A_t) = \sqrt{t}$, for a classical simple group S we define $d(S)$ to be the dimension of the projective space on which S acts; for all other simple groups S, $d(S) = 1$.* □

(For the alternating groups, the weaker bound c/t was quoted in **[BaB]**. This would suffice for the polynomial-time claim but not for the claim stated in Remark 4.7.)

Analysis of the algorithm. It is obvious that the order of g mod S will divide the order of each gs_i so if $g \notin S$ then the output will always be correct ("probably not member").

Let us now assume $g \in S$. In this case, each gs_i is a random member of S. Let $p_1, \dots, p_k$ be the set of primes dividing the order of S; clearly, $k < n$ (in fact, $k < n/\log n$) where n is the encoding length of our black box group.

Suppose now that we chose m random elements s_i. By Theorem 4.9, the probability that a particular p_j divides the orders of each gs_i is less than $(1 - c/d)^m < \exp(-cm/d)$. The probability that the g. c. d. of these orders is greater than 1 is therefore less than $k\exp(-cm/d) \leq \epsilon$ assuming $m \geq (d/c)\log(k/\epsilon)$.

Note that $d = O(\sqrt{\log|S|}) = O(\sqrt{n})$; therefore sampling $m = O(\sqrt{n}\log n)$ random elements s_i suffices for any constant margin of error. □

Remark 4.10. As remarked in Section 4.1, we cannot compute the order of the elements in polynomial time. Hence, strictly speaking, we did not complete the proof of Theorem 4.6. This can be fixed by copying the methods of **[BaB**, Section 8.4**]** where a notion of "pseudo-order" with respect to a set of "pretend-primes" is introduced. The "pretend-primes" are the "small" primes ($\leq n$) plus a set of numbers obtained as cyclotomic polynomials and semi-cyclotomic polynomials of the characteristic p of our black-box group. (These numbers occur as the orders of certain maximal tori in simple groups of characteristic p.) More precisely, we divide these numbers by any small prime divisors; this way we end up with a set of pairwise relatively prime "pretend-primes."

The "pseudo-order" is the smallest multiple of the order which factors as a product of our "pretend-primes." The set of pretend-primes needed for computation in a black-box group of given characteristic is computable in polynomial time, and so is the pseudo-order of elements with respect to a given set of pretend-primes.

If we apply the above algorithm to black-box groups of known characteristic, we can replace "order" by "pseudo-order" of an element. As in **[BaB**, Section 8.4], this transition requires a more detailed version of Theorem **[BaPS]** (see **[BaB, Theorem 8.7]**). (Note: omit the word "cyclic" from the statement of **[BaB**, Theorem 8.7]. This correction does not affect any of the known applications of the theorem.)

4.5. Reduction to perfect, solvable-by-simple groups. A group G is perfect if $G' = G$. It is *solvable-by-simple* if it has a solvable normal subgroup N such that the quotient G/N is simple nonabelian. If S is a nonabelian simple group then we say that G is a solvable-by-simple extension of S or simply G is solvable-by-S if G has a solvable normal subgroup N such that $G/N \cong S$.

Theorem 4.11. *Let S be a finite simple group of Lie type of characteristic p. If S can be recognized in Monte-Carlo polynomial time among its perfect, solvable-by-simple extensions, given as black-box groups of characteristic p, then S can be recognized in Monte-Carlo polynomial time among all finite groups given as black-box groups of any known characteristic (not necessarily p and p need not be known).*

Proof. Let G be a black-box group of characteristic r; we wish to decide whether or not $G \cong S$, where S is also given as a black-box group of some (known) characteristic (not necessarily r or p).

Applying Corollary 4.4 to the given representation of S, we find the standard name of S and learn the value p. Applying Corollary 4.4 to G we are able to refute the hypothesis $G \cong S$ unless G has exactly one nonabelian composition factor, L, and this composition factor is a simple group of Lie type of characteristic p and r simultaneously. Since a finite number of groups can be treated as sporadic (using ad hoc methods), we infer that $r = p$. We may further assume that $L \cong S$. The question that remains is to decide under these conditions whether or not G is simple.

Let H be the stable derivative of G, i. e., the smallest normal subgroup of G such that G/H is solvable. This can be found in Monte Carlo polynomial time **[BaCFLS]**.

Clearly, H is perfect and solvable-by-simple, given as a black-box group of the right characteristic. Therefore, by assumption, we are able to decide whether or not H is simple. We may assume it is, so $H \cong S$.

Now an application of Theorem 4.6 completes the proof: we pick some random elements of G and test whether or not they belong to H. If any of them does not, then $G \neq H$; otherwise we can "bet" that $G = H$ (the probability of error is at most 2^{-m} where m is the number of random elements selected). □

Remark 4.12. We did not estimate the overall probability of error in this algorithm; 2^{-m} is only the error probability attributable to our failure to catch an element outside the proper subgroup H. All other Monte Carlo subroutines used contribute to the error (most notably the routine to recognize the standard names of the composition factors (Corollary 4.4), and on a more fundamental level, the routines used for generating nearly uniform random elements and normal closures). These routines must run long enough to make their contribution to the error probability negligible. Exponentially small error probabilities are achievable in polynomial time by the definition of polynomial-time Monte Carlo algorithms (cf. **[Ba2]**).

4.6. Recognizing classical groups over tiny fields. "Tiny" is a well-defined technical term: an input parameter q is *tiny* if it is given in *unary* (q tally marks). Since we measure the complexity of an algorithm as a function of the length of the input, a polynomial-time algorithm with a tiny field of order q as input is allowed to take time $q^{O(1)}$, as opposed to $(\log q)^{O(1)}$ under the ordinary (binary) encoding of the field order.

Theorem 4.13. *Every finite simple group of Lie type of a given characteristic over a tiny field can be recognized among all finite groups, given as black-box groups of any characteristic, by a polynomial-time Monte Carlo algorithm.*

The proof is based on the recent monumental work by Kantor and Seress **[KS1]** which proves the following main result.

Theorem 4.14 (**[KS1]**). *Let $\mathcal{C}_p$ denote the class of classical finite simple groups of characteristic p. Then one can, in Monte-Carlo polynomial time, constructively recognize black-box members of $\mathcal{C}_p$ within $\mathcal{C}_p$ assuming the field of definition is tiny.* □

In this result, p is an input variable, not assumed to be constant.

"Constructive recognition" means that an explicit isomorphism is constructed with a standard matrix representation of the group in question. As pointed out in **[KS1]**, it follows, using Steinberg's presentations, that a presentation in terms of generators and relations can also be constructed in Monte Carlo polynomial time. This observation immediately upgrades the Monte Carlo algorithm of Theorem 4.14 to a *Las Vegas algorithm*, i. e., a Monte Carlo algorithm which does not err but is allowed to report failure with a small probability. (If the presentation does not work out as expected, report failure; else we have proof that the group is what we believe it is.) This conceptual upgrade is especially significant if we generate random elements using unproven heuristics (as we always do in practice). (See **[Ba2]** for a discussion of related issues.)

Unfortunately, a Las Vegas upgrade of Theorem 4.13 would seem to require entirely new ideas. Kantor and Seress **[KS1]** express pessimism about the possibility of such upgrade (in a paragraph following their Corollary 1.8) even when both groups are simple classical and one of them is over a tiny field. In fact, even a Las Vegas separation (proof of non-isomorphism) of $\mathrm{PSL}(2m, p)$ and $\mathrm{PSL}(2, p^m)$ in time $m^{O(1)}$ seems to be out of reach, where $\mathrm{PSL}(2, p^m)$ is given as a black-box group of characteristic p, while $\mathrm{PSL}(2m, p)$ is given explicitly in its natural projective representation and p is a fixed small prime. The trouble is that we are unable to recover the geometry behind a linear group over a very large field, and in particular we are unable to construct a presentation of PSL$(2, p^m)$, given as a black-box group, when m is moderately large.

Even bigger problems seem to loom when looking for a Las Vegas comparison of the unitary groups PSU$(3, p^m)$ and PSU$(3m, p)$: *short presentations* are not even known to exist for PSU$(3, p^m)$ (even for fixed small p), let alone constructed efficiently. (A "short presentation" of a group G is a presentation that can be stated using only $(\log |G|)^{O(1)}$ symbols. In the case of PSU$(3, p^m)$ this would mean length $(m \log p)^{O(1)}$.) Short presentations are known to exist for all finite simple groups with the possible exception of the rank-1 twisted Lie-type groups, viz. PSU$(3, q)$, ${}^2B_2(q)$ (Suzuki) and ${}^2G_2(q)$ (Ree) **[BaGKLP]**.

Proof of Theorem 4.13. According to Theorem 4.11, we may assume that G is solvable-by-S, where S is a finite simple group of Lie type of characteristic p over a tiny field $\mathbb{F}_q$ and G is given as a black-box group of characteristic p; the task is to decide whether or not $G \cong S$.

First we dispose of the case when S is an exceptional group (this case is not considered in **[KS1]**). Indeed, in this case $|S| < q^{248}$ so the order of S is polynomially bounded as a function of the length of the input (because $\mathbb{F}_q$ is *tiny*), and we can decide in polynomial time whether or not $|G| > |S|$ by exhaustive search. (The permutation group methods of **[BaB]** reviewed in Section 4.8 will considerably reduce the revolting exponent 248 but substantial further work will be required to produce a manageable exponent.)

Let us now assume that S is classical and work from Theorem 4.14.

Let R denote the solvable radical of G (largest solvable normal subgroup). Membership in R can be decided in Monte Carlo polynomial time (check solvability of the normal closure of the given element, **[BaCFLS]**). Note that $G/R \cong S$.

All we need to do is decide whether or not $R = 1$. This is done by the standard method of "sifting" the defining relations of S as follows: let $\varphi : G \to S$ be the epimorphism to which our explicit $G/R \to S$ isomorphism lifts. Let $S = \langle s_i \,:\, i \leq a \mid R_j \,:\, j \leq b \rangle$; let $g_i \in G$ be a lifting of s_i to G. Let G be given by a set $\{h_\ell \,:\, \ell \leq f\}$ of generators. Let $t_\ell = \varphi(h_\ell)$. Express t_ℓ as a word $w_\ell(\ldots, s_i, \ldots)$. Then R is the normal closure of the elements $R_j(\ldots, g_i, \ldots)$ and the elements $h_\ell^{-1} w_\ell(\ldots, g_i, \ldots)$. □

4.7. Central-by-simple groups

Theorem 4.15. *Let G be a black-box group of characteristic p. Assume $G/Z(G)$ is simple nonabelian. Then we can find $Z(G)$ in Monte Carlo polynomial time.*

Proof. The algorithm will again be based on Theorem 4.9.

For simplicity, we again pretend that we know a superset of primes dividing the order of G and so we can compute the order of elements. (We commented on this assumption in Section 4.1. We can avoid this assumption with the same technique as for Theorem 4.6. For the details, see Remark 4.10.)

Let r be a prime and let R be a Sylow r-subgroup of $Z(G)$. We shall collect generators for R in a set $\mathcal{S}$.

Algorithm

initialize the set $\mathcal{S} := \emptyset$
repeat a reasonable number of times:
  pick a random element $g \in G$
  write the order of g as $r^{\ell}b$ where r does not divide b
  let $h = g^b$
  if $h \in Z(G)$ **then** add h to $\mathcal{S}$

Remark 4.16. We shall see that $O(n^{3/2})$ is a "reasonable number," where n is the encoding length of the black-box group G.

Analysis. Let $S = G/Z(G)$ and let $\varphi : G \to S$ be the natural epimorphism.

Let us imagine that the random choice of elements of G is performed in two stages: first we pick a random element $s \in S$, and then a random member $g \in \varphi^{-1}(s)$. Clearly, if both choices are uniform, so is their composition.

By Theorem 4.9, with reasonable frequency we are likely to pick an r-regular $s \in S$. If this is the case, it is easy to see that the resulting element h will be a uniformly distributed random member of R. Indeed, let T denote the r'-Hall subgroup of $Z(G)$ (comprising all r-regular elements of $Z(G)$). Let further $g_0 \in \varphi^{-1}(s)$. Then $g = g_0 z_1 z_2$ where z_1 is a random member of T and z_2 is a random member of R. Now $h = g^b$ has order r^{ℓ}; on the other hand, s is r-regular; therefore $h \in Z(G)$. It follows that $h \in R$. The same is true for $h_0 := g_0^b$. Therefore $z_1^b z_2^b \in R$; consequently, $z_1^b = 1$. Finally, z_2^b is uniformly distributed over R, therefore so is h.

To estimate what is a "reasonable number" of repetitions, let $d = d(S)$ be the quantity defined in Theorem 4.9. $C_1 d \log |R|)$ repetitions are likely to turn out $\geq C_2 \log |R|$ random elements of R; for $C_2 > 1$, these are likely to generate R. We can use the crude estimates $d < \sqrt{n}$ and $\log |R| < n$ to obtain an $O(n^{3/2})$ bound on the number of repetitions (n is the encoding length). □

4.8. Reduction to perfect, p-by-simple groups. In this section we further narrow the obstacle exhibited in Theorem 4.11.

A group G is *p-by-simple* if it has a normal p-subgroup N such that the quotient G/N is simple nonabelian. If S is a nonabelian simple group then we say that G is a p-by-simple extension of S if G is p-by-simple and $G/N \cong S$. We say that G is *non-central minimal p-by-simple* if N is a non-central minimal normal subgroup of G.

Theorem 4.17. *Let S be a finite simple group of Lie type of characteristic p. If S can be distinguished in Monte-Carlo polynomial time from its non-central minimal p-by-simple extensions, given as black-box groups of characteristic p, then S can be recognized in Monte-Carlo polynomial time among all finite groups given as black-box groups of any characteristic (not necessarily p and p need not be known).*

Proof. In view of Theorem 4.11, we may assume that we are given a perfect, solvable-by-simple group G such that $G/N \cong S$ where N is the solvable radical of G; and moreover that G is given as a black-box group of characteristic p. The task is to decide whether or not $N = 1$.

We note that membership in N is decidable in Monte Carlo polynomial time (check solvability of the normal closure). As soon as a nonidentity element of N is found, we are done.

First we perform the algorithm "PERM(G, m)" of **[BaB**, Theorem 6.1**]** with a judiciously chosen value m. ($m = n$ will suffice.) This algorithm carries the following guarantee:

Theorem 4.18 ([BaB]). *Assume G is a black-box group of given characteristic. Assume further that G has a nontrivial permutation representation of degree $\leq m$. Then Algorithm* PERM(G, m) *either returns a nonidentity element which belongs to a proper normal subgroup of G or returns a faithful permutation representation of G of degree $\leq m^c$. The algorithm works in Monte Carlo time $(m+n)^{O(1)}$.* □

If the algorithm produces a permutation representation, the fact that it is faithful can be verified in deterministic polynomial time. In this case we know virtually everything about G from the theory of permutation group algorithms (cf. **[Lu3]**). Specifically, an algorithm of Luks can be used to decide simplicity **[Lu1]**.

If PERM fails to produce either of the desired objects, we perform the algorithm of Theorem 4.15. With each element h generated, we check whether $h \in N$.

If no element of N is found, we pretend that either $N = 1$ or N is an elementary abelian p-group which is non-central, minimal normal in G, and run the postulated algorithm to decide whether or not $N = 1$, checking along the way every element generated for membership in N. If none is found, we decide that $N = 1$, i. e., G is simple.

Analysis. We conclude that $N \neq 1$ only if we find a nonidentity element of N, so if $N = 1$ then our decision is certainly correct (except for possible errors made by the Monte Carlo algorithm for normal closures).

Suppose now that $N \neq 1$ and let M be a proper subgroup of N, maximal with respect to the condition that $M \triangleleft G$. Then N/M is a minimal normal subgroup of G/M and therefore it is an elementary abelian r-group for some r; let $|N/M| = r^k$. Note that $k \leq \log_2 |G| \leq n$.

If $N/M = Z(G/M)$ then it is easy to see that the second phase of the algorithm will find a nonidentity element of N. Indeed, pretend we compute in G/M. The only difference occurs when some element $u \in G$ is tested for being the identity. If $u = 1$, we report that u is the identity and continue. If $u \in N$ and $u \neq 1$ then we terminate: we found what we wanted. If $u \notin N$, we report that $u \neq 1$ and continue. As long as no nonidentity element of N is encountered, the computations in G and in G/M proceed in the exact same steps.

Assume now that N/M is not central in G/M. Now S acts faithfully (via conjugation) on N/M.

If $r \neq p$ then this implies that S has a permutation representation of small degree ($\leq k^c \leq n^c$ for a small constant c). This follows from a theorem of Landazuri and Seitz **[LaS]** (cf. **[BaB**, Theorem 4.1]). But then, the first phase of the algorithm must have produced the required result.

Finally if $r = p$ then the last phase of the algorithm succeeds for a reason similar to why the algorithm to find the center worked when pretending to compute over G/M. □

4.9. Recognizing unisingular simple groups. Finally we are in the position to tackle some classes of Lie-type simple groups over large fields.

Theorem 4.19. *If S is a finite simple group of Lie type of characteristic p and S is unisingular then S can be recognized in Monte Carlo polynomial time among all black-box groups of known characteristic (not necessarily p and p need not be known).*

Proof. Let q be the order of the field of definition of S. For $q < 13$ we refer to the work of Kantor and Seress for groups over tiny fields and specifically to Theorem 4.13 above. Henceforth we assume $q \geq 13$. As a consequence, we have $\rho_p(S) < 1/3$. Indeed, by the quoted result of Guralnick and Lübeck (Theorem 3.2), we have

$$\rho_p(S) \leq \rho_p'(S) < \frac{3}{q-1} + \frac{2}{(q-1)^2} < \frac{1}{3}.$$

By Theorem 4.17, the proof of Theorem 4.19 is reduced to the following situation: we are given a black-box group G of characteristic p known to have a normal subgroup N such that $G/N \cong S$ and either $N = 1$ or N is a non-central minimal normal p-subgroup of G.

The task is to decide whether or not $N = 1$.

Let us call $g \in G$ a "witness" of the assertion "$N \neq 1$" if g is p-singular (in G) but gN/N is p-regular (in S). The algorithm tries to find such a witness.

If a witness is found, we declare "$N \neq 1$." Otherwise we declare "probably $N = 1$."

Algorithm

repeat a reasonable number of times:
  pick a random element $g \in G$
  let b be the (pseudo)order of g

if $p|b$ **then** let $h = g^{b/p}$
if the normal closure $\langle h^G \rangle$ is solvable **then** output h, **halt**
end(repeat)
output "probably $N = 1$."

Assuming the normal closure and solvability routines do not fail, no error can occur if $N = 1$. Assume now that $N \neq 1$. Then the probability that a random $g \in G$ is a witness is $\rho_p(G) - \rho_p(S)$. Therefore the probability that we fail to find a witness is $\left(1 - \rho_p(G) + \rho_p(S)\right)^m$ where m is the number of repetitions.

Combining Corollary 3.8 and Observation 3.10 we obtain that for unisingular simple groups, this quantity is always less than $(\rho_p(S)+1/p)^m$. Under our assumption of $q \geq 13$ we have $\rho_p(S) < 1/3$ and therefore the probability of failing to detect $N \neq 1$ is less than $(5/6)^m$. So a modest constant number of repetitions is "reasonable."

For large p we can use the better estimate $(4/q + 1/p)^m$ to see that in that case, already $m = 1$ suffices with large probability. □

Remark 4.20. Note that this algorithm not only decides whether or not $N = 1$ but it is also finds a nontrivial element of N if $N \neq 1$. This demonstrates a special case of a general principle discovered by Beals [**Be**, Lemma 5.2]: *any* black-box algorithm which distinguishes a simple group S from its extension G ($G/N \cong S$, $N \neq 1$) can be modified to find a nontrivial element of a proper normal subgroup of G. (The Monte Carlo complexity of the modified algorithm is polynomially bounded in the input length and the complexity of the original algorithm.)

5. Open questions

1. (*p-core problem.*) The big open question is to decide whether or not $O_p(G) = 1$ where $O_p(G)$ is the p-core of G (the largest normal p-subgroup) and G is a matrix group over $\mathbb{F}_p$ or more generally, a black-box group of characteristic p (Section 4.1). – Is the p-core problem any easier for matrix groups than for black-box groups of characteristic p?

2. (*Affine case.*) The special case of the p-core problem described in the Abstract deserves particular attention. Its solution would resolve the simplicity problem and would constitute a major step toward solving the general p-core problem.

3. (*Non-unisingular groups.*) The question is, how bad are those simple groups S of Lie-type of characteristic p which are not unisingular. The measure of badness is the quantity

$$\lambda(S) = \min \lambda(A_S) \tag{4}$$

where the minimum is taken over all irreducible S-modules A_S of characteristic p. The smaller this value, the worse the group. The running time of our naive algorithm for the affine case is proportional to $1/\lambda(S)$. (Recall that $\lambda(A_S)$ is the proportion in S of those p-regular elements which act as unisingular transformations on A. We called a linear transformation *unisingular* if 1 is among its eigenvalues. We called the group S unisingular if all elements of S act as unisingular transformations on all nontrivial S-modules.)

Let $\mathbb{F}_q$ be the field of definition of S. We are concerned with the cases of large q.

For unisingular S, we have $\lambda(S) = 1 - O(1/q)$, i. e., $\lambda(S)$ is close to 1 (see equation (2)).

On the other hand, for $S = \mathrm{PSL}(2, q)$, $q = p^f$ $(f \geq 2)$, Proposition 3.14 asserts that

$$\lambda(S) \leq p/q. \tag{5}$$

Does this upper bound generalize to all simple groups of Lie type that are not unisingular? Guralnick, Saxl, and Tiep suggest **[GST]** that a bound of the form

$$\lambda(S) \leq cpd^C/q \tag{6}$$

might hold, where d is the Lie-rank, p is the characteristic, and $q = p^f$ is the order of the field of definition; c and C are absolute constants. Such a bound would show that the naïve algorithm does not work for any of the non-unisingular groups with $f \geq 2$ when q is large.

The estimation of $\lambda(S)$ for the non-unisingular simple groups with $f = 1$ $(q = p)$ is a further relevant question. The groups $\mathrm{PSL}(3, p)$ with $p \not\equiv 1 \pmod 3$ would be first class to consider.

4. Recognize (test membership in) the derived subgroup G' of a matrix group, or more generally, of a black box group of characteristic p.

Acknowledgments The first named author wishes to thank Jonathan Alperin, George Glauberman, Charles Leedham-Green, Péter P. Pálfy and Ákos Seress for valuable discussions. The authors are indebted to Bill Kantor and Ákos Seress for their meticulous editorial work and in particular, for catching a number of inaccuracies in the paper. The authors are also indebted to Bill and Ákos for the opportunity to present this work at the OSU meeting and an update at the May 2000 meeting in Milan. Last but not least, we wish to thank Bob Guralnick, Jan Saxl and Pham Huu Tiep for providing a quick and complete solution, within days of the Milan meeting, to the problem of characterizing the unisingular simple groups of characteristic p. Their fast work allowed a last-minute update of the paper before being sent to print.

References

[AloS] N. Alon, J. H. Spencer, The Probabilistic Method, Wiley, 1992.

[AltB] C. Altseimer, A. V. Borovik, Probabilistic recognition of orthogonal and symplectic groups, in: [G&C3], 1–20.

[Ba1] L. Babai, Local expansion of vertex-transitive graphs and random generation in finite groups, in: Proc. 23rd ACM Symp. on Theory of Computing (STOC'91), 1991, 164–174.

[Ba2] L. Babai, Randomization in group algorithms: conceptual questions, in: [G&C2], 1–16.

[BaB] L. Babai, R. Beals, A polynomial-time theory of black-box groups I, in: Groups St Andrews 1997 in Bath, I (C. M. Campbell, E. F. Robertson, N. Ruskuc, G. C. Smith, eds.), London Math. Soc. Lecture Note Ser. 260, Cambridge University Press, 1999, 30–64.

[BaCFLS] L. Babai, G. Cooperman, L. Finkelstein, E. M. Luks, Á. Seress, Fast Monte Carlo algorithms for permutation groups, J. Comput. System Sci. 50 (1995), 296–308.

[BaGKLP] L. Babai, A. J. Goodman, W. M. Kantor, E. M. Luks, P. P. Pálfy, Short presentations for finite groups, J. Algebra 194 (1997), 79-112.

[BaKPS] L. Babai, W. M. Kantor, P. P. Pálfy, Á. Seress, Black-box recognition of finite simple groups of Lie type by statistics of element orders, manuscript, in preparation.

[BaPS] L. Babai, P. P. Pálfy, J. Saxl, On the number of p-regular elements in simple groups, manuscript.

[BaS] L. Babai, E. Szemerédi, On the complexity of matrix group problems I, in: Proc. 25th IEEE Symp. Found. Comp. Sci. (FOCS'84), 1984, 229–240.

[Be] R. Beals, Towards polynomial time algorithms for matrix groups, in: [G&C2], 31–54.

[Ber] Á. Bereczky, On the proportion of p-singular elements in finite simple groups of Lie type of characteristic p, private communication, 1999.

[BrN] R. Brauer, C. Nesbitt, On the modular characters of groups, Ann. of Math. 42 (1941) 556–590.

[CLMNO] F. Celler, C. R. Leedham-Green, S. Murray, A. Niemeyer and E. A. O'Brien, Generating random elements of a finite group, Comm. Algebra 23 (1995), 4931–4948.

[G&C1] Groups and Computation (L. Finkelstein and W. M. Kantor, eds.), DIMACS Ser. Discrete Math. Theoret. Comput. Sci. 11, Amer. Math. Soc., Providence, RI, 1993.

[G&C2] Groups and Computation II (L. Finkelstein and W. M. Kantor, eds.), DIMACS Ser. Discrete Math. Theoret. Comput. Sci. 28, Amer. Math. Soc., Providence, RI, 1997.

[G&C3] Groups and Computation III (W. M. Kantor and Á. Seress, eds.), Ohio State Univ. Math. Res. Inst. Publ. 8, Walter de Gruyter, Berlin–New York, 2001.

[GL] R. M. Guralnick, F. Lübeck, On p-singular elements in Chevalley groups in characteristic p, in: [G&C3], 169–182.

[GST] R. M. Guralnick, Jan Saxl, Pham Huu Tiep, private communication, May 2000.

[Ka] W. M. Kantor, Sylow's theorem in polynomial time, J. Comput. System Sci. 30 (1985), 359–394.

[KS1] W. M. Kantor and Á. Seress, Black box classical groups, Mem. Amer. Math. Soc. 149, Nr. 708, Amer. Math. Soc., Providence, RI, 2001.

[KS2] W. M. Kantor, Á. Seress, Prime power graphs for groups of Lie type, in preparation.

[LaS] V. Landazuri, G. M. Seitz, On the minimal degrees of projective representations of the finite Chevalley groups, J. Algebra 32 (1974), 418–443.

[Lu1] E. M. Luks, Computing the composition factors of a permutation group in polynomial time, Combinatorica 7 (1987), 87–99.

[Lu2] E. M. Luks, Computing in solvable matrix groups, in: Proc. 33rd IEEE Symp. Found. Comp. Sci. (FOCS'92), 1992, 111–120.

[Lu3] E. M. Luks, Permutation groups and polynomial-time computation, in: [G&C1], 139–175.

Department of Computer Science
University of Chicago
Chicago, IL 60637, U.S.A.
E-mail: laci@cs.uchicago.edu

Department of Mathematics
Hebrew University
Jerusalem, Israel
E-mail: shalev@math.huji.ac.il

Improved algorithms for the Tits alternative

*Robert Beals**

Abstract. Let G be a symbolically given matrix group over an algebraic number field. Our main result is a new algorithm to test whether the subgroup of scalar matrices has finite index in G. This algorithm is Monte Carlo (randomized with one-sided error) and runs in polynomial time. Its advantage over previous Las Vegas (randomized, zero error) and deterministic polynomial time algorithms is in its efficiency: the previous approaches required squaring the dimension of the matrices, the current method avoids this, and is easily implemented in GAP using the tools of the GRIM package.

Our eventual goal is to provide a similar improvement for algorithms to decide the Tits alternative: does G has a solvable normal subgroup of finite index. The existing algorithm, while polynomial time, involves squaring the dimension twice in succession. We describe a new method which performs all computations in the original dimension (or smaller), making extensive use of the finiteness test for G modulo scalars. This new method, while not yet tested, overcomes the major bottleneck in a practical implementation of the Tits alternative algorithm.

1991 Mathematics Subject Classification: primary 20H20; secondary 68W30.

1. Introduction

Let $G = \langle S \rangle \leq \mathrm{GL}(n, F)$, where F is an algebraic number field. A theorem of Tits states that either G has a solvable normal subgroup of finite index, or G contains a nonabelian free subgroup. This dichotomy, known as the Tits alternative, is intimately related to computability: several fundamental computational problems for matrix groups are decidable exactly for the class of solvable-by-finite groups.

In a previous paper **[Be]**, we gave a polynomial time algorithm to determine, given $S \subseteq \mathrm{GL}(n, F)$, whether or not $G = \langle S \rangle$ is solvable-by-finite. Unfortunately, this algorithm is impractical, as it involves solving systems of linear equations with n^4 unknowns. We describe here a new approach, which seeks to perform all computations in dimension n.

There are at least two schools of thought concerning computational efficiency. In theory of computing, it is customary to simply define efficiency as polynomial time.

*This research was conducted while the author was with the University of Arizona Department of Mathematics.

On the other hand, for those who wish to actually run the algorithms in order to compute answers, an algorithm is efficient if it allows typical instances of reasonable size to be solved using a reasonable amount of computational resources (time, memory, disk space). The theory of computing definition is precise and allows theorems to be proved concerning the efficiency of various algorithms, but it allows various absurd algorithms to be classified as efficient. For example, our Tits alternative algorithm **[Be]** is unusable except for very small dimension. We justify the polynomial time criterion by noting that the mathematical structure discovered in designing and analyzing a polynomial time algorithm is often useful in designing practical algorithms. This paper represents a step in that direction for the problem of deciding the Tits alternative.

However, we do not wish to abandon asymptotic analysis of worst-case running time. A theoretically proven performance guarantee is very valuable, especially for Monte Carlo algorithms: if an algorithm outputs "probably yes", then the user would want to know a proven bound on the error probability (the conditional probability of a "yes" output given that the correct output is "no"). And a worst-case performance bound overcomes concern about the credibility of claims of typical case performance. I may know how the algorithm performs on *my* typical cases, but I would hesitate to make claims about *general* typical cases.

Our goal therefore is to provide algorithms which have been implemented and tested, and which are backed by a theoretical analysis. An example is the **[BBR]** finiteness test for matrix groups, which has been implemented as part of the GRIM package for GAP **[GAP]**.

We emphasize that we are viewing the dimension n as something that may be in the hundreds; the special case that n is quite small is not dealt with. One approach, for various problems concerning the existence of finite index normal subgroups of a particular type, is to consider the kernel of the homomorphism to $\mathrm{GL}(n, p)$ obtained by reducing matrix entries modulo p. (The prime p is chosen so that it does not divide the denominators of the entries of the generators or their inverses.) For example, G is finite iff this kernel is trivial for $p > 2$. Ostheimer **[Os96, Os99]** characterizes the solvable-by-finite groups in a similar manner. We are not aware of any successful implementation of this which works in large dimension. There are two main difficulties. First, computing the kernel involves extensive computation with the image, which is typically handled by treating the finite vector space as a permutation domain, so the cardinalities of the vector orbits (exponentially large in general) are a factor in the running time. Second, even if the image could be handled efficiently, finding the kernel will involve computations with the original group. A sequence of k group operations in characteristic zero may increase the bit lengths of the matrix entries by a factor of 2^k. This is achieved by repeated squaring, which may seem contrived, but coefficient explosion is a real practical difficulty when computing with large matrices in characteristic zero.

The dimension squaring in the original **[Be]** Tits alternative algorithm occurs when finding irreducible constituents of the conjugation action of $G \leq \mathrm{GL}(n, F)$ on the ring of $n \times n$ matrices. This is an n^2 dimensional representation of G, and the Friedl–

Rónyai **[FR]** algorithm for finding irreducible constituents squares the dimension again. The algorithm relies upon identification of those constituents on which G act as a finite group. Finiteness testing for the conjugation action turns out to be the real issue; and our main result is an algorithm for this problem which works in dimension n.

Theorem 1.1. *Let F be a symbolically given algebraic number field, let $G \leq \mathrm{GL}(n, F)$ be given by a set S of generators, and let Z denote subgroup of scalar matrices in G. The finiteness question for G/Z reduces in Monte Carlo polynomial time to the finiteness question for subgroups of $\mathrm{SL}(n, F)$. The reduction has one-sided error: if the reduction produces an infinite subgroup H of $\mathrm{SL}(n, F)$, then G/Z is certainly infinite.*

The conjugation action is finite iff the subgroup of scalar matrices in G has finite index in G. A natural approach to this subproblem would be to test finiteness directly of the conjugation representation, using an algorithm of Babai, Beals, and Rockmore. Unfortunately, this requires working in dimension n^2, which is excessive. In Section 3, we prove Theorem 1.1, showing how this problem may be solved by testing finiteness of a subgroup of $\mathrm{SL}(n, F)$.

In Section 4, we indicate how the general Tits alternative problem reduces to finiteness testing of the conjugation action. The reduction is complicated and has not yet been tested. In restricting the computations to dimension n, we believe that we have removed most, if not all, of the practical difficulty in deciding the Tits alternative.

Section 2 presents background material on the algorithmic and group theoretic tools we bring to bear on the problem.

2. Preliminaries and background

For an ordered set S of group elements, a *random subproduct* of S is the product, in order, of a uniformly chosen subset of S. Cooperman and Finkelstein **[BCFLS]** established the following useful properties of random subproducts:

Lemma 2.1. *Let $G = \langle S \rangle$, and let H be a proper subgroup of G. Then H contains at most half of the random subproducts of S.* □

Theorem 2.2. *Let $G = \langle S \rangle$ and $H = \langle T \rangle$ be subgroups of a group K, with G not contained in the normalizer of H. Then at most $3/4$ of the elements y^x lie in H, where x is a random subproduct of G and y is a random subproduct of H.* □

This is the key in analyzing the following "black-box" (cf. **[BSz]**) normal closure algorithm: add the element y^x to T, and repeat. If L is a bound on lengths of subgroup chains in K, and ϵ is an error parameter, then $O(L + \sqrt{L \log(1/\epsilon)})$ iterations will suffice to generate $\langle H^G \rangle$ with error $< \epsilon$. We will use this in a different context, where we seek not the full normal closure but a subgroup which is at the top of a chain of at least L strictly increasing subgroups. Morris Newman's bound **[New]** on the orders of finite matrix groups will provide our L:

Theorem 2.3. *A finite subgroup of* $\mathrm{GL}(n, \mathbb{Q})$ *has order dividing* $(2n)!$. *A finite subgroup of* $\mathrm{GL}(n, F)$ *has order dividing* $(2nk)!$, *where* k *is the degree of* F *over* $\mathbb{Q}$. □

Corollary 2.4. *Suppose* $H_0 \leq H_1 \leq \cdots \leq H_L$ *are distinct subgroups of* $GL(n, F)$, *and* $L > \log_2((2nk)!)$ *where* k *is the degree of* F *over* $\mathbb{Q}$. *Then* H_L *is infinite.* □

For a matrix group G, we will denote by $\mathrm{Env}(G)$ the *enveloping algebra* of G, which is the linear span of the matrices in G.

For any group $G = \langle S \rangle$, and subgroup H, let $T \subseteq G$ be a system of coset representatives for H. Then

$$H = \langle \tau g \tau'^{-1} \mid \tau \in T, g^{\pm 1} \in S, \tau' \text{ such that } \tau g \in H\tau' \rangle.$$

These are the Schreier generators for H. Note that, given S and T, to compute the Schreier generators it suffices to be able to recognize when two elements of G lie in the same coset of H. In typical algorithmic applications, G acts naturally on a small set, and H is the stabilizer of a point.

Our algorithms work with algebraic number fields. The input consists of matrices over some field F; in the reduction, we may use extensions of F (we do gain by this; if we work over an extension of degree ℓ, then we are able to reduce the dimension by a factor of ℓ). For convenience, we will think of all fields as being subfields of the complex field $\mathbb{C}$, so that we may speak, for example, of complex roots of polynomials in $F[x]$, and floating point approximations to these roots.

3. Testing finiteness of the projective action

Let $G = \langle S \rangle \subseteq \mathrm{GL}(n, F)$, where F is a symbolically given algebraic number field. G acts on lines of F^n through the origin; this is called the projective action. The kernel Z of this action is the subgroup of scalar matrices in G. In this section, we present an algorithm to test whether or not G/Z is finite, i.e., whether or not G acts projectively as a finite group. For brevity, we will say that G is *projectively finite* if G/Z is finite.

As described above, one method to test projective finiteness is to represent G as an $n^2 \times n^2$ matrix group: G acts by conjugation on the set of $n \times n$ matrices over F, with kernel Z. The **[BBR]** algorithm can then be used to test finiteness of the image. Here we avoid this squaring of dimension: we are able to test projective finiteness of G by using the **[BBR]** algorithm in dimension n.

Before describing the algorithm, we introduce some notation. Let

$$\tilde{S} = \{(\det(a))^{-1/n} \cdot a \mid a \in S\} \subseteq \mathrm{SL}(n, \mathbb{C}),$$

and let $\tilde{G} = \langle \tilde{S} \rangle \leq \mathrm{SL}(n, \mathbb{C})$ (the exponent $-1/n$ is a source of ambiguity, however, for what follows it does not matter which nth roots are chosen). Since the generators of $\tilde{G}$ are scalar multiples of the generators of G, both groups have the same projective action. Also, since $\tilde{G}$ is in the special linear group, the subgroup of scalar matrices in $\tilde{G}$ is finite (of order dividing n). Therefore, G is projectively finite iff $\tilde{G}$ is finite.

Unfortunately, we cannot simply give $\tilde{G}$ to the **[BBR]** algorithm; all those nth roots may require a field extension of degree $n^{|S|}$, which is unacceptable. The good news is that $\tilde{G}$ has a finite index subgroup H which is representable in dimension n over the original field F. Let

$$H = \langle \tilde{G}' \cup \{a^n \mid a \in \tilde{S}\} \rangle.$$

By construction, $\tilde{G}/H$ is a finitely generated abelian group of exponent dividing n, and therefore finite. So $\tilde{G}$ is finite iff H is finite. Also, G and $\tilde{G}$ have the same set of commutators, and the same conjugation action, and therefore the same commutator subgroups. So we may write H equivalently as

$$H = \langle G' \cup \{\det(a)^{-1} \cdot a^n \mid a \in S\} \rangle,$$

from which it is clear that $H \subseteq \mathrm{SL}(n, F)$.

We still do not have an algorithm. We cannot give H to the **[BBR]** algorithm, since we cannot efficiently compute a generating set for H. Indeed, if $\tilde{G}$ is free on $\tilde{S}$, then H is free on the Schreier generators, so the number of Schreier generators, which is basically $|S| \cdot |\tilde{G} : H| = |S|n^{|S|}$, is a lower bound for the cardinality of *any* generating set for H. We solve this problem probabilistically: for any prescribed error bound ϵ, we can in time polynomial in $\log 1/\epsilon$ and the input length compute a subset T of H with the property that if H is infinite, then the probability that $\langle T \rangle$ is finite is less than ϵ. This yields a one-sided error algorithm; it may (with probability $< \epsilon$) incorrectly state that a projectively infinite group is projectively finite, but if it states that a group is projectively infinite, it does so with certainty.

We describe the selection of T. We begin by including normal generators for H: initially $T = [S, S] \cup \{\det(a)^{-1}a^n \mid a \in S\}$. Then we add elements to T using the Cooperman–Finkelstein black-box normal closure algorithm (cf. Theorem 2.2) with subgroup chain bound $L = \log_2((2nk)!)$, where k is the degree of F, and error parameter ϵ. By the Newman bound (cf. Theorem 2.3), if H is infinite we will with probability $> 1 - \epsilon$ have generators for an infinite subgroup of H. We have proven Theorem 1.1. □

A comment on the **[BBR]** algorithm: let $K = \langle T \rangle$. If **[BBR]** produces an element of infinite order, we are certain that H is infinite. Otherwise, it produces a finiteness certificate in the form of a lattice X and a positive definite quadratic form B, both K-invariant. (The lattice X is in $\mathbb{Z}^{nk}$, where k is the degree of F; $\mathrm{GL}(n, F)$ acts naturally on $\mathbb{Q}^{nk}$.) We could certify that H is finite if we could somehow certify that $\langle K^G \rangle$ preserves both X and B. This is an interesting and apparently difficult open problem; a solution would make our algorithm Las Vegas. Of course, if membership testing in K is possible, then we may check that K is normalized by G.

4. The Tits alternative

Let $G \leq \mathrm{GL}(V)$ be given by generators, where $V = F^n$ and F is a symbolically given algebraic number field. We wish to decide the Tits alternative for G while working only in dimension n. To do this, we must depart significantly from the **[Be]** algorithm. We cannot afford to use the **[FR]** algorithm, which was used to reduce the problem to the case that G acts irreducibly and also to decompose the conjugation representation of G into irreducible constituents. A key step of the new approach is the algorithm of the previous section, which tests if the scalars Z in G are a finite index subgroup. If they are, then G is solvable-by-finite. If not, then a matrix a is found such that some scalar multiple of a is in G, and no power of a is a scalar matrix. In this section, we show how to use a to aid in the testing of G.

We begin in Section 4.1 with some necessary structure theory. Section 4.2 gives algorithms for certain computations with polynomials which are necessary for the reduction. The main reduction is given in Section 4.3.

4.1. Structure theory. By a theorem of Mal'cev (cf. **[We]**, Theorem 3.6), G is solvable-by-finite iff it has a triangularizable normal subgroup T of finite index.

For Lemmas 4.1–4.2, we suppose that G is solvable-by-finite, $a \in T$, and we let $E = \mathrm{Env}(\langle a^G \rangle) \subseteq \mathrm{Env}(T)$. Then:

Lemma 4.1. *The algebra E is either a direct sum of fields or has a nontrivial radical.*

Proof. Since T is triangularizable, any nontrivial nilpotent element of E is in the radical. If a is not diagonalizable, then $\mathrm{Env}(\langle a \rangle)$ contains nilpotent elements. If E is noncommutative, then for any noncommuting $x, y \in E$, the element $xy - yx$ is nilpotent. So we may suppose E is commutative and a is diagonalizable. In this case, E is generated by commuting diagonalizable matrices, which are simultaneously diagonalizable, so E is a direct sum of fields. □

Lemma 4.2. *Suppose that E is a direct sum of fields $E_1, \ldots, E_r$. Let W_i denote the subspace $E_i V$. Then G permutes the W_i; this action is either transitive, or some subset of the W_i generates a nontrivial proper G-invariant subspace.*

Proof. The conjugation action of G permutes the E_i. Suppose that $g^{-1} E_i g = E_j$. Then

$$gW_j = gE_j V = E_i gV = E_i V = W_i,$$

so G permutes the W_i. If this action is not transitive, then the sum of a G-orbit of the W_i is a nontrivial proper G-invariant subspace. □

Lemma 4.3. *Suppose that E contains a nontrivial nilpotent element x. Then the nullspace of x contains a nontrivial proper G-invariant subspace.*

Proof. The nullspace of the radical of E is a nontrivial proper G-invariant subspace contained in the nullspace of x. □

For Lemma 4.4, we no longer assume G is solvable-by-finite.

Lemma 4.4. *Let $G \leq \mathrm{GL}(V)$, where $V = F^n$, be given by a generating set S. Then*

(a) *If G acts transitively on the subspaces $W_1, \ldots, W_r$ of a G-invariant direct sum decomposition, then G is solvable-by-finite iff the stabilizer H of W_1 acts as a solvable-by-finite group on W_1.*

(b) *If a subalgebra E of $\mathrm{Env}(G)$ exists such that E is a field and G normalizes E, then G is solvable-by-finite iff the centralizer H of E is. Let $\ell = |E : F|$. Then H has index at most ℓ, and is represented faithfully in $\mathrm{GL}(n/\ell, E)$.*

(c) *If G preserves a nontrivial proper subspace W, then G is solvable-by-finite iff G acts as a solvable-by-finite group on both W and V/W.*

Proof. For (a), let K denote the permutation action of G on the W_i, and let L denote the action of H on W_1. Then G is isomorphic to a subgroup of the wreath product $L \wr K$, so G is solvable-by-finite iff L is.

For (b), first note that the index of H is at most the order of the group Γ of F-automorphisms of E. Let $\alpha \in E$ be such that $E = F(\alpha)$, and let $p \in F[x]$ be the minimal polynomial of α. Then each $\gamma \in \Gamma$ is determined by $\gamma(\alpha)$, which must be one of the $\leq \ell$ roots of p in E. So $|G : H| \leq |\Gamma| \leq \ell$.

The algebra E is a subalgebra of $\mathrm{End}(V)$, so V is a vector space over E. Since $\dim_F(V) = |E : F| \dim_E(V)$, we have $\dim_E(V) = n/\ell$. The group H acts E-linearly on V; so by picking an E-basis for V we represent H in $\mathrm{GL}(n/\ell, E)$. This is a faithful representation, since H acts faithfully on V (the action is the same whether we view V as an E-vector space or an F-vector space).

Since H has finite index in G, G is solvable-by-finite iff H is.

For (c), the kernels of the actions of G on W and V/W intersect in a solvable (indeed nilpotent) group. □

4.2. Manipulation of polynomials. Let $m(x) \in F[x]$ be a given square-free polynomial. We wish to find a polynomial $f(x)$ satisfying a particular property which we describe below. Later we will use this in the case that m is the minimal polynomial of an element a of G, some power of which lies in a subgroup T. The polynomial f will have the property that $f(a)$ is in $\mathrm{Env}(T)$.

Let R be the F-algebra $F[x]/(m)$, and for natural numbers k, let S_k denote the F-subalgebra of R generated by $x^k + (m)$. With $S = \cap_k S_k$, we want to find an $f \in F[x]$ with $f + (m)$ a non-scalar element of S, or to report that S has dimension 1 over F.

Such an f has the property that if for any k, $a^k \in T$, then $f(a) \in \mathrm{Env}(T)$, since $f(a)$ is a linear combination of powers of a^k. In particular, if T has finite index in G, then $f(a) \in \mathrm{Env}(T)$.

We consider what it means for $f + (m)$ to be in S_k. The algebra R has dimension equal to the degree n of $m(x)$; we can think of elements $r + (m)$ of R as corresponding to the n-tuple of values $r(\alpha_1), \ldots, r(\alpha_n)$, where the α_i are the complex roots of m. To say that $f + (m) \in S_k$ is to say that $f(\alpha_i) = f(\alpha_j)$ whenever $\alpha_i^k = \alpha_j^k$. We will say that k is *relevant* if for some i, j, $\alpha_i \alpha_j^{-1}$ is a primitive kth root of unity; clearly only the relevant k's are important in the definition of S, since for k not relevant, S_k is the intersection of S_ℓ for relevant ℓ dividing k. We have shown:

Lemma 4.5. *Let R, S_k, and S be as above. Then*

$$S = \bigcap_{k \text{ relevant}} S_k$$

□

Our first step in finding f is to determine the set of relevant k. Once this is done, we will search for a polynomial f by solving linear equations in the coefficients of f. We look for f of degree less than $\deg(m)$, so f has n unknown coefficients. The condition $f \in S_k$ is equivalent to a system of linear equations in these coefficients. These equations are given by requiring $f(x)$ to be in the span of $\{x^{k\ell} \bmod m(x) \mid 0 \leq \ell < n\}$. We compute an f by solving simultaneously the equations coming from all relevant k.

Before detailing how we determine the set of relevant k, we investigate a slightly more general problem:

For square-free polynomials $p, q \in F[x]$, we wish to know whether or not p and q have roots α and β such that $\omega = \alpha\beta^{-1}$ is a root of unity, and, if so, we wish to know the orders of all such ω.

The first step is to compute the polynomial $r(x)$ with roots

$$\{\alpha^{-1}\beta \mid p(\alpha) = q(\beta) = 0\}.$$

We compute r as follows: let ℓ be the degree of p, and let n be the degree of q, so

$$p(x) = \sum_{i=0}^{\ell} a_i x^i$$

and

$$q(x) = \sum_{i=0}^{n} b_i x^i.$$

Let $r(x)$ be the determinant of the following $(\ell+n)\times(\ell+n)$ matrix:

$$\begin{pmatrix} a_\ell & a_{\ell-1} & \dots & a_0 & 0 & \dots & 0 \\ 0 & a_\ell & \dots & a_1 & a_0 & \dots & 0 \\ \vdots & \ddots & \ddots & \ddots & \ddots & \ddots & \\ 0 & 0 & \dots & a_\ell & a_{\ell-1} & \dots & a_0 \\ b_n & b_{n-1}x & \dots & b_1x^{n-1} & b_0x^n & \dots & 0 \\ \vdots & \ddots & \ddots & \ddots & \ddots & \ddots & \\ 0 & 0 & \dots & b_n & b_{n-1}x & \dots & b_0x^n \end{pmatrix}$$

We claim:

Lemma 4.6. *Let $p(x)$, $q(x)$, and $r(x)$ be as above. Then the roots of $r(x)$ are exactly the $\alpha\beta^{-1}$, where α ranges over roots of $p(x)$ and β ranges over roots of $q(x)$.*

Proof. First note that $r(1)$ is the familiar resultant of $p(x)$ and $q(x)$. That is, $r(1)=0$ iff p and q have a common root. Indeed, for any number λ, $r(\lambda)$ is λ^n times the resultant of $p(x)$ and $q(x\lambda^{-1})$. So the roots of r are as specified. □

Corollary 4.7. *If polynomials $p(x), q(x) \in F[x]$ have complex roots α and β such that $\alpha\beta^{-1}$ is a primitive kth root of unity, then $\phi(k) \le \deg(p)\deg(q)\deg_{\mathbb{Q}}(F)$ and $k \le 2(\deg(p)\deg(q)\deg_{\mathbb{Q}}(F))^2$.*

Proof. Let $r(x)$ be as above, so $\deg(r) = \deg(p)\deg(q)$. Let ω be a primitive kth root of unity such that $r(\omega) = 0$. Then ω satisfies a polynomial of degree $\deg(p)\deg(q)\deg_{\mathbb{Q}}(F)$ over $\mathbb{Q}$, which must be at least the degree of the minimal polynomial of ω over $\mathbb{Q}$. As $\phi(k)$ is the degree of the minimal polynomial of ω over $\mathbb{Q}$, we have the first inequality. For the second, we need $k \le 2\phi(k)^2$, which follows from the multiplicativity of ϕ and the fact that for primes $s > 2$, $s^t \le (\phi(s^t))^2$, and for $s = 2$ we have $s^t = 2\phi(s^t)$). □

Given the bound for k from Corollary 4.7 we can simply compute the GCD of $r(x)$ and all cyclotomic polynomials of sufficiently low degree. This yields a polynomial time algorithm to compute a basis for S. We have:

Theorem 4.8. *Let $p(x), q(x) \in F[x]$ be given polynomials. In polynomial time, we can determine the set of k such that $\alpha\beta^{-1}$ is a primitive kth root of unity for some complex α, β with $p(\alpha) = q(\beta) = 0$.* □

Letting $p(x) = q(x) = m(x)$, Theorem 4.8 implies we can compute the set of relevant k in polynomial time. By Lemma 4.5, this reduces the computation of f to computing the intersection of a polynomial number of linear subspaces (the S_k). We have:

Corollary 4.9. *Let $m(x) \in F[x]$ be a given square-free polynomial of degree n. In polynomial time, we can either compute a non-constant polynomial $f(x)$ of degree $< n$ such that for all k, $f(x) \in \mathrm{Span}\{x^{k\ell} \bmod m(x) \mid 0 \leq \ell < n\}$, or determine that no such f exists.* □

Note: Bradford and Davenport **[BD]** give efficient algorithms for finding cyclotomic factors of polynomials, including a sharper bound on k in terms of $\phi(k)$. However, actually computing $r(x)$, as in the proof of Lemma 4.6, is probably not the most practical solution to our problem. Corollary 4.7 gives us a small set of candidate k's; we can narrow the choice further by an examination of complex floating point approximations to the roots of p and q, sorted by magnitude. As long as we compute a superset of the relevant k, we will produce an $f(x)$ with $f + (m) \in S$.

Since $m(x)$ is square-free, the algebras R and S are direct sums of fields. In the case that S is not a field, it is desirable to compute an f such that $f + (m)$ is an idempotent. To do this, we let $p_1, \ldots, p_t$ be the irreducible factors of m. We apply Theorem 4.8 to all pairs p_i, p_j. Let $\mathcal{I}$ be the set of those i such that p_i has a root which is the product of a root of p_1 and a root of unity.

Suppose $\mathcal{I} \neq \{1, \ldots, t\}$. Let f be a polynomial which is congruent to 1 modulo p_i for $i \in \mathcal{I}$ and congruent to 0 modulo p_i for $i \notin \mathcal{I}$. The polynomial f exists by the Chinese Remainder Theorem, we find it with the Euclidean Algorithm. We have $f + (m)$ is an idempotent, since $f(\alpha) \in \{0, 1\}$ for all roots α of m. Also, for any relevant k, and for any roots α and β of m such that $\alpha\beta^{-1}$ is a primitive kth root of unity, we have $f(\alpha) = f(\beta)$, so $f + (m) \in S$.

Conversely, if $\mathcal{I} = \{1, \ldots, t\}$, then S is a field. Otherwise, there would be a non-scalar idempotent $f + (m)$. Such an f would have to satisfy $f(\alpha) \in \{0, 1\}$ for all α roots of m. So for all i, f or $f - 1$ has a nontrivial GCD with p_i, so f is congruent to 0 or 1 mod p_i. For any $i \in \mathcal{I}$, let α, β be roots of p_1 and p_i such that $\alpha\beta^{-1}$ is a root of unity. Then $f + (m) \in S$ guarantees that $f(\alpha) = f(\beta)$, so f is congruent to the same value (0 or 1) modulo both p_1 and p_i. Since i was arbitrary, we see that f is a constant polynomial. We have shown:

Lemma 4.10. *Let $m(x) \in F[x]$ be a given square-free polynomial of degree n. In polynomial time we can either compute a non-constant polynomial $f(x)$ of degree $< n$ such that $f(x)^2 \equiv f(x) \bmod m(x)$ and for all k, $f(x) \in \mathrm{Span}\{x^{k\ell} \bmod m(x) \mid 0 \leq \ell < n\}$, or determine that no such f exists.* □

The set $\mathcal{I}$ is an equivalence class of the relation $\sim$, where $i \sim j$ iff p_j has a root which is the product of a root of p_i and a root of unity. We can work just as well with a weaker equivalence relation (i.e., one which gives a coarser partition into equivalence classes), as long as we have an equivalence class which is a proper subset of $\{1, \ldots, t\}$. We suggest a weaker equivalence relation which seems likely to be easier to compute than $\sim$, leading to a speedup in some cases. Let $i \approx j$ denote that $p_i(0)^{\deg(p_j)} p_j(0)^{-\deg(p_i)}$ is a root of unity. The following lemma implies that $\approx$ is a weaker equivalence relation than $\sim$:

Lemma 4.11. *Let $p, q \in F[x]$ be irreducible polynomials and suppose that for some complex α, β with $p(\alpha) = q(\beta) = 0$, $\alpha\beta^{-1}$ is a root of unity. Then $p(0)^{\deg(q)} q(0)^{-\deg(p)}$ is a root of unity.*

Proof. Let Γ be the Galois group of the splitting field of pq. Let $\cong$ denote the following equivalence relation on the roots of pq: we say $\alpha \cong \beta$ if $\alpha\beta^{-1}$ is a root of unity. As Γ permutes the roots of pq, it preserves the partitioning of roots into $\cong$-equivalence classes. The action on the set of classes is transitive, since the roots of p are permuted transitively and there is at least one root of p in each class (since *some* pair of roots α, β with $p(\alpha) = q(\beta) = 0$ satisfies $\alpha \cong \beta$, *each* root α' of p is in an equivalence class with some root β' of q, and vice versa). Therefore there are integers d, e such that each $\cong$-equivalence class has d roots of p and e roots of q; by counting we see that $\deg(p)/d = \deg(q)/e$ is the number ℓ of equivalence classes. Let T be a set of representatives for the equivalence classes of $\cong$. Then

$$p(0) = \omega \prod_{\alpha \in T} \alpha^d,$$

for some root of unity ω. A similar argument applies to $q(0)$, so $p(0)^e q(0)^{-d}$ is a root of unity. Raising this to the power ℓ, we have $p(0)^{\deg(q)} q(0)^{-\deg(p)}$ is a root of unity as well. □

We summarize our algorithms for manipulation of polynomials:

Lemma 4.12. *Let $m(x) \in F[x]$ be a given square-free polynomial of degree n. In polynomial time, we can either compute a non-constant polynomial $f(x)$ of degree $< n$ such that for all k, $f(x) \in \operatorname{Span}\{x^{k\ell} \bmod m(x) \mid 0 \le \ell < n\}$, or determine that no such f exists. In addition, either $f + (m)$ is an idempotent in $F[x]/(m)$, or, for some k, $x^k + (m)$ generates a field in $F[x]/(m)$.* □

4.3. Main reduction. We will proceed as if G is solvable-by-finite, looking for a non-scalar element of $\operatorname{Env}(T)$. This may run into trouble, in which case we deduce that G is not solvable-by-finite. We use the element a returned by the projectively finite test, together with the algorithms of Section 4.2, to help in finding an element of $\operatorname{Env}(T)$. We hope to reduce the solvable-by-finiteness question as in Lemma 4.4.

Note that, instead of reducing the problem, we may encounter a contradiction to our assumption that G is solvable-by-finite, in which case the algorithm halts with a negative answer. For example, we may find a nilpotent element which, assuming G is solvable-by-finite, must be in $\operatorname{Env}(T)$, yet whose nullspace contains no nontrivial proper G-invariant subspace. Then the algorithm terminates; G is not solvable-by-finite by Lemma 4.3.

We now describe the algorithm. By **[BBCIL]**, we may compute the Jordan decomposition of a: we can compute matrices a_d, a_u such that $a = a_d a_u$, both a_d and a_u are in the matrix algebra generated by a, a_d is diagonalizable, and a_u is unipotent.

Suppose first that a_u is not the identity. In this case no power of a is diagonalizable, so by Lemma 4.1 Env(T) has a nontrivial radical. We have $a^k = a_d{}^k a_u{}^k \in \text{Env}(T)$, and $W = \text{Nullspace}(a_u{}^k - I)$ contains a nontrivial proper G-invariant subspace. W is easy to find, since $a_u - I$ and $a_u{}^k - I$ have the same nullspace (consider the rational canonical form). We can then "spin" W: while W is not G-invariant, replace W by $W \cap gW$ for some $g \in G$ with $gW \neq W$. If the resulting subspace is trivial, we know by Lemma 4.3 that G cannot be solvable-by-finite. On the other hand, if we find a proper G-invariant subspace W, then by case (c) of Lemma 4.4 it suffices to test the action of G on W and the action of G on the quotient V/W.

Now suppose that $a_u = I$, so $a = a_d$ is diagonalizable. Let $m(x)$ be the minimal polynomial of a. By Lemma 4.12, we can compute a polynomial f so that $f(a) \in \text{Env}(T)$, and $f(a)$ is either an idempotent or has an irreducible minimal polynomial (this occurs when the algebra generated by $f(a)$, which is isomorphic to the algebra generated by $f + (m)$ in $F[x]/m(x)$, is a field). (Since no power of a is scalar, there are non-constant polynomials f satisfying the criteria of Lemma 4.12.)

Suppose first that $f(a)$ is an idempotent. Let $W = \text{Nullspace}(f(a))$. If all G-conjugates of $f(a)$ commute, then by Lemma 4.1, they generate a direct sum of fields. Since $f(a)$ is a non-scalar idempotent, by Lemma 4.2 we have that W is a direct sum of subspaces from a G-invariant direct sum decomposition. We search for this decomposition assuming that it exists; if our search fails then we will be able to find non-commuting conjugates of $f(a)$.

Algorithmically, we find the direct sum decomposition by a different "spinning" technique. Let $V = W_1 + \cdots + W_r$ be a G-invariant direct sum, with $W = \sum_{i \in I} W_i$ for some (unknown) nonempty $I \subseteq \{1, \ldots, r\}$. As G permutes the W_i, G acts on $\{1, \ldots, r\}$. There are two cases. It may happen that I is a block of imprimitivity for this action of G, in which case W and its G-images lie in direct sum with each other and we are done. Otherwise, for some $g \in G$, $\emptyset \neq I \cap g(I) \neq I$, in which case $0 \neq W \cap gW \neq W$. We can determine which of these cases holds by enumerating G-images of W until we either have a G-invariant direct sum or we have two distinct images $g_1 W$, $g_2 W$ with nontrivial intersection; in the latter case we have $0 \neq W \cap gW \neq W$ for $g = g_1^{-1} g_2$.

We give the spinning algorithm: while $0 \neq W \cap gW \neq W$ for some $g \in G$, replace W by $W \cap gW$. At each step, W is a direct sum of some of the W_i. At the end, W and its G-images will lie in direct sum with each other. G permutes these subspaces.

If W and its images do not span V, then they span a G-invariant subspace U and we proceed as in case (c) of Lemma 4.4; G is solvable by finite iff it acts as a solvable-by-finite group on both U and V/U. Otherwise, standard permutation group techniques (cf. **[Si, FHL]**) allow us to compute Schreier generators for the stabilizer H of W. By case (a) of Lemma 4.4, we may reduce the solvable-by-finite problem to the action $H|_W$ of H on W.

On the other hand, we may find, during the spinning algorithm, nonintersecting G-images of W which do not lie in direct sum with each other. By bookkeeping, we may find two elements x and y of the algebra generated by T which do not commute. At each step, we are working with subspaces of the form Nullspace(x). Initially, $x = f(a)$. We wish to keep track of the x's as we perform the spinning algorithm. We can do this easily, taking advantage of the fact that x is an idempotent: If x and x^g commute, the intersection of Nullspace(x) and g^{-1} Nullspace(x) = Nullspace(x^g) is Nullspace($x + x^g - x \cdot x^g$), where $x + x^g - x \cdot x^g$ is again an idempotent.

Once the spinning algorithm has failed, we will have the desired non-commuting elements x, y. (In practice this should be easy, since *most* pairs of $f(a)$ conjugates should fail to commute.) The element $xy - yx$ is a nilpotent element of the algebra generated by T, and we proceed as in the $a_u \neq I$ case: Nullspace($xy - yx$) contains a nontrivial proper G-invariant subspace, which we find (or report that G is not solvable-by-finite).

Now suppose that $f(a)$ generates a field over F. By construction, $f(a)$ lies in the algebra generated by T. Let E be the algebra generated by G-conjugates of $f(a)$ (this may be bigger than the field generated by $f(a)$). If E is a field, we are in case (b) of Lemma 4.4. We have that G acts as F-automorphisms of E; the kernel is a low-index subgroup H for which we can find Schreier generators. The problem is reduced to the subgroup H (which may be all of G!). This reduction really accomplishes something, since H is E-linear, so we represent H in smaller dimension over E. In this representation, the notion of a scalar matrix is more general, so when the algorithms of the previous sections are run, they will return different results (in particular $f(a)$ is now considered a scalar).

Finally, if E is not a field, then either it contains non-commuting elements x, y, or it is a direct sum of at least two fields by Lemma 4.1. If we have noncommuting x, y, we proceed as above, "spinning" the nullspace of $xy - yx$ to find a nontrivial proper G-invariant subspace W, and reducing the problem as in case (c) of Lemma 4.4. If E is a direct sum of at least two fields, then any basis for E over F contains an element b with a reducible minimal polynomial. Let $p(x)$ be a proper divisor of the minimal polynomial of b, and let W be the nullspace of $p(b)$. Then by Lemma 4.2, W is the direct sum of subspaces from a G-invariant direct sum decomposition, and we proceed as in the case that $f(a)$ is an idempotent.

5. Concluding remarks

The problem of certifying finiteness of a group given by normal generators is open. Success would mean a Las Vegas algorithm for testing finiteness of the conjugation action. We restate the problem: given $S, T \subseteq \mathrm{GL}(n, F)$, let $G = \langle S \rangle$ and let $H = \langle T^G \rangle$; decide if H is finite. The algorithm used in Section 3 yields a Monte Carlo algorithm for this problem with one-sided error—it may incorrectly state that H is finite. A method of proving that H is finite would be desirable. For our application, we are interested in the case that H contains G' and the nth power of any element of G centralizes H.

A practical implementation of the reduction of the previous section would probably require an improved state-of-the-art in the area of working with matrix algebras and polynomials over number fields. However, the n^4 bottleneck has been removed, and a practical implementation seems within reach.

Acknowledgements. This work was stimulated by the 1997 Dagstuhl workshop on Computational Methods for Permutation and Matrix Groups and the 1997 Oberwolfach workshop on Computational Group Theory. Much thanks to the organizers and sponsors of those meetings as well as the 1999 Groups and Computation meeting at the Ohio State University. As I was too sick at the OSU meeting to present this work, I was robbed of much constructive feedback from my colleagues; I thank the referee for providing an extra helping of this. I also thank Gretchen Ostheimer for introducing the additive commutator to me as a tool for handling triangularizable linear groups.

References

[Ba] L. Babai, Deciding finiteness of matrix groups in Las Vegas polynomial time, Proc. 3rd ACM-SIAM SODA (Symp. on Discrete Algorithms), ACM Press, New York, 1992, 33–40.

[BBCIL] L. Babai, R. Beals, J-y. Cai, G. Ivanyos, E. M. Luks, Multiplicative equations over commuting matrices, Proc. 7th ACM-SIAM SODA, ACM Press, New York, 1996, 498–507.

[BBR] L. Babai, R. Beals, D. Rockmore, Deciding finiteness of matrix groups in deterministic polynomial time, in preparation, preliminary version in Proc. ISSAC'93, ACM Press, New York, 1993, 117–126.

[BCFLS] L. Babai, G. Cooperman, L. Finkelstein, E. M. Luks, Á. Seress, Fast Monte Carlo algorithms for permutation groups, J. Comput. System Sci. 50 (1995), 296–308.

[BSz] L. Babai, E. Szemerédi, On the complexity of matrix group problems I, Proc. 25th IEEE FOCS, IEEE Computer Society Press, Los Alamitos, CA, 1984, 229–240.

[Be] R. Beals, Algorithms for matrix groups and the Tits alternative, J. Comput. System Sci. 58 (1999) 260–279.

[BD] R. J. Bradford, J. H. Davenport, Effective tests for cyclotomic polynomials, in: Symbolic and algebraic computation (Rome, 1988) Lecture Notes in Comput. Sci. 358, Springer-Verlag, Berlin–New York, 1989, 244–251.

[FR] K. Friedl, L. Rónyai, Polynomial time solutions of some problems in abstract algebra, Proc. 17th ACM STOC, ACM Press, New York, 1985, 153–162.

[FHL] M. L. Furst, J. Hopcroft, E. M. Luks, Polynomial-time algorithms for permutation groups, in: 21st IEEE FOCS, IEEE Computer Society Press, Los Alamitos, CA, 1980, 36–41.

[GAP] The GAP Group, GAP — Groups, Algorithms, and Programming, Version 4.2; Aachen, St Andrews, 2000 (`http://www-gap.dcs.st-and.ac.uk/~gap`).

[Le] A. K. Lenstra, Factoring polynomials over algebraic number fields, Proc. EUROCAL'83, Lecture Notes in Comput. Sci. 162, Springer-Verlag, Berlin–New York, 1983, 245–254.

[LLL] A. K. Lenstra, H. W. Lenstra, L. Lovász, Factoring polynomials with rational coefficients, Math. Ann. 261 (1982), 515–534.

[Lu] E. M. Luks, Computing in solvable matrix groups, Proc. 33rd IEEE FOCS, IEEE Computer Society Press, Los Alamitos, CA, 1992, 111–120.

[New] M. Newman, Integral Matrices, Pure Appl. Math. 45, Academic Press, New York, 1972.

[Os96] G. Ostheimer, PhD thesis, Dept. of Mathematics, Rutgers University, 1996.

[Os99] G. Ostheimer, Practical algorithms for polycyclic matrix groups, J. Symbolic Comput. 28 (1999), 361–379.

[Si] C. C. Sims, Computation with Permutation Groups, in: Proc. Second Symposium on Symbolic and Algebraic Manipulation (S. R. Petrick, ed.), ACM Press, New York, 1971, 23–28.

[Su63] D. Suprunenko, Soluble and nilpotent linear groups, Transl. Math. Monogr. 9, Amer. Math. Soc., Providence, RI, 1963.

[Su76] D. Suprunenko, Matrix Groups, Transl. Math. Monogr. 45, Amer. Math. Soc., Providence, RI, 1976.

[Ti] J. Tits, Free subgroups in linear groups, J. Algebra 20 (1972), 250–270.

[We] B. A. F. Wehrfritz, Infinite linear groups, Ergeb. Math. Grenzgeb. 76, Springer-Verlag, Berlin, 1973.

IDA Center for Communications Research
29 Thanet Road
Princeton, NJ 08540, U.S.A.
`beals@ccr-p.ida.org`

A constructive recognition algorithm for the matrix group $\Omega(d, q)$

*Peter A. Brooksbank**

Abstract. We present an efficient algorithm to recognise constructively the orthogonal groups $\Omega(d, q)$ in their natural representation.

1991 Mathematics Subject Classification: primary 20C40, 20G40, 20P06; secondary 51F25.

1. Introduction

The matrix group recognition project is a broad and complex area of research, incorporating the work of many different authors (cf. **[LG]**). Among the key ingredients is the need to recognise when a matrix group G, given by arbitrary generators, is a classical group, and, if it is, to use this fact to compute efficiently inside G using the given generators. This latter requirement leads to the notion of a *constructive* recognition algorithm. Namely, suppose that the matrix group $G = \langle \mathcal{S} \rangle \leq \mathrm{GL}(d, q)$ is a classical group. The objective, for a given element g of G, is to write g as a word in $\mathcal{S}$. However, such a word may be very long, so instead we will find a *straightline program* from $\mathcal{S}$ to g.

A straightline program, or *SLP*, from $\mathcal{S}$ is a sequence of elements of $G = \langle \mathcal{S} \rangle$, where each element appearing in the sequence is either in $\mathcal{S}$, or is the inverse of an earlier element in the sequence, or else is the product of two earlier elements. If $g \in G$ is the last element of the SLP, then we say that g has been *constructed using an SLP from* $\mathcal{S}$; we will call the sequence an *SLP from* $\mathcal{S}$ *to* g.

Celler and Leedham-Green **[CeLG]** provided the first example of a constructive recognition algorithm when G is known to contain $\mathrm{SL}(d, q)$, and Celler **[Ce]** later dealt with $G = \mathrm{Sp}(d, q)$. Both algorithms have been implemented, and are available in GAP3.

This type of problem arises in various algorithmic settings (for example, for quotients of matrix groups and for finding Sylow subgroups in permutation groups **[Ka,**

*This research was supported in part by the National Science Foundation.

KLM, Ma, Mo]). In an attempt to solve many such problems at once, constructive recognition has been addressed in the much more general framework of *black box* groups. In [**Bt, BrK, CFL**] constructive recognition algorithms are given for black box groups isomorphic to perfect central extensions of $\mathrm{PSL}(d, q)$, while [**KS**] gives algorithms for all black box classical groups.

In this note, we remain within the context of matrix groups, and outline an algorithm for constructive recognition of all classical groups in their natural representation. In this setting, we are able to use linear algebra more efficiently than is possible inside a black box group, and this gives rise to faster algorithms (cf. Remark (i) below). Furthermore, with a little work, these matrix group algorithms can be inserted into versions of black box algorithms in [**KS**], obviating the need for a recursive call.

Although this paper only deals with the groups $\Omega(d, q)$, the algorithm is easily modified to deal with all classical groups. Moreover, when $G = \mathrm{Sp}(d, q)$ our algorithm has improved asymptotic running time over the algorithm in [**Ce**]. The author has implemented this algorithm for the orthogonal groups in GAP4, and the implementation of the remaining classical groups will be completed soon. We will discuss the main modifications required for the latter in Section 5.

Our algorithm employs both Lie theory and linear algebra and we often change our perspective in order to obtain a clearer understanding of different segments of the algorithm. Moreover, we bring together ideas from several algorithmic sources. For example, the central technique of the algorithm can be viewed as a synthesis of the point-stabiliser methods in [**KS**] and the m-space stabiliser methods in [**Ce**].

Primitive prime divisor methods akin to those in [**KS**] are employed to remove all explicit occurrences of q in the running time. In particular, we show that the polynomial time bottleneck is now the efficient treatment of the low dimensional cases. The recent advance of Conder and Leedham-Green [**CoLG**], reported at this conference, emphasises the practical importance of this seemingly theoretical observation. They present a fast recognition algorithm for $\mathrm{SL}(2, q)$ as a matrix group in its natural representation using an oracle for solving discrete logarithm problems in $\mathrm{GF}(q)^*$. Here, we assume the availability of an oracle for constructive recognition of $\mathrm{SL}(2, q)$, implicitly assuming that this means a discrete logarithm oracle is also available.

Our main result can be stated as follows.

Theorem 1.1. *Let $G = \langle \mathcal{S} \rangle \leq \mathrm{GL}(d, q) = \mathrm{GL}(V)$, where $q = p^k$ for a prime p and a positive integer k, and G is known to be $\Omega(d, q)$, preserving some (unknown) quadratic form on V. In Las Vegas time*

$$O(d^3 \log d \log^4 q + \xi d + \mu d^2 \log q + \chi d \log q),$$

the elements of a new generating set $\mathcal{T}$, of size $O(kd^2)$, can be constructed using straightline programs (SLPs) from $\mathcal{S}$, such that there is a deterministic algorithm to write an SLP of length $O(d^2 \log q)$ from $\mathcal{T}$ to any given element g of G in time $O(d^3 \log q + \chi)$.

Here, ξ is an upper bound on the time requirement per element for the construction of independent, (nearly) uniformly distributed random elements of G, μ is an upper bound on the time required to perform each group operation in G, and χ represents the cost, per call, to SL(2, q) *and discrete logarithm oracles.*

Remarks. **(i)** The timing in **[KS]** for the black box recognition of the groups $P\Omega^{\epsilon}(d, q)$, even without verifying a presentation, is $O(\xi q d \log d(d+\log q)+\mu q d^4 \log^2 q \log^3 d)$. **[KS]**, 3.6.1 gives a black box algorithm for constructive recognition of $SL(2, q)$ in time $\chi = O(\xi q \log q + \mu q \log^2 q)$. Even using this estimate of χ and estimating discrete logs as $O(q)$, the complexity of our algorithm is $O(d^3 \log d \log^4 q + \xi(d + q \log q) + \mu(d^2 \log q + q \log^2 q))$: our algorithm for $\Omega^{\epsilon}(d, q)$ in its natural representation runs significantly faster.

(ii) The basic idea of the algorithm, as in **[CeLG, Ce]**, is to construct a set $\mathcal{T}$ from which it is easy to write an SLP, of modest length, to any given element of G. In **[CeLG]**, $\mathcal{T}$ consists of sufficiently many transvections to perform Gaussian elimination in $SL(d, q)$. *We will describe a more involved analogue of Gaussian elimination for $\Omega(d, q)$, and the set $\mathcal{T}$ will consist of the elements necessary to execute this procedure.*

(iii) We use the parameter μ in order to keep a count of the number of group operations in G that are required, but we note that, since $G \leq GL(d, q)$, $\mu = O(d^3 \log q)$. As in **[KS]**, we assume that $\xi > \mu|\mathcal{S}|$.

(iv) The definition of straightline program given here implies that each element in the sequence is obtained as a matrix by computing inside G. In practice, however, it is not always necessary to write each entry in the sequence as a group element. We will often know the element (matrix) to which we are writing the SLP in advance, and simply wish to record how the element is *constructed* from a certain set of elements. In such instances, the time required to write the SLP will not involve the parameter μ.

(v) The result is stated for all groups $\Omega(d, q)$, but here we only give a proof for d odd. Of course this means that we need not consider $p = 2$, since in this case, $\Omega(d, q) \cong PSp(d - 1, q)$. Nevertheless, we still encounter most of the technical difficulties apparent in the general case while keeping the exposition as simple as possible.

(vi) Either by using Meataxe methods **[HR]**, or alternatively by a polynomial time deterministic algorithm due to Luks, one can construct a G-module isomorphism between the natural module of G and its dual. *We therefore assume that we can find a matrix Φ corresponding to a quadratic form on V preserved by G.*

(vii) The fast Monte Carlo algorithm in **[BCFLS]** for computing the derived group of a matrix group G allows us to extend this result to groups G with $\Omega(d, q) \leq G \leq N_{\Gamma L(d,q)}(\Omega(d, q))$.

2. Background

We assume a basic familiarity with the classical groups and the geometries associated with them. We generally adhere to the definitions and terminology of **[Ta]**. Since we only prove Theorem 1.1 in the case when d and q are odd, we restrict our summary of the necessary background material to this case alone.

Let $G = \Omega(d, q)$, where d is odd, and $q = p^k$ for some odd prime p and positive integer k. Let V denote the underlying vector space of dimension d over $\mathbb{F} = \mathrm{GF}(q)$. Let $\varphi : V \to \mathbb{F}$ denote a nonsingular quadratic form on V, preserved by G. We say that the subspace W of V is *totally singular* (t.s.) in case $\varphi(W) = 0$.

For the remainder of the paper, ρ will always denote a fixed generator of $\mathbb{F}^*$, and $\mathbb{F}_p$ will denote the prime subfield of $\mathbb{F}$.

2.1. Standard bases. Let $m = \lfloor \frac{d}{2} \rfloor$. Replacing φ with $\lambda\varphi$ for non-square $\lambda \in \mathbb{F}$ if necessary, there exists an ordered basis for V of the form $\mathfrak{B} = \{e_1, \dots, e_m, v, f_1, \dots, f_m\}$, where $\varphi(v) = 1$ and $\varphi(e_i) = \varphi(f_i) = (e_i, e_j) = (f_i, f_j) = (e_i, v) = (f_i, v) = 0$, $(e_i, f_j) = \delta_{ij}$ for $1 \le i, j \le m$. Such a basis $\mathfrak{B}$ is called a *standard basis* for V.

Fix a standard basis $\mathfrak{B}$ for V. Relative to $\mathfrak{B}$, all elements of G have the form

$$g = \begin{pmatrix} A & z_1^t & B \\ z_2 & \lambda & z_3 \\ C & z_4^t & D \end{pmatrix}, \tag{2.1}$$

where $A, B, C, D \in \mathbb{M}_{m \times m}(\mathbb{F})$, $z_i \in \mathbb{F}^m$ for $1 \le i \le 4$ and $\lambda \in \mathbb{F}$. Simple matrix relations exist between the entries of g, arising from the fact that g preserves φ; we will use some of these relations later to study the structure of certain key subgroups of G.

Let $\mathfrak{B}_\rho$ be the $\mathbb{F}_p$-basis obtained from $\mathfrak{B}$ by replacing each element w of $\mathfrak{B}$ with the k vectors $\rho^l w$ for $0 \le l < k$. $\mathfrak{B}_\rho$ is ordered lexicographically, by the ordering of $\mathfrak{B}$ first, then by increasing values of l.

2.2. Point stabilisers. Let $x = \langle e_1 \rangle$ and $y = \langle f_1 \rangle$. Then the point stabiliser

$$G_x = Q(x) \rtimes G_{x,y}, \tag{2.2}$$

where $Q(x) = O_p(G_x)$ is the group of isometries inducing 1 on x and on $x^\perp/x$. $Q(x)$ has order q^{d-2} and inherits the structure of an orthogonal space of the same isometry type as V. In fact, $Q(x)$ is isometric to the $d-2$-space $\langle e_1, f_1 \rangle^\perp$ and we have the following $G_{x,y}$-invariant correspondences:

$$\left\{ \begin{array}{c} \text{singular points} \\ \langle w \rangle \text{ in } \langle e_1, f_1 \rangle^\perp \end{array} \right\} \longleftrightarrow \left\{ \begin{array}{c} \text{`long root groups'} \\ \text{of } Q(x) \text{ associated to} \\ \text{the t.s. lines } \langle e_1, w \rangle \end{array} \right\},$$

$$\left\{\begin{array}{c}\text{nonsingular points}\\ \langle w\rangle \text{ in } \langle e_1, f_1\rangle^{\perp}\end{array}\right\} \longleftrightarrow \left\{\begin{array}{c}\text{`root groups'}\\ \text{of } Q(x) \text{ associated to}\\ \text{the lines } \langle e_1, w\rangle\end{array}\right\}. \tag{2.3}$$

The elements of certain root groups will play the rôle of elementary transvections in our analogue of Gaussian elimination.

Let $w \in \langle e_1, f_1\rangle^{\perp}$ be singular. The long root group $R_{e_1 w}$ associated to the t.s. line $\langle e_1, w\rangle$ consists of the linear transformations

$$r_{e_1 w}(\lambda) : u \mapsto u - \lambda(u, w)e_1 + \lambda(u, e_1)w, \tag{2.4}$$

for $\lambda \in \mathbb{F}$. Next, let $w \in \langle e_1, f_1\rangle^{\perp}$ be nonsingular. The root group $R_{e_1 w}$ associated to the line $\langle e_1, w\rangle$ consists of the linear transformations

$$r_{e_1 w}(\lambda) : u \mapsto u - \lambda(u, w)e_1 + \lambda(u, e_1)w - \lambda^2\varphi(w)(u, e_1)e_1. \tag{2.5}$$

The map $\psi_x : \langle e_1, f_1\rangle^{\perp} \to Q(x)$ sending $w \mapsto r_{e_1 w}(1)$ defines an isometry giving rise to the correspondences in (2.3). We similarly define an isometry $\psi_y : \langle e_1, f_1\rangle^{\perp} \to Q(y)$. Starting with the standard $\mathbb{F}_p$-basis $\mathfrak{B}_\rho$ (cf. 2.1), we use these isometries to obtain generating sets $B_x = \psi_x(\mathfrak{B}_\rho)$ and $B_y = \psi_y(\mathfrak{B}_\rho)$ for $Q(x)$ and $Q(y)$ respectively. We refer to the pair (B_x, B_y) as a *standard opposite pair* for $Q(x)$ and $Q(y)$.

2.3. *m*-space stabilisers. Let E and F be the t.s. m-spaces $E = \langle e_1, \dots, e_m\rangle$ and $F = \langle f_1, \dots, f_m\rangle$. Then the subspace stabilisers G_E and G_F are such that

$$G_E = U(E) \rtimes L \quad\text{and}\quad G_F = U(F) \rtimes L.$$

Here $U(E)$ (resp. $U(F)$) is the p-core of G_E (resp. G_F) and $L = G_{E,F}$, the subgroup of G stabilising E and F. Consider the embedding $\iota : \mathrm{GL}(m, q) \hookrightarrow \mathrm{O}(d, q)$ via $\iota : A \mapsto \operatorname{diag}(A, 1, A^{-t})$. Let H denote the subgroup of index 2 of $\mathrm{GL}(m, q)$ consisting of all matrices having square determinant. Then $L = \iota(H)$.

$U = U(F)$ consists of all matrices of the form $u(z, M) = \begin{pmatrix} I & z^t & M \\ & 1 & -2z \\ & & I \end{pmatrix}$, where $M + M^t + 2z^t z = 0$. Hence $Z(U) = \{\, u(0, M) \mid M + M^t = 0 \,\}$ is isomorphic to the $\mathbb{F}$-space of all skew-symmetric matrices. $U/Z(U) \cong \mathbb{F}^m$ and $U(E) = U(F)^t$.

Generators for U: $Z(U)$ is generated by the long root groups $R_{f_i f_j}$. Observe that $F \cong U/Z(U)$ via $f \mapsto r_{fv}(1)Z(U)$, where $r_{fv}(1)$ is an element of the short root group R_{fv} associated to the line $\langle f, v\rangle$. It follows that U is generated by the set

$$\mathcal{T}_F = \{\, r_{f_i f_j}(\rho^l),\ r_{f_i v}(\rho^l) \mid 1 \le i < j \le m,\ 0 \le l < k \,\}. \tag{2.6}$$

Similarly, $U(E)$ is generated by the set

$$\mathcal{T}_E = \{\, r_{e_i e_j}(\rho^l),\ r_{e_i v}(\rho^l) \mid 1 \le i < j \le m,\ 0 \le l < k \,\}. \tag{2.7}$$

Generators for L: $L = \iota(H)$ and $H/\mathrm{SL}(m,q) \cong \langle \mathrm{diag}(\rho^2, 1, \ldots, 1)\rangle$. Let $a = \iota(\mathrm{diag}(\rho^2, 1, \ldots, 1)) \in L$. The image, under ι, of the elementary transvection $x_{ij}(\lambda) = I + E_{ij}(\lambda)$ for $i \neq j$, is the element $r_{f_i e_j}(\lambda)$ in the long root group $R_{f_i e_j}$ associated to the t.s. line $\langle f_i, e_j\rangle$. Hence, L is generated by the set

$$\mathcal{T}_L = \{\, a,\ r_{f_i e_j}(\rho^l) \mid 1 \le i \neq j \le m,\ 0 \le l < k \,\}. \tag{2.8}$$

The generating set $\mathcal{T}$ of Theorem 1.1 *will be the union of the generating sets $\mathcal{T}_E$, $\mathcal{T}_F$ and $\mathcal{T}_L$ for $U(E)$, $U(F)$ and L.*

2.4. Commutator relations in $\Omega(d,q)$. It is difficult, algorithmically, to construct the generating sets $\mathcal{T}_E$, $\mathcal{T}_F$ and $\mathcal{T}_L$ directly. *Instead, we will construct a standard opposite pair (cf.* 2.2*) for groups $Q(\langle e_1\rangle)$ and $Q(\langle f_1\rangle)$ and then use familiar commutator relations among long root elements.* The next lemma summarises these commutator relations.

Lemma 2.1. *Let $G = \Omega(d,q)$, let $r_{e_1 f_j}(\lambda)$, $r_{e_1 e_j}(\lambda)$, for $2 \le j \le m$ and $\lambda \in \mathbb{F}$, be the long root elements of $Q(\langle e_1\rangle)$ defined as linear transformations in* (2.4)*, and let $r_{f_1 e_j}(\lambda)$, $r_{f_1 f_j}(\lambda)$ be the analogous long root elements in $Q(\langle f_1\rangle)$. Then, for $2 \le i \neq j \le m$, we have*

1. $r_{e_1 w}(\lambda)^g = r_{(e_1 g)(wg)}(\lambda)$ *for all* $g \in G$,
2. $[r_{f_1 f_j}(\mu), r_{e_1 e_i}(\lambda)] = r_{e_i f_j}(\lambda\mu) \in L$,
3. $[r_{f_1 f_j}(\mu), r_{e_1 f_i}(\lambda)] = r_{f_i f_j}(\lambda\mu) \in U(F)$,
4. $[r_{f_1 e_j}(\mu), r_{e_1 e_i}(\lambda)] = r_{e_i e_j}(\lambda\mu) \in U(E)$.

2.5. Primitive prime divisors. By a fundamental theorem of Zsigmondy **[Zs]**, if p is prime and $n \ge 2$ then there is a prime dividing $p^n - 1$ but not $p^i - 1$ for $1 \le i < n$, except when either $p = 2$, $n = 6$, or $n = 2$ and p is a Mersenne prime. Such a prime is called a *primitive prime divisor* of $p^n - 1$.

We will call an integer j a $\mathrm{ppd}^\#(p; n)$ if $j \mid p^n - 1$ and either $n = 2$, p is a Mersenne prime, and $4 \mid j$; or j is divisible by a primitive prime divisor of $p^n - 1$. We will call an element g of $\mathrm{GL}(d, p^k)$ a $\mathrm{ppd}^\#(p; n)$-element if $|g|$ is a $\mathrm{ppd}^\#(p; n)$.

Lemma 2.2. *Let $G = \Omega(d,q)$. Then the number of* $\mathrm{ppd}^\#(p; ed)$ *elements in G is at least $\frac{|G|}{2d}$.*

Proof. See **[KS]** or **[NiP]**.

We require a means of testing whether or not a given element of G is a $\mathrm{ppd}^\#(p; n)$-element. The following test is found in **[NP]**.

Lemma 2.3. *Following a one time integer computation taking time $O(n^3 \log n \log^4 p)$, one can test whether or not any given element of G is a $\mathrm{ppd}^{\#}(p; n)$-element in time $O(\mu n \log p)$.*

3. The main algorithm

In the next two sections we present the two algorithms which together prove Theorem 1.1 when $d \geq 5$ is odd. In this section we give a Las Vegas algorithm to construct the desired generating set $\mathcal{T}$. We also discuss the low dimensional groups $\Omega(3, q)$ and $\Omega^+(4, q)$.

3.1. Low dimensional groups. It is well known that $\Omega(3, q) \cong \mathrm{PSL}(2, q)$, and that $\Omega^+(4, q) \cong \mathrm{SL}(2, q) \circ \mathrm{SL}(2, q)$. If $G = \langle \mathcal{S} \rangle = \Omega(3, q)$, then one can find an isomorphism with $\mathrm{PSL}(2, q)$ by specifying images in $\mathrm{SL}(2, q)$, modulo scalars, of the elements of $\mathcal{S}$. Similarly, given $G = \langle \mathcal{S} \rangle = \Omega^+(4, q)$, one can find generators for each of the $\mathrm{SL}(2, q)$ factors in the isomorphic group. In each case, we may then use our hypothesised oracle to complete the recognition of G. The complexity is dominated by the call(s) to the $\mathrm{SL}(2, q)$-oracle, namely $O(\chi)$.

In the proof of Theorem 1.1 for d odd, we encounter the groups $\Omega^+(4, q)$ in the following algorithmic setting. Given $G = \langle \mathcal{S} \rangle = \Omega(d, q) = \Omega(V)$ for $d \geq 5$ odd, and a subset $\mathcal{U}$ of G such that $[V, \mathcal{U}]$ is a nonsingular 4-space of maximal Witt index, we need to recognise, constructively, when $\langle \mathcal{U} \rangle \cong \Omega^+(4, q)$. One can easily compute the action of the elements of $\mathcal{U}$ on $[V, \mathcal{U}]$ to get the 4-dimensional representation, find the two $\mathrm{SL}(2, q)$ factors, and then call the oracle to constructively recognise each factor. Once the isomorphism $\langle \mathcal{U} \rangle \cong \Omega^+(4, q)$ has been established, we will also need to write an SLP from $\mathcal{U}$ to any given element $g \in \Omega^+(4, q)$. This is done using the $\mathrm{SL}(2, q)$-oracle at a cost of $O(\chi)$ per element.

We may assume that whenever $\langle \mathcal{U} \rangle \cong \Omega^+(4, q)$, a call to the constructive test above will confirm this with probability at least 3/4.

3.2. Constructing $\mathcal{T}$. We do this in several stages. First we construct generators for groups $Q(x)$ and $Q(y)$ corresponding to some pair x, y of singular points with $y \notin x^\perp$. We then obtain the elements of a standard opposite pair (B_x, B_y) for $Q(x)$ and $Q(y)$ using SLPs from these constructed generators. Finally, we show how to construct $\mathcal{T}$ using SLPs from $B_x \cup B_y$.

Standard bases. By Remark (vi) following Theorem 1.1, we may assume that we have constructed a $d \times d$ matrix Φ representing a nonsingular quadratic form φ on V, preserved by G.

Let W be any subspace of V. Then we can use linear algebra to compute the perp space $W^{\perp}$ of W with respect to the bilinear form associated to φ. Also, if W is a nonsingular subspace of dimension $n \leq d$, of the same isometry type as V, then it is not hard to see that there is a Las Vegas algorithm to construct a standard basis for W. This involves recursively finding perp-spaces inside W, together with $\lfloor n/2 \rfloor$ calls to the discrete log oracle. The latter computations dominate the complexity, giving a running time of $O(n\chi)$.

Constructing $Q(x)$. This part of the routine is a brief commentary on the methods of **[KS]** for finding $Q(x)$ in the case that we are interested in. In fact, we assume that $q \geq 17$, since we can find $Q(x)$ for smaller q, within the required time constraints, using **[KS]**, 4.2.1, Case 1.

Step 1. Our first task is to find generators for a naturally embedded $\Omega^+(4, q)$ subgroup of G. Choose up to $16(d-3)$ random elements of G, and use Lemma 2.3 to find one, τ, such that $|\tau|$ is divisible by a $\mathrm{ppd}^{\#}(p; k(d-3))$ *and* by a $\mathrm{ppd}^{\#}(p; k)$. Set $z \leftarrow q^{(d-3)/2} + 1$ and $\sigma \leftarrow \tau^z$. Choose up to 2^{10} random elements g of G, and for each use 3.1 and the $\mathrm{SL}(2, q)$-oracle to test whether the group $\langle \sigma, \sigma^g \rangle$ is isomorphic to $\Omega^+(4, q)$. If so, then set $J \leftarrow \langle \sigma, \sigma^g \rangle$. Then J is a naturally embedded $\Omega^+(4, q)$ subgroup.

Remark 3.1. In practice, it would be inefficient to make 2^{10} calls to a constructive oracle. Instead, one could use a much faster implicit recognition algorithm for $\Omega^+(4, q)$, and run the constructive algorithm only when we are fairly sure we have the group we require.

Correctness. Each element τ of order divisible by a $\mathrm{ppd}^{\#}(p; k)$ and a $\mathrm{ppd}^{\#}(p; k(d-3))$ must split V as $V = V_1 \perp V_2^+ \perp V_{d-3}^-$ of V, inducing an element of order dividing $q-1$ on V_2^+ and of order dividing $z = q^{\frac{d-3}{2}} + 1$ on V_{d-3}. Since $(q-1, z) = 2$, $\sigma = \tau^z$ induces an element of $\mathrm{ppd}^{\#}(p; e)$-order on V_2^+ and 1 on $V_2^{+\perp}$. In particular, if $J \cong \Omega^+(4, q)$, then J has 4-dimensional support $[V, J] = V_2^+ \oplus V_2^{+g}$, so J is a naturally embedded $\Omega^+(4, q)$ subgroup.

Reliability. By Lemma 2.2, there are at least $\frac{|G|}{4(d-3)}$ elements of G having the desired order, so that at least one of our random choices succeeds with probability $> 1 - e^{-4}$. By **[KS]**, Lemma 4.12, two $\mathrm{ppd}^{\#}(p; k)$ elements, having two dimensional support, generate a naturally embedded $\Omega^+(4, q)$ with probability at least 1/640. For such a group, the test in 3.1 will confirm this constructively with probability at least 3/4. Therefore, for each choice of g, with probability at least $\frac{3}{4} \times \frac{1}{640}$, we constructively recognise a group $\langle \sigma, \sigma^g \rangle \cong \Omega^+(4, q)$, so that at least one of our 2^{10} choices σ^g gives the desired group with probability $\geq 1 - (1 - 3/2560)^{2^{10}} > 1 - 1/e$.

Complexity. $O(d^3 \log d \log^4 q + \xi d + \mu d^2 \log q)$ for the $O(d)$ ppd tests (using Lemma 2.3) together with $O(\chi)$ for the 2^{10} calls to the test in 3.1.

Step 2. Compute $V_2^+ \leftarrow [V,\sigma]$ and $V_4^+ \leftarrow \langle V_2^+, [V,\sigma^g]\rangle$. Find the two singular points x and y of V_2^+ (these are the unique points of V_2^+ fixed by σ). Use 3.1 and the SL(2, q)-oracle to find a $2k$-element (where $k = \log_p q$) generating set $\mathcal{A}_J$ for the group $Q_J(x) = O_p(J_x)$. Compute a standard basis $\{e_1, e_2, f_1, f_2\}$ for V_4^+ such that $x = \langle e_1\rangle$ and $y = \langle f_1\rangle$. Use the oracle again to construct the elements $j(y) : e_i \mapsto f_i,\ f_i \mapsto e_i$ for $i = 1, 2$, and $a : e_1 \mapsto \rho^2 e_1,\ e_2 \mapsto e_2,\ f_1 \mapsto \rho^{-2} f_1,\ f_2 \mapsto f_2$ of J. Note that the oracle constructs each element of $\mathcal{A}_J \cup \{a, j(y)\}$ using an SLP from $\{\sigma, g\}$.

Complexity. $O(\chi \log q)$ to construct $\mathcal{A}_J$ and the elements a and $j(y)$.

Step 3. We now give a Monte Carlo algorithm to construct the group $Q(x) = O_p(G_x)$: set $\mathcal{A} \leftarrow \bigcup_{i=0}^{d-2} \mathcal{A}_J^{\tau^{i(q-1)}}$ and return $\langle \mathcal{A}\rangle$.

Correctness. We need to show that with high probability, $\langle \mathcal{A}\rangle$ is the target group $Q(x)$.

Note that $\langle \mathcal{A}_J\rangle = Q_J(x)$ is a hyperbolic line of the target group $Q(x)$ and that τ acts irreducibly on a hyperplane of $Q(x)$. If $Q_J(x)$ is contained in this hyperplane, then the normal closure $\langle Q_J(x)^{\langle\tau\rangle}\rangle$ is precisely the hyperplane; otherwise it is all of $Q(x)$. The probability that the hyperbolic line lies inside a given hyperplane is at most $\frac{1}{q-1} \le \frac{1}{16}$. If $Q_J(x)$ is not contained in the hyperplane of τ, then by an argument similar to **[KS]** Lemma 3.9, $\langle Q_J(x)^{\langle\tau\rangle}\rangle = \langle \mathcal{A}\rangle$. Hence, with probability at least $\frac{15}{16}$, $Q(x) = \langle \mathcal{A}\rangle$.

Complexity. $O(\mu d k)$ for the construction of $\mathcal{A}$.

Remark 3.2. In the unlikely event that $\langle \mathcal{A}\rangle$ is only a hyperplane of $Q(x)$, this failure will be uncovered when we construct the opposite pair (B_x, B_y) below. Hence, we will soon upgrade the algorithm for $Q(x)$ to Las Vegas.

Total Reliability. With probability at least $1 - (\frac{1}{e} + \frac{1}{e^4} + \frac{1}{16}) > \frac{1}{2}$, we constructed a set $\mathcal{A}$, and elements $j(y)$ and a, such that $Q(x) = \langle \mathcal{A}\rangle$, $Q(y) = \langle \mathcal{A}^{j(y)}\rangle$ and a acts as an element of order $(q-1)/2$ on V_2^+, and is the identity on the perp of this space.

The remainder of the algorithm is deterministic.

Changing basis. Compute a standard basis for the $d-4$-space $V_4^{+\perp}$. Augment it with the one found earlier for V_4^+ to obtain a standard basis $\mathfrak{B} = \{e_1, \ldots, e_m, v, f_1, \ldots, f_m\}$ for V. Write each element of $\mathcal{S} \cup \mathcal{A}$, together with the elements $j(y)$ and a, relative to $\mathfrak{B}$.

Complexity. $O(\mu(|\mathcal{S}| + d\log q) + d\chi)$.

The opposite pair (B_x, B_y). In the next part of the algorithm, we will construct the nice $\mathbb{F}_p$-bases B_x and B_y for the groups $Q(x)$ and $Q(y)$ described in Section 2.2. Observe that we need only find B_x, since $B_y = B_x^{j(y)}$, which is easily seen by taking an element of B_x and conjugating with $j(y)$ using Lemma 2.1(1).

Step 0. We first describe simple, fast, procedures for computing with the isometry $\psi_x : \langle e_1, f_1\rangle^\perp \to Q(x)$ (cf. Section 2.2), thereby demonstrating that linear algebra in $Q(x)$ relative to $B_x = \psi_x(\mathfrak{B}_\rho)$ is algorithmically equivalent to linear algebra in $\langle e_1, f_1\rangle^\perp$ relative to $\mathfrak{B}_\rho$.

Given any $w \in \langle e_1, f_1\rangle^\perp$, it is easy to write down the matrix representing the linear transformation $\psi_x(w) = r_{e_1 w}(1) \in Q(x)$. Conversely, let $u \in Q(x)$. Compute $f_1 u$ and ignore coefficients of e_1 and f_1 in order to obtain a vector w in $\langle e_1, f_1\rangle^\perp$.

Step 1. Using linear algebra in $Q(x)$, find an $\mathbb{F}_p$-basis for $Q(x)$ inside $\mathcal{A}$, and discard the remaining elements of $\mathcal{A}$.

Note: If $\mathcal{A}$ generates only a hyperplane of the target group $Q(x)$, then we will uncover this failure at this point in the algorithm. Hence, we have now upgraded the algorithm for finding $Q(x)$ to Las Vegas (cf. Remark 3.2).

Step 2. We know the elements of our target standard basis B_x but we now need to *construct* them using SLPs from $\mathcal{A}$. Write each element of $\mathcal{A}$ as an $\mathbb{F}_p$-linear combination of elements from B_x. This gives rise to a $k(d-2) \times k(d-2)$ base change matrix C from $\mathcal{A}$ to B_x. Set $D \leftarrow C^{-1}$. Then the rows of D are $\mathbb{F}_p$-vectors of length $k(d-2)$ giving each element of B_x as a linear combination of the $\mathbb{F}_p$-basis $\mathcal{A}$ for $Q(x)$. Use these $\mathbb{F}_p$-vectors to write down SLPs of length $O(d \log q)$ from $\mathcal{A}$ to each element of B_x.

Complexity. The only thing being used here is linear algebra on a $k(d-2)$ dimensional vector space over $\mathbb{F}_p$. Each of the $O(kd)$ SLPs has length $O(d \log q)$, giving a total time of $O(d^2 \log^2 q)$.

From (B_x, B_y) to $\mathcal{T}$. It is well known that G is generated by the groups $Q(x)$ and $Q(y)$. Let $\mathcal{T}$ be the union

$$\mathcal{T} = \mathcal{T}_L \cup \mathcal{T}_E \cup \mathcal{T}_F,$$

where $\mathcal{T}_L$, $\mathcal{T}_E$ and $\mathcal{T}_F$ are respectively the generating sets for the groups $U(F)$, $U(E)$ and L, defined in (2.6),(2.7) and (2.8) respectively. We now show how to construct the elements of these three sets using SLPs from $B_x \cup B_y$.

Constructing $\mathcal{T}_L$: By Lemma 2.1(2), for $0 \le l < k$ and $2 \le i < j \le m$, $[r_{f_1 f_j}(\rho^l), r_{e_1 e_i}(1)] = r_{e_i f_j}(\rho^l)$, so that each root element of $\mathcal{T}_L$ is obtained using an SLP of length 4 from $B_x \cup B_y$. *Time:* $O(d^2 \log q)$, since we need not compute each commutator inside G (cf. Remark (iv) following Theorem 1.1).

Constructing $\mathcal{T}_E$ and $\mathcal{T}_F$: We do this in three stages. First, use Lemma 2.1(3,4) to write SLPs of length 4 from $B_x \cup B_y$ to the long root elements $r_{f_i f_j}(\rho^l) \in U(F)$ and $r_{e_i e_j}(\rho^l) \in U(E)$ for $0 \le l < k$ and $1 \le i < j \le m$. *Time:* $O(d^2 \log q)$.

Next, let c be any element of L such that $c : e_1 \mapsto e_2 \mapsto \ldots \mapsto e_m$. Use 4.1 to write an SLP of length $O(d)$ from $\mathcal{T}_L$ to such a c. *Time:* $O(d^2 \log q)$, since only $O(d)$ row/column operations are required.

Finally, we use c to construct the remaining short root elements of $\mathcal{T}_E$ and $\mathcal{T}_F$. Since $c \in L$, it follows that for each short root element $r_{f_1 v}(\rho^l)$ $(0 \le l < k)$,

$r_{f_1 v}(\rho^l)^{c^i} = r_{(f_1 c^i)(v c^i)}(\rho^l) = r_{f_{i+1} v}(\rho^l)$ for $1 \le i < m$. Hence, we obtain SLPs of length $O(\log m)$ from $B_x \cup B_y \cup \{c\}$ to the short root elements of $\mathcal{T}_E$ and $\mathcal{T}_F$. *Time:* $O(kd \log d)$.

This completes the construction of the generating sets $\mathcal{T}_E$ and $\mathcal{T}_F$, and hence the construction of $\mathcal{T}$. The total cost of this construction is $O(d^2 \log q)$.

Remark 3.3. The generating sets $\mathcal{T}_L$, $\mathcal{T}_E$ and $\mathcal{T}_F$ are the analogues of the generating sets S, W_u and W_l used in **[Ce]** when $G = \mathrm{Sp}(d, q)$. Observe, however, that the groups $U(E)$ and $U(F)$ are nonabelian here.

Some useful elements. We conclude this section by constructing elements $w_0, \ldots, w_{m-1}$ of $U(E)$ needed in the next section. These elements are products of short root elements, and we define them recursively as follows:

$$w_{m-1} = r_{e_m v}(1), \quad \text{and} \quad w_i = w_{i+1} r_{e_{i+1} v}(1), \quad \text{for } 0 \le i < m-1.$$

Each w_i can therefore be expressed using SLPs of length at most m from $\mathcal{T}$.

Total time to construct $\mathcal{T}$. Adding together all of the complexity calculations from the various parts of 3.2, we arrive at the timing stated in Theorem 1.1.

4. Straightline programs

In this section, we give a deterministic algorithm which will write an SLP of length $O(d^2 \log q)$ from the constructed generating set $\mathcal{T}$ to any given element $g \in G$. This will complete the proof of Theorem 1.1. Our approach is analogous to that used in **[Ce]** for $G = \mathrm{Sp}(d, q)$. We omit many technical details involving matrix relations satisfied by elements of our orthogonal group, preferring rather to give an overview of the procedure.

We may assume that a given element $g \in G$ is written relative to $\mathfrak{B}$. In particular, we may assume that g has the block matrix form of (2.1). Our strategy is to pre- or post-multiply g by certain elements from our key subgroups L, $U(E)$, $U(F)$ in order to filter g down the following short chain of subgroups:

$$G > G_E > G_{E,F} = L > \{I\}.$$

We therefore begin by giving subroutines to write an SLP to any given element $g \in L$ (resp. $U(E)$, $U(F)$) from $\mathcal{T}_L$ (resp. $\mathcal{T}_E$, $\mathcal{T}_F$).

4.1. SLPs from $\mathcal{T}_L$. Let $g \in L$ be given. Then $g = \mathrm{diag}(A, 1, A^{-t})$ for some $A \in H < GL(m, q)$ (cf. 2.3). Use Gaussian elimination simultaneously on the upper and lower blocks to write an SLP of length $O(d^2 \log q)$ from the set of root elements of $\mathcal{T}_L$ to an element g_1 such that $gg_1 = \iota(\lambda, 1, \ldots, 1)$. Since $gg_1 \in L$, it follows that λ

is a square; use the discrete log oracle to find $0 \le n < (q-1)/2$ such that $\lambda = (\rho^2)^n$. Hence $gg_1 = a^n$. Write an SLP of length $\log q$ from $\{a\}$ to a^n, and concatenate with the inverse of the SLP for g_1, to obtain the desired SLP from $\mathcal{T}_L$ to g.

Complexity. $O(d^3 \log q + \chi)$ for Gaussian elimination and one call to the discrete log oracle.

4.2. SLPs from $\mathcal{T}_E$ and $\mathcal{T}_F$. Since $U(E)$ is simply the transpose of $U(F)$, we will just discuss the latter. We again remark that the group $U(F)$ is nonabelian.

Let $u \in U = U(F)$ be given. Then $u = u(z, M)$ for some z, M with $M + M^t + 2z^t z = 0$ (cf. 2.3). Identify z with a vector in the t.s. m-space F and compute the $\mathbb{F}_p$-vector $(\lambda_i^{(l)} \mid 1 \le i \le m, 0 \le l < k)$ representing z relative to the $\mathbb{F}_p$-basis $\{\rho^l f_i \mid 0 \le l < k, 1 \le i \le m\}$ for F. This is an ordered basis, ordered lexicographically by i first, then l. Write an SLP of length $O(d \log q)$ from $\mathcal{T}_F$ to the element

$$w = \prod_{i=1}^{m} \prod_{l=0}^{k-1} r_{f_i v}(\rho^l)^{\lambda_i^{(l)}},$$

where the order of the product coincides with that of the basis. Then $uw^{-1} = u(0, M') \in Z(U)$, for some skew-symmetric matrix M' (note that the matrix of w is easily computed without evaluating the SLP, and that using a different ordering for the product would still give rise to a central element, but would yield a different M'). It is now an easy matter to write an SLP of length $O(d^2 \log q)$ from $\mathcal{T}_F$ to uw^{-1}, since the elements of $Z(U)$ in $\mathcal{T}_F$ have the form $u(0, E_{ij}(\rho^l) - E_{ji}(\rho^l))$, for $1 \le i < j \le m$, $0 \le l < k$.

Complexity. $O(d^2 \log q + \mu)$ for finding the matrix of w and computing uw^{-1}.

4.3. The general case. Let $g \in G$ be given, as in (2.1).

Step 1. *Reduce to* $g \in G_E$. There are two cases to consider.
Case 1: A is nonsingular. Since g preserves (,), we obtain the matrix relation $AB^t + BA^t + 2z_1^t z_1 = 0$, from which it follows that

$$u = u(z_1 A^{-t}, A^{-1}B) = \begin{pmatrix} I & A^{-1}z_1^t & A^{-1}B \\ & 1 & -2z_1 A^{-t} \\ & & 1 \end{pmatrix} \in U(F).$$

Use 4.2 to write an SLP from $\mathcal{T}_F$ to u. Then $g \leftarrow gu^{-1} \in G_E$, as required.

Case 2: $\operatorname{rank}(A) = r < m$. We reduce to case 1. Observe that pre- and post-multiplying g by elements of $\mathcal{T}_L$ simultaneously effects elementary row and column operations on each of the entries A, B, C, D. Using two-sided Gaussian elimination, find SLPs from $\mathcal{T}_L$ to elements $g_1, g_2 \in L$ such that $g_1 g g_2$ has entry 'A' in (2.1) of the form $\left(\begin{smallmatrix} I_r & 0 \\ 0 & 0 \end{smallmatrix}\right)$.

Note that we do not need to evaluate these SLPs (i.e. find the matrices g_1 and g_2) in order to write down $g_1 g g_2$; we simply perform each elementary row/column operation on g.

It is not hard to check that $g_1 g g_2 w_r$, where w_r is one of the useful elements constructed at the end of Section 3, has entry 'A' in (2.1) of the form $\begin{pmatrix} I_r & 0 \\ 0 & A_{m-r} \end{pmatrix}$, where A_{m-r} is a nonsingular $(m-r) \times (m-r)$ matrix. In particular, this 'A' is now nonsingular. Set $g \leftarrow g_1 g g_2 w_r$, and use case 1.

Case 1 can be thought of as an algorithmic demonstration of the well known transitivity of $U(F)$ on the set of t.s. m-spaces opposite F (i.e., those having trivial intersection with F). More precisely, in case 1, Eg is opposite F, and we constructed $u \in U(F)$ such that $Egu = E$.

Step 2. *Reduce to* $g \in L$. Since $g \in G_E$, Fg is opposite E. As above, we can find an element $u' \in U(E)$, directly from the matrix of g, such that $Fgu' = F$. Use 4.2 to write an SLP from $\mathcal{T}_E$ to u' and set $g \leftarrow gu'$.

Step 3. *Completion.* Since $g \in L$, we can use 4.1 to write an SLP from $\mathcal{T}_L$ to g.

By concatenating the various SLPs obtained during the three stages of the filtering process, we obtain the desired SLP of length $O(d^2 \log q)$ from $\mathcal{T}$ to g.

Complexity. The entire routine involves only multiplying matrices, computing the inverses of matrices, and calls to 4.1 and 4.2: time $O(d^3 \log q + \mu + \chi) = O(d^3 \log q + \chi)$.

Together with the timing estimate for Section 3, this completes the proof of Theorem 1.1.

5. Other classical groups

We mentioned earlier that the algorithm presented here is easily modified to deal with the other classical groups. In this section, we discuss the most significant alterations, including those needed for the remaining isometry types of orthogonal groups.

Constructing $Q(x)$. In our routine to construct $Q(x)$, we first found an element τ acting irreducibly on a subspace of large dimension, while centralising an element σ having two dimensional support. In order to achieve this goal for each of $G = \Omega^{\pm}(2n, q)$, $\mathrm{SU}(2n, q)$, $\mathrm{SU}(2n+1, q)$, $\mathrm{Sp}(2n, q)$, we use elements τ having slightly different orders than the element employed here.

We then found a naturally embedded $\Omega^+(4, q)$ subgroup J; when $G = \mathrm{Sp}(d, q)$ (respectively $\mathrm{SU}(d, q)$) we instead use a naturally embedded $\mathrm{Sp}(4, q)$ (respectively $\mathrm{SU}(4, q)$) subgroup. Once again, the key to polynomial time efficiency lies in the treatment of the low dimensional groups $\mathrm{Sp}(4, q)$, $\mathrm{SU}(3, q)$ and $\mathrm{SU}(4, q)$.

Finally, when $G = \mathrm{Sp}(d, q)$ or $\mathrm{SU}(d, q)$, we confront the additional problem that $Q(x)$ is usually nonabelian. However, in these cases, we always have $Z(Q) = \Phi(Q) = T$ (where $\Phi(Q)$ is the Frattini subgroup and T is the group of $(x, x^{\perp})$-transvections), and we use the action of τ on Q/T to generate Q. In some sense, we dealt with the most difficult case in Section 3.2, since there, τ did not act irreducibly on Q.

Suppose that the algorithm has reached the stage where it has constructively recognised a naturally embedded four dimensional subgroup J, and has extracted generators for the analogue Q_J of Q inside J. Then, in the cases where τ acts irreducibly on Q or on Q/T, the normal closure $\langle Q_J^{\langle\tau\rangle}\rangle$ is *guaranteed* to be all of Q; this was only *probably* true in 3.2. Hence, the subroutine constructing Q which follows the construction of the subgroup J is deterministic in many of the remaining cases.

Constructing $\mathcal{T}$. Again, $\mathcal{T}$ is comprised of generators for the analogues of the groups $L, U(E), U(F)$. The structure of these groups in each case is similar to that studied here.

There are also versions of the commutator relations which allow us to pass from point stabilisers to m-space stabilisers, where m is the Witt index. Together with generators for appropriate transvection groups (when G is $\mathrm{Sp}(d, q)$ or $\mathrm{SU}(d, q)$), we again use these relations to directly construct $\mathcal{T}$ from an opposite pair (B_x, B_y).

Straightline programs. Decomposing a given element $g \in G$ as a product of elements from the subgroups $L, U(E), U(F)$, as we did in Section 4, is once again similar for the other classical groups, and is somewhat simpler for $\Omega^+(d, q)$ and $\mathrm{Sp}(d, q)$.

References

[BCFLS] L. Babai, G. Cooperman, L. Finkelstein, E. M. Luks and Á. Seress, Fast Monte Carlo algorithms for permutation groups, J. Comput. System Sci. 50 (1995), 296–308.

[Bt] S. Bratus, Recognition of finite black box groups, Ph.D. Thesis, Northeastern Univ. 1999.

[BrK] P. A. Brooksbank and W. M. Kantor, On Constructive Recognition of a Black Box PSL(d, q), in: Groups and Computation III (W. M. Kantor and Á. Seress, eds.), Ohio State Univ. Math. Res. Inst. Publ. 8, Walter de Gruyter, Berlin–New York, 2001, 95–111.

[Ce] F. Celler, Matrixgruppenalgorithmen in GAP. Ph. D. thesis, RWTH Aachen 1997.

[CeLG] F. Celler and C. R. Leedham-Green, A constructive recognition algorithm for the special linear group, in: The Atlas of Finite Groups: Ten Years On (Birmingham 1995), London Math. Soc. Lecture Note Ser. 249, Cambridge Univ. Press, 1998, 11–26.

[CoLG] M. Conder, C. R. Leedham-Green, Fast recognition of classical groups over large fields, in: Groups and Computation III (W. M. Kantor and Á. Seress, eds.), Ohio State Univ. Math. Res. Inst. Publ. 8, Walter de Gruyter, Berlin–New York, 2001, 113–121.

[CFL] G. Cooperman, L. Finkelstein and S. Linton, Constructive recognition of a black box group isomorphic to $\mathrm{GL}(n, 2)$, in: Groups and Computation II (L. Finkelstein and W. M. Kantor, eds.), DIMACS Ser. Discrete Math. Theoret. Comput. Sci. 28, Amer. Math. Soc., Providence, RI, 1997, 85–100.

[HR] D. F. Holt and S. Rees, Testing modules for irreducibility, J. Austral. Math. Soc. (Ser. A) 57 (1994) 1–16.

[Ka] W. M. Kantor, Sylow's theorem in polynomial time, J. Comp. Syst. Sci. 30 (1985) 359–394.

[KLM] W. M. Kantor, E. M. Luks and P. D. Mark, Sylow subgroups in parallel, J. Algorithms 31 (1999) 132–195.

[KS] W. M. Kantor and Á. Seress, Black box classical groups, Mem. Amer. Math. Soc. 149, Nr. 708, Amer. Math. Soc., Providence, RI, 2001.

[LG] C. R. Leedham-Green, The Computational Matrix Group Project, in: Groups and Computation III (W. M. Kantor and Á. Seress, eds.), Ohio State Univ. Math. Res. Inst. Publ. 8, Walter de Gruyter, Berlin–New York, 2001, 229–247.

[Ma] P. D. Mark, Sylow's theorem and parallel computation, Ph. D. thesis, Univ. of Oregon, 1993.

[Mo] P. Morje, A nearly linear algorithm for Sylow subgroups of permutation groups, in: Proc. Internat. Symp. Symbolic and Algebraic Computation, ACM Press, 1995, 270–277.

[NP] P. M. Neumann and C. E. Praeger, A recognition algorithm for special linear groups, Proc. London Math. Soc. (3) 65 (1992), 555–603.

[NiP] A. C. Niemeyer and C. E. Praeger, A recognition algorithm for classical groups over finite fields, Proc. London Math. Soc. (1) 77 (1998), 117-169.

[Ta] D. E. Taylor, The geometry of the classical groups, Heldermann, Berlin 1992.

[Zs] K. Zsigmondy, Zur Theorie der Potenzreste, Monatsh. Math. Phys. 3 (1892), 265–284.

Department of Mathematics
University of Oregon
Eugene, OR 97403, U.S.A.
brooks@geometry.uoregon.edu

On constructive recognition of a black box $\mathrm{PSL}(d, q)$*

Peter A. Brooksbank and William M. Kantor

Abstract. Assuming the availability of an oracle for handling black box groups isomorphic to $\mathrm{SL}(2, q)$, we present an algorithm that constructively recognises a group G known to be isomorphic to $\mathrm{PSL}(d, q)$ for known $d \geq 3$ and q. We also outline an analogous result for symplectic groups.

1991 Mathematics Subject Classification: primary 20C40, 20G40, 20P06; secondary 51F25.

1. Introduction

The presently known algorithms for constructive recognition of black box classical simple groups have asymptotic running times that are not polynomial **[Br, KS]**: the complexity contains factors of q (the field size), whereas the input size involves only $\log q$. In this paper, we will take a step toward the polynomial time paradigm for the $\mathrm{PSL}(d, q)$ case, removing all factors of q in the timing at the expense of making calls to an oracle that allows us to work inside subgroups isomorphic to $\mathrm{SL}(2, q)$; thus, $\mathrm{SL}(2, q)$ is the polynomial-time bottleneck.

Consequently, we pay close attention to the cost χ of each call to our hypothesised $\mathrm{SL}(2, q)$-oracle, in addition to the more standard timing parameters used in **[KS]** (namely, the input length, which is greater than $(d^2 \log q)/2$, together with the time μ required to perform group operations in G, and the time ξ required per element for the construction of independent, (nearly) uniformly distributed random elements of G). The motivation for this oracle hypothesis is provided by a recent advance by Leedham-Green and Conder reported at this conference **[CLG]**: an algorithm that deals with the groups $\mathrm{SL}(2, q)$ as 2×2 matrix groups assuming the availability of an oracle for the discrete log problem. This advance prompted Leedham-Green to ask us the question answered by the following result (where SLP stands for "straight-line program"; cf. **[KS]**, 2.2.5):

*This research was supported in part by the National Science Foundation.

Theorem 1.1. *Let* $G = \langle \mathcal{S} \rangle$ *be a black box group, believed to be isomorphic to a nontrivial homomorphic image of* $\mathrm{SL}(d, q)$ *for known* $q \geq 17$ *and* $d \geq 3$*, and equipped with an oracle for handling black box groups isomorphic to* $\mathrm{SL}(2, q)$*. Then the following hold.*

(1) *There is a Las Vegas algorithm which, with probability* $\geq \frac{1}{2}$*, in time polynomial in* ξ, χ, μ, d *and* $\log q$*, namely in*

$$O(d^2 \log(d \log q) \log q\{\xi + \chi d \log q + \mu d^2 \log^2 q\} + |\mathcal{S}| \log(d|\mathcal{S}|)\{\xi + \chi d^2 \log q + \mu d^2 \log^2 q + d^5 \log q\})$$

time, verifies that G *is a homomorphic image of* $\mathrm{SL}(d, q)$ *and constructs a new generating set* $\mathcal{S}^*$ *for* G*, a generating set* X *for* $\mathrm{SL}(d, q)$*, and a bijection* $\Psi\colon X \to \mathcal{S}^*$ *that extends to an epimorphism* $\Psi\colon \mathrm{SL}(d, q) \to G$*.*

(2) X*,* $\mathcal{S}^*$ *and* Ψ *have the following properties.*

(a) *In* $O(d\chi + \mu d^2 \log q)$ *time, when given* $A \in \mathrm{SL}(d, q)$ *one can construct an SLP of length* $O(d^2 \log q)$ *from* X *to* A*, and then mimic it in* G *to obtain* $\Psi(A)$*.*

(b) *In Las Vegas* $O(\log d\{\xi + \chi(d + \log q) + \mu d \log^2 q\} + d^5 \log q)$ *time, with probability* $\geq 1 - \frac{1}{8d^2}$ *an SLP of length* $O(d^2 \log q)$ *can be found from* $\mathcal{S}^*$ *to any given* $g \in G$*, and then mimicked in* $\mathrm{SL}(d, q)$ *to obtain* $\Psi^{-1}(g)$ *modulo scalars.*

Note that failure in (1) could be due to bad luck with random selections, or to the fact that G is not isomorphic to the stated type of quotient group; but if there is an output then it is guaranteed to be correct (since the algorithm is Las Vegas). The lower bound on q is designed to simplify our arguments; the algorithm of **[KS]** applies for the remaining small q. This paper is intended, primarily, as a commentary on the $\mathrm{PSL}(d, q)$ algorithm presented in **[KS]**. We frequently refer to results and routines contained therein with little further comment, and will generally adhere to notational conventions used there.

The case $d = 3$ is the hardest one we consider (cf. Section 3). The general case, in Section 2, needs only a few non-obvious modifications of **[KS]**. We note that we use our $\mathrm{SL}(2, q)$-oracle for various subgroups isomorphic to $\mathrm{SL}(2, q)$, many of which involve somewhat random generators; it would have been far better to have used the oracle only a few times. However, it is hard to imagine not needing at least $|\mathcal{S}|$ oracle calls in order to verify that $\langle \mathcal{S} \rangle = \langle \mathcal{S}^* \rangle$.

We have only dealt with homomorphic images of $\mathrm{SL}(d, q)$, but homomorphic images of other subgroups of $\mathrm{Aut}\,\mathrm{SL}(d, q)$ containing $\mathrm{SL}(d, q)$ can be reduced to this case using a Monte Carlo algorithm for finding the derived group **[BCFLS]**.

In Section 4 we outline an analogue of this theorem for symplectic groups and comment on the other classical groups. We have continued to follow the recursive approach in **[KS]**. There is some hope that this can be avoided. In particular, we suspect that the methods used in **[Br]** for the group PSL(d, q) also can be modified in the same way that **[KS]** has been in this paper.

The SL(2, q)-oracle. We assume that our hypothesised oracle is able to perform various tasks in a black box group $L = \langle \mathcal{S} \rangle$ (believed to be) isomorphic to SL$(2, q)$ for known q. We list these tasks together with the time required to perform them.

(i) The oracle provides us with sets $\mathcal{S}^* \subset L$ and $X = \left\{ \begin{pmatrix} 1 & \rho^k \\ 0 & 1 \end{pmatrix}, \begin{pmatrix} 1 & 0 \\ \rho^k & 1 \end{pmatrix} \mid 0 \le k < e \right\} \subset \mathrm{SL}(2, q)$ for a known generator ρ of GF$(q)^*$, and a bijection $\Psi \colon X \to \mathcal{S}^*$, behaving as in Theorem 1.1. Wc denote by χ the time required for Theorem 1.1(1) as well as for each application of Theorem 1.1(2).

(ii) The oracle is also equipped with routines for discrete logarithms in GF$(q)^*$. Let T denote the transvection group of L generated by the image under Ψ of $\left\{ \begin{pmatrix} 1 & 0 \\ \rho^k & 1 \end{pmatrix} \mid 0 \le k < e \right\}$ for ρ in (i), and suppose we are given $t \in T$. We can use the oracle to find $\Psi^{-1}(t) = \begin{pmatrix} 1 & 0 \\ \lambda & 1 \end{pmatrix}$ for some $\lambda \in \mathrm{GF}(q)$. In addition, the oracle can compute the integer n such that $0 \le n < q - 1$ and $\lambda = \rho^n$. We assume that the cost per call to the discrete log oracle is χ.

We will find task (ii) particularly useful in the following algorithmic setting. If we have elements $g, h \in N_G(T)$ (but not necessarily in L) both inducing scalars of the same order on T, then h can be replaced by a specific power of itself to ensure that g and h induce the same scalar on T. The discrete log oracle tells us which power.

We view this oracle as a black box. However, in practice it will presumably be replaced by a Las Vegas algorithm for Theorem 1.1(1): the algorithms in **[KS]** and **[CLG]** for this is are, indeed, Las Vegas. On the other hand, we assume that our discrete log oracle is deterministic in practice.

2. The general case: $d \ge 4$

Somewhat as in **[KS]**, the methods we employ for dimension 3 differ widely from the larger dimensions. In this section, we assume that $d \ge 4$, and whenever possible we follow closely the algorithm in **[KS]**. In particular, the algorithm we present here is recursive. We wish to ensure that the probability of the routine failing up to the point where the recursive call is made in Section 2.6 is no more than $\frac{1}{d^2}$. In any non-deterministic step in the main routine, this is achieved by making an extra factor of (at least) $\lceil \log d \rceil$ random choices.

2.1. Background. We recall some background material discussed at greater length in **[KS]**.

(1) Groups of transvections. We assume a familiarity with the basic properties of transvections and transvection subgroups of $\mathrm{SL}(d, q)$. For a hyperplane H of V and a point $\alpha \in H$, we denote by $T_{\alpha,H}$ the group of transvections with centre α and axis H. As in **[KS]**, we denote by $Q(\alpha)$ the elementary abelian group of order q^{d-1} consisting of all transvections with centre α, and by $Q = Q(H)$ the group of all transvections with axis H. In this way, we may view points (resp. hyperplanes) as existing within the group G as subgroups $Q(\alpha)$ (resp. $Q(H)$).

(2) Primitive prime divisors. By a fundamental theorem of Zsigmondy **[Zs]**, if p is prime and $m \geq 2$ then there is a prime dividing $p^m - 1$ but not $p^i - 1$ for $1 \leq i < m$, except when either $p = 2, m = 6$, or $m = 2$ and p is a Mersenne prime. Such a prime is called a *primitive prime divisor* of $p^m - 1$. We will call an integer j a ppd$^\#(p; m)$*-number* if $j | p^m - 1$ and either $p = 2, m = 6$ and $21 | j$; $m = 2$, p is a Mersenne prime, and $8 | j$; $m = 1$ and $j > 4$; or j is divisible by a primitive prime divisor of $p^m - 1$ (this is slight departure from **[KS]**, taking advantage of the fact that here we will always have $p^m \geq 7$).

We test whether the order of an element of G is a ppd$^\#(p; m)$-number using a deterministic $O(\mu m^2 \log p)$ time algorithm in **[KS]**, 2.6.

(3) Field computations. These are discussed in **[KS]**, 2.3. We will have at our disposal the hypothesised $\mathrm{SL}(2, q)$ and discrete log oracles in place of the Zech table used there.

2.2. Time involving factors of q in [KS]. Since our goal here is only to remove occurrences of factors of q in the timing in **[KS]**, we begin by pinpointing those occurrences, which were of five types when $d \geq 4$:

(i) finding certain elements of G whose probability of occurrence is bounded below only by $\frac{1}{qM}$ for some integer M;

(ii) testing for isomorphism to $\mathrm{SL}(3, q)$ and finding such an isomorphism;

(iii) performing a test using all conjugates of the form A^b, $b \in B$, for a transvection group B and an element or subgroup A;

(iv) listing a full transvection group of order q, primarily in order to perform linear algebra within that group; and

(v) a discrete log computation in $\mathrm{GF}(q)^*$ (using a Zech table for $\mathrm{GF}(q)$).

Type (i). This was used to obtain transvections when $d \geq 4$ (**[KS]**, 3.2.1), and here is replaced by Sections 2.3 for $d \geq 5$ and 2.4 when $d = 4$.

In addition to finding transvections, by using the probability estimate in (i), **[KS]**, 3.2.1, obtained an element of suitable ppd order that is not available here when d is even. Hence we use substitute procedures to find certain subgroups Q, $Q(\alpha)$, H, L and $\mathcal{S}^*$ when $d \geq 5$ (cf. **[KS]**, 3.3.1). However, in the present situation we only find these subgroups by a Monte Carlo algorithm (in Section 2.5), verifying correctness later in Section 2.8 and hence only then converting this to a Las Vegas algorithm.

Type (ii). In Section 3 we give such a test.

Type (iii). This occurred only in **[KS]**, Lemma 3.11 (effective transitivity of the groups Q and $Q(\alpha)$), which here is replaced by Lemma 2.1. This transitivity was then used several times. First it was used in **[KS]**, Corollary 3.12 to find a subgroup $L \cong \mathrm{SL}(d-1, q)$; this is replaced in Section 2.5. Then it was used in **[KS]**, 3.5.1, to label the points of the target vector space, which in turn was needed in **[KS]**, 3.5.2, to construct the desired homomorphism $G \to \mathrm{PSL}(d, q)$; see Section 2.8. Also it was used in **[KS]**, Proposition 3.18, for the analogue of our Theorem 1.1(2b); again see Section 2.8.

Type (iv). This occurred in **[KS]**, 3.4.3(1),(4),(5): finding and computing with $\mathrm{GF}(p)$- and $\mathrm{GF}(q)$-bases of Q and $Q(\alpha)$; expressing any element of Q or $Q(\alpha)$ as linear combinations or SLPs using these bases; and finding matrices of the linear transformations induced on Q or $Q(\alpha)$. See Section 2.7.

Type (v). This occurred in **[KS]**, Propositions 3.17 and 3.18, where a discrete log calculation ensured that a certain $(d-1) \times (d-1)$ matrix had determinant 1. Of course our discrete log oracle is used here.

[KS], 7.3.2. Verifying a presentation is the only other place where a factor q appears in a timing. This verification uses all of the above procedures, and these already have had q removed from their timings.

2.3. Constructing transvection groups and J when $d \geq 5$.

Calls to $d = 3$ or 4. We will make frequent use of the cases $d = 3, 4$ of Theorem 1.1 for subgroups generated by very special subgroups of G, although we will not use the additional time required to verify a presentation. With small probability, the subgroups generated by our chosen elements might not be $\mathrm{SL}(3, q)$ (or $\mathrm{SL}(4, q)$) and the failure goes undetected. However, in this case, the algorithm will eventually fail, and we will not succeed in our goal of finding and verifying a presentation for G. We will see, in Sections 2.8 and 3, that a call to Theorem 1.1(1), without the additional presentation verification, costs $O(\xi+\chi e+\mu \log^2 q)$ when $d = 3$ and $O(e \log e(\xi+\chi e+\mu \log^2 q))$ when $d = 4$.

The elements a_1, a_2. We first present an $O(d \log d(\xi + \mu e d^2 \log q))$-time Las Vegas algorithm which, with probability $\geq 1 - \frac{1}{8d^2}$, constructs an element $a \in G$ which, in its action on the sought-after vector space, has 2-dimensional support and is the identity on a $d-2$-space (for even d we construct two such elements). We also construct $\sigma \in G$ centralising a and having $d-3$- or $d-2$-dimensional support.

Let $r = d-2$ for d odd and $d-3$ for d even. Choose up to $\lceil 32(d-2)\log(2d)\rceil$ elements τ of G, and for each test whether $|\tau|$ is divisible by both a $\mathrm{ppd}^{\#}(p; er)$- and $\mathrm{ppd}^{\#}(p; 2e)$-number. If so let $a = \tau^{q^r-1}$ and $\sigma = \tau^{q^2-1}$. Then $|a|$ is a $\mathrm{ppd}^{\#}(p; 2e)$-divisor of $q+1$.

Let $\tau_1 = \tau_2 = \tau$ if d is odd. When d is even, repeat the above a second time in order to obtain two pairs (a_1, σ_1) and (a_2, σ_2), where $a = a_1$.

The proofs of correctness, timing and probability of success of this procedure are similar to those in **[KS]**, 4.2.1 and 5.2.1. Note that no oracle calls are needed here.

Constructing $K \cong SL(4,q)$. We next present an $O((\log d)\cdot(e\log e)(\xi+\chi e+\mu\log^2 q))$-time Las Vegas algorithm which, with probability $\geq 1-\frac{1}{8d^2}$, constructs a naturally embedded subgroup $K\cong SL(4,q)$ together with a generating set $\mathcal{S}_K^*$ of K, a field $F=GF(q)$, a vector space V_K and an isomorphism $SL(V_K)=SL(4,q)\to K$ behaving as in Theorem 1.1.

Choose up to $\lceil 16\log(2d)\rceil$ elements $g\in G$, and for each use the case $d=4$ of Theorem 1.1 to test whether $K=\langle a_1,a_2^g\rangle\cong SL(4,q)$, in which case also obtain an isomorphism $\Psi_K\colon SL(4,q)\to K$.

Note that, in view of the subgroup structure of $SL(4,q)$, two elements having 2-dimensional supports and whose orders are $\mathrm{ppd}^{\#}(p;2e)$-divisors of $q+1$ generate $SL(4,q)$ with probability at least $\frac{1}{2}$ (this result is the concatenation of many papers, summarized in **[KL]**, Theorem 5.1). The stated timing arises from the $\Theta(\log d)$ calls to the case $d=4$ of the theorem.

At this point we have a field $F=GF(q)$; fix a generator ρ of F^* throughout the remainder of the proof (observe that this generator is available from just one call to the $SL(2,q)$-oracle).

Constructing $J\cong SL(3,q)$. We prefer, for two reasons, to work with a subgroup $SL(3,q)$ rather than the $SL(4,q)$ that we have constructed. Firstly, using an $SL(3,q)$ subgroup matches up better with **[KS]**, making it easier for the reader to translate. Also, we will need to construct an $SL(3,q)$ subgroup when we deal with $SL(4,q)$ in Section 2.4.

Therefore we now construct a subgroup $J\cong SL(3,q)$ of K, succeeding with probability $\geq 1-\frac{1}{8d^2}$.

Use Theorem 1.1(2b) $O(\log d)$ times for Ψ_K to find matrices $\Psi_K^{-1}(a_1)$ and $\Psi_K^{-1}(a_2^g)$ in $SL(V_K)$. Use linear algebra to compute the 2-spaces $W_1=[V_K,\Psi_K^{-1}(a_1)]$ and $W_2=[V_K,\Psi_K^{-1}(a_2^g)]$, and find $A\in SL(V_K)$ such that $W_1\cap W_2A$ is 1-dimensional. Use Theorem 1.1(2a) to find $\Psi_K(A)$. Replace $a_2\leftarrow a_2^{g\Psi_K(A)}$, $\sigma_2\leftarrow\sigma_2^{g\Psi_K(A)}$ and set $J=\langle a_1,a_2\rangle$.

Next, use Section 3 $O(\log d)$ times to find a generating set $\mathcal{S}_J^*$ of J, and an isomorphism $\Psi_J\colon SL(V_J)\to J$ behaving as in Theorem 1.1. Use Theorem 1.1(2b) $O(\log d)$ times for Ψ_J to find the matrices $\Psi_J^{-1}(a_1)$ and $\Psi_J^{-1}(a_2)$ in $SL(V_J)$.

The purpose of each of the $O(\log d)$ repetitions above is to ensure that we correctly find all of the matrices $\Psi_K^{-1}(a_1)$, $\Psi_K^{-1}(a_2^g)$, $\Psi_J^{-1}(a_1)$, $\Psi_J^{-1}(a_2)$ and the isomorphism Ψ_J with probability $\geq 1-\frac{1}{8d^2}$. All of the remaining uses for Ψ_J will be to compute images of matrices in the group J, using the deterministic Theorem 1.1(2a), so there will be no more randomness needed in this section.

Some elements and subgroups of J. Use linear algebra in V_J to compute the point α of intersection of the 2-spaces $V_{2,i}=[V_J,\Psi_J^{-1}(a_i)]$, $i=1,2$. Find $B,C\in SL(3,q)$

such that $T^B_{\alpha, V_{2,2}} = T_{\alpha, V_{2,1}}$ and $C = \text{diag}(1, \rho, \rho^{-1})$, using a basis of V_J starting with a vector in α.

Use Ψ_J to find the following: $f = \Psi_J(B)$, $T = \Psi_J(T_{\alpha, V_{2,1}})$ (centralised by both σ_1 and σ_2^f); $j = \Psi_J(C)$ (inducing on T an automorphism of order $q-1$); the two subgroups X_1, X_2 of J having order q^2, consisting of transvections and containing T; any $j(\gamma) \in J$ such that $[[X_1, X_2^{j(\gamma)}], X_1] \neq 1$ (i.e., X_1 and $X_2^{j(\gamma)}$ are not "incident" within J); and the subgroup $D \cong \text{SL}(2, q)$ of J normalising both X_1 and $X_2^{j(\gamma)}$. For each of the above elements of J we have an $O(\log q)$-length SLP from $\mathcal{S}_J^*$, found in the time $O(\chi + \mu \log q)$ to use Theorem 1.1(2a) for J.

The total time in Section 2.3 is $O(d \log d(\xi + \mu e d^2 \log q) + (\log d)(e \log e)(\xi + \chi e + \mu \log^2 q))$, and it succeeds with probability at least $1 - \frac{1}{4d^2}$.

2.4. Constructing transvection groups and J when $d = 4$. Next we also construct a naturally embedded subgroup $J \cong \text{SL}(3, q)$ when $d = 4$.

As in Section 2.3, find an element τ whose order is divisible by both a ppd$^\#(p; e)$- and ppd$^\#(p; 3e)$-number, and set $a = \tau^{3(q^2+q+1)}$. Consider conjugates a^{g_1}, a^{g_2}, let $A = \langle a, a^{g_1}, a^{g_2} \rangle$, find the derived subgroup $J = A'$ using **[BCFLS]**, and use Section 3 to test whether $J \cong \text{SL}(3, q)$. If this isomorphism holds, we obtain a field, an isomorphism $\Psi_J \colon \text{SL}(3, q) = \text{SL}(V_J) \to J$ and a generating set $\mathcal{S}_J^*$ for J.

By an argument similar to that in **[KS]**, Lemma 3.7, we have $A \cong \text{SL}(3, q) \circ \langle a \rangle$ with probability $\geq \frac{1}{2^5}$, while Section 3 succeeds with probability $\geq \frac{1}{2}$. Hence, we obtain J and then Ψ_J with probability $\geq \frac{1}{2^6}$. Finding A' uses the $O(\mu \log^2 q)$-time black box Monte Carlo algorithm in **[BCFLS]**.

Note that there are no elements σ, σ_i here; and f will not be needed. Use the isomorphism Ψ_J and linear algebra to find the following: the 2-dimensional eigenspace H and the 1-dimensional eigenspace γ of $\Psi_J^{-1}(a)$ in V_J; the transvection group $T = \Psi_J(T_{\alpha,H})$ for some point $\alpha \in H$; $B' \in \text{SL}(V_J)$ sending α to γ and $j(\gamma) = \Psi_J(B')$; a diagonal matrix C inducing ρ on $T_{\alpha,H}$ and $j = \Psi_J(C)$; and X_1, X_2, D as in Section 2.3. This takes $O(\xi + \chi e + \mu \log^2 q)$ time.

2.5. The subgroups Q, $Q(\alpha)$, $Q(\gamma)$, H and L. If $d \geq 5$, let

$$Q = \langle X_1^{\sigma_1^i}, X_1^{(\sigma_2^f)^i} \mid 0 \leq i \leq d-2 \rangle, \quad Q(\alpha) = \langle X_2^{\sigma_1^i}, X_2^{(\sigma_2^f)^i} \mid 0 \leq i \leq d-2 \rangle$$

and $Q(\gamma) = Q(\alpha)^{j(\gamma)}$. We claim that, with probability $> 1 - \frac{4}{q^{d-4}}$, Q and $Q(\alpha)$ are the subgroups of G, of order q^{d-1}, consisting of all transvections having the same centre (or axis) as T (cf. **[KS]**, 3.1.3). We will only prove this for the group Q and the case of even $d \geq 6$. Up to duality in the target vector space V, the stated conjugates of X_1 all have the same axis W. With probability $> 1 - \frac{2}{q^{d-4}}$ the codimension 1 or 2 subspaces of W preserved by $\sigma_1, \sigma_2^f \in C_G(T)$ generate W, in which case

$\langle X_1^{\langle\sigma_1\rangle}, X_1^{\langle\sigma_2^f\rangle}\rangle$ is the desired group of order q^{d-1}. Finally, the stated conjugates of X_1 generate these normal closures by **[KS]**, Lemma 2.7.

We emphasise that *we do not* yet *know for certain that the output groups Q and $Q(\alpha)$ actually have the desired order q^{d-1}*. This is somewhat similar to what occurred in **[KS]**, 5.3.1. Until we have verified this in Section 2.8 we merely *assume* that these groups are correct. We have seen that this assumption is correct with high probability.

When $d = 4$ we proceed more simply, as well as deterministically, by defining $Q = \langle X_1, X_1^\tau\rangle$, $Q(\alpha) = \langle X_2, X_2^{\tau^{j(\gamma)^{-1}}}\rangle$ and $Q(\gamma) = Q(\alpha)^{j(\gamma)}$. Since the methods for constructing X_1, X_2, τ and $j(\gamma)$ differ from **[KS]** in this case, we supply a proof that Q and $Q(\alpha)$ behave as desired.

Let $\Psi : \mathrm{SL}(4, q) = \mathrm{SL}(V) \to G$ denote the target isomorphism and identify V_J with the 3-dimensional support of $\Psi^{-1}(J)$ in V. Then $\Psi^{-1}(T) = T_{\alpha,W}$ where $W \cap V_J = H$. Let δ denote the 1-space of V centralised by $\Psi^{-1}(J)$, so that $W = \langle H, \delta\rangle$ and δ is in the 3-dimensional eigenspace of $\Psi^{-1}(a)$. Hence $\Psi^{-1}(\tau)$ acts irreducibly on W and $\langle X_1, X_1^\tau\rangle$ is the image under Ψ of the group of transvections with axis W. Also, $\Psi^{-1}(\tau)$ fixes γ so that $(\Psi^{-1}(\tau))^{(B')^{-1}}$ fixes α: $Q(\alpha)$ is the image under Ψ of the group of transvections with centre α.

The subgroup *H*. Let $H = \langle (X_1D)^{\sigma_1^i}, (X_1D)^{(\sigma_2^f)^i} \mid 0 \le i \le d\rangle$ if $d > 4$ and $\langle X_1D, (X_1D)^\tau\rangle$ if $d = 4$. The argument in **[KS]**, 3.3.1, shows that $H = N_G(Q)'$ *provided that Q is the correct group, of order q^{d-1}* (as is already the case when $d = 4$). Thus, $H = N_G(Q)'$ with high probability (and this is verified in Section 2.8).

Transitivity of *Q* and *Q*(α). The following variation of **[KS]**, Lemma 3.10, is not deterministic: the restriction imposed by our timing goals seems to force a departure from **[KS]** in this regard. While our use of randomization only takes place inside elementary abelian groups, we cannot be sure of the correctness of the output of this lemma until we are sure of the correctness of Q (cf. Section 2.8).

Note that **[KS]**, 3.3.2, contains simple and fast procedures for deciding whether or not two conjugates $Q(\alpha)^x$ and $Q(\alpha)^y$ (or Q^x and Q^y) are equal, or whether the "point" $Q(\alpha)^x$ is "on" the "hyperplane" Q^y. There is also a simple test for whether two conjugates T_1, T_2 of T are opposite (i.e., generate an $\mathrm{SL}(2, q)$): $[[t_1, t_2,]t_2] \ne 1$ for just one choice of $1 \ne t_i \in T_i$.

Lemma 2.1. *There are $O(\xi + \chi e + \mu(d + \log^2 q))$-time Las Vegas algorithms which, with probability $\ge \frac{1}{26}$, solve the following problems.*

(i) *Given conjugates $Q(\alpha)^x$ and $Q(\alpha)^y$ not on Q, find the unique $u \in Q$ such that $Q(\alpha)^{xu} = Q(\alpha)^y$.*

(ii) *Given conjugates Q^x and Q^y not on $Q(\alpha)$, find the unique $u \in Q(\alpha)$ such that $Q^{xu} = Q^y$.*

Proof. We only prove (i). We may assume that $Q(\alpha)^x \ne Q(\alpha)^y$. Let $T_0 = T$ and let T_1 be one of the conjugates of T generating $Q(\alpha)^x$ and opposite to T. Choose (generators for) a random transvection group $T_2 < Q(\alpha)^y$, use Section 3 to test whether $K = \langle T_0, T_1, T_2 \rangle \cong \mathrm{SL}(3, q)$, and if so to find an isomorphism $\Psi_K\colon \mathrm{SL}(3, q) \to K$. Use Ψ_K to find the subgroup Q_{Ki} of K that contains T_i, consists of q^2 transvections (for $i = 0, 1, 2$), and satisfies $[Q, Q_{K0}] = [Q(\alpha)^x, Q_{K1}] = [Q(\alpha)^y, Q_{K2}] = 1$. Finally, use Lemma 3.1 to find the element $u \in Q_{K0}$ such that $Q_{K1}^u = Q_{K2}$. Output u.

We claim that u behaves as required with the stated probability. First note that $K \cong \mathrm{SL}(3, q)$ with probability$\geq \frac{1}{2^5}$ (as in **[KS]**, Lemma 3.7), while Section 3 verifies this with probability $\geq \frac{1}{2}$, so we obtain K behaving as stated, together with the needed isomorphism, with probability $\geq \frac{1}{2^6}$. Since K acts on the target vector space V by preserving a decomposition as a direct sum of a 3- and a $d-3$-space, each Q_{Ki} consists of transvections of V, $Q_{K1} \leq Q(\alpha)^x$, $Q_{K2} \leq Q(\alpha)^y$ and $u \in Q_{K0} \leq Q$. The timing is dominated by the use of the above commutator tests and the use of Theorem 1.1(1) with $d = 3$. □

We note that the corresponding time in **[KS]** was $O(\mu(qe+qd))$ for a *deterministic* algorithm. However, we will always need to repeat the lemma at least $\lceil 2^6 \log d \rceil$ times in order to ensure that it produces an output with probability $> 1 - \frac{1}{d^2}$; we will see this very soon below, as well as later in Section 2.8. We note also that the probability $\frac{1}{2^6}$ can be increased considerably here, since we are generating with full transvection groups, and we are assuming that $q \geq 17$.

The subgroup L. Next we (probably) find a subgroup L of G such that $N_G(Q)' = Q \rtimes L$, essentially as in **[KS]**, Corollary 3.12. For each choice of $i = 1, 2$ and $0 \leq j \leq d-2$, use up to $\lceil 2^6 \log(16d^3) \rceil$ calls to Lemma 2.1 to find $u_{i,j} \in Q$ such that $Q(\gamma)^{\sigma_i^j u_{i,j}} = Q(\gamma)$. Let $L = \langle D^{\sigma_i^j u_{i,j}} \mid i = 1, 2;\ 0 \leq j \leq d-2 \rangle$.

That L behaves as stated is clear (cf. **[KS]**, Corollary 3.12). For each i, j, with probability $\geq 1 - \frac{1}{16d^3}$ at least one of our calls succeeds, and hence all $u_{i,j}$ are found with probability $\geq 1 - \frac{1}{8d^2}$. The procedure takes $O(d \log d(\xi + \chi e + \mu d \log^2 q))$ time in view of the $\Theta(d \log d)$ calls to Lemma 2.1. □

2.6. Recursion. Recursively we find an isomorphism $\Psi_L\colon \mathrm{SL}(d-1, q) = \mathrm{SL}(V_{d-1}) \to L$; however, as in Section 2.3 and 2.5 we do not include the verification part of the algorithm in this call, postponing verification until Section 2.8. If this recursive call fails then the algorithm terminates. Failure can occur for two very different reasons: a group Q, $Q(\alpha)$ or L constructed in Section 2.5 was not the desired group; or these groups were correct but the recursive call failed to give an output. The probability of any of these events is small, and hence we can ignore them: we have an isomorphism Ψ_L.

We may need to modify the isomorphism Ψ_L in order to ensure that the stabiliser of a 1-space in the natural $\mathrm{SL}(d-1, q)$-module is sent to the stabiliser of a 1-space of Q; see **[KS]**, 3.3.3.

2.7. Linear algebra. We now need analogues of **[KS]**, Lemma 3.13 and Section 3.4.3, which deal with workable bases for Q and for $Q(\alpha)$. Proceeding as in **[KS]**, Lemma 3.13, we obtain a direct sum decomposition $Q = A_1 \oplus \ldots \oplus A_{d-1}$; we obtain (using Ψ_L) an element $c \in L$ such that $A_i = A_1^{c^{i-1}}$ for $1 \le i \le d-1$; and we have a deterministic $O(\mu d)$ algorithm to express any given $u \in Q$ in the form $u = \prod_1^{d-1} a_i$ with $a_i \in A_i$. While **[KS]**, 3.4.2, goes through without additional change, we need replacement routines for those found in **[KS]**, 3.4.3, since we wish to avoid storing an entire transvection group. This is easily achieved, with the aid of our hypothesised oracle, by noting that A_1 lies inside an SL$(2, q)$ subgroup L_1 of J, where we already have a field to work with.

As in **[KS]**, 3.4.3(1), let $\mathcal{B} = \{b_i\}_{i=1}^{d-1}$ be an $\mathbb{F}$-basis for Q, where $b_i = b_1^{c^{i-1}}$ for $1 \le i \le d-1$. The SL$(2, q)$ oracle gives an $\mathbb{F}_p$-basis for A_1, namely the image of $\left\{\left(\begin{smallmatrix} 1 & 0 \\ \rho^k & 1 \end{smallmatrix}\right) \mid 0 \le k < e\right\}$, where we may assume that b_1 corresponds to $\left(\begin{smallmatrix} 1 & 0 \\ 1 & 1 \end{smallmatrix}\right)$. We now obtain the $\mathbb{F}_p$-basis $\mathcal{B}_p$ of **[KS]**, 3.4.3(1), as the union of conjugates of the $\mathbb{F}_p$-basis of A_1 by the various powers of c.

Next, if we are given $u \in Q$ as in **[KS]**, 3.4.3(4), then use **[KS]**, Lemma 3.13 to write $u = \prod_{i=1}^{d-1} a_i$, with $a_i \in A_i$. For each $1 \le i \le d-1$, compute $a_i^{c^{1-i}} \in A_1$ and find the $\lambda_i \in \mathbb{F}$ such that the oracle for L_1 assigns $\left(\begin{smallmatrix} 1 & 0 \\ \lambda_i & 1 \end{smallmatrix}\right)$ to $a_i^{c^{1-i}}$. The coefficient of each b_i can therefore be found in $O(\chi + \mu)$ time, so that the $\mathbb{F}$-vector of u relative to $\mathcal{B}$ is found in $O(d(\chi + \mu))$ time. It is now an easy matter to obtain an SLP of length $O(d \log q)$ from $\mathcal{B}_p$ to u. The time for **[KS]**, 3.4.3(4), is now $O(\chi d + \mu d \log q)$ compared with $O(\mu q d)$ in **[KS]**.

Note that the same procedures apply even when $d = 3$ once we have obtained Q and $Q(\alpha)$ together with suitable direct sum decompositions.

2.8. Conclusion. As in **[KS]**, 3.4.4, we extend Ψ to QL, but here we use Section 2.7 for linear algebra in Q. We label conjugates of $Q(\alpha)$ as in **[KS]**, 3.5.1, replacing **[KS]**, Lemma 3.13, with our Lemma 2.1. Hence, we label a single given conjugate $Q(\alpha)^x$, with probability at least $\frac{1}{2^6}$ in time $O(\xi + \chi e + \mu(d + \log^2 q))$.

We next define $O(d^2 \log q)$-element generating sets $\mathcal{S}^*$ of G and X of SL(d, q) and a bijection $X \to \mathcal{S}^*$ that extends to an epimorphism Ψ: SL$(d, q) \to G$, as in **[KS]**, 3.5.2. Computing $\Psi^{-1}(\mathcal{S}^*)$ goes through without change (note that in order to compute $\Psi^{-1}(g)$ for each of the $O(e)$ elements $g \in \mathcal{S}^* \cap J$ we must find the labels of $d + 1$ conjugates of $Q(\alpha)$). We require that *all* of the images are correct with probability $\ge 1 - \frac{1}{16d^2}$, so we repeat Lemma 2.1 $\lceil 2^6 \log(16ed^3) \rceil$ times in each of our $O(ed)$ applications of 3.5.1. Hence, the total time for **[KS]**, 3.5.2, is now $O(d \log(d \log q)\{\xi + \mu d^2 \log q + \chi d \log q\})$.

As part of our recursion we call the subroutines of this and previous Sections $O(d)$ times. It follows that, if we omit the verification of a presentation for G (see below, as well as Section 2.6), then the algorithm runs in time $O(d^2 \log q \log(d \log q)\{\xi + \mu d^2 \log^2 q + \chi d \log q\})$.

We also need procedures for finding SLPs from $\mathcal{S}^*$ to any given element of G, or from X to any given element of SL(d, q). The latter is already in $O(d\chi + \mu d^2 \log q)$ time in **[KS]**, Proposition 3.17, modulo the use of discrete logs, as indicated above in (**v**); while the following is essentially **[KS]**, Proposition 3.18:

Proof of Theorem 1.1(2b). Use the proof of Proposition 3.18 in **[KS]**, substituting our procedures wherever the timing there involves q. As above, the only care required concerns probability. Namely, the argument in **[KS]** uses Lemma 2.1 to find up to 3 elements of Q or $Q(\alpha)$ and expresses each using an SLP from $\mathcal{S}^*$; finds the matrix A of an element of H as a linear transformation of the $\mathbb{F}$-space Q by using $d-1$ calls to Section 2.7; forces A to have determinant 1 by using a discrete log call; and applies Theorem 1.1(2a) to A. We repeat Lemma 2.1 $\Theta(\log d)$ times in order to ensure the stated probability, so that the algorithm takes $O(\log d\{\xi+\chi(e+d^2)+\mu d \log q(\log q+d)\} + d^5 \log q)$ time. □

Ψ is an epimorphism. Finally, we need to confirm this by checking the relations for a suitable short presentation inside G for a quotient of SL(V), which takes time polynomial in all of the preceding procedures (cf. **[KS]**, 7.2.2); and to check that $G = \langle \mathcal{S} \rangle$ is $\langle \mathcal{S}^* \rangle$, by checking that each member of the original generating set $\mathcal{S}$ is in $\langle \mathcal{S}^* \rangle$. We follow the timing calculation in **[KS]**, 7.2.2, keeping in mind that the probability of success of each Las Vegas routine must be large enough so that the probability that any routine fails is small: the total time is $O(\{ed^2 \log(ed)\{\xi + \chi(e+d)+\mu d^2 \log^2 q\}+|\mathcal{S}| \log(d|\mathcal{S}|)\{\xi+\chi(d^2+e)+\mu d \log q(d+\log q)+d^5 \log q\}$, where e is $O(\log q)$. (Note that we repeat the algorithm in Theorem 1.1(2b) $\Theta(\log(d|\mathcal{S}|))$ times for each element of $\mathcal{S}$ in order to make the probabilities behave correctly.) □

3. PSL(3, q)

Throughout this section we assume that $G \cong$ PSL$(3, q)$ or SL$(3, q)$ for known $q \geq 17$. The occurrences of factors q in the timing in **[KS]**, 3.6.3, are as follows:

(**i**) testing isomorphism with SL$(2, q)$ and finding such an isomorphism;

(**ii**) performing a test using all conjugates of the form A^b, $b \in B$, for a transvection group B and an element or subgroup A;

(**iii**) listing a transvection group, primarily in order to perform linear algebra within that group; or

(**iv**) a discrete log computation in GF$(q)^*$.

Occurrences of these in **[KS]**, 3.6.3, resemble those indicated in Section 2.

Finding τ, a, A and $L \cong$ SL(2, q). As in **[KS]**, 3.6.3, find an element τ whose order is divisible by a ppd$^\#(p;\, 2e)$- and a ppd$^\#(p;\, e)$-number, as well as by a ppd$^\#(p;\, e/2)$-number if e is even. Let $a = \tau^{2(q+1)}$.

Choose a conjugate b of a, let $A = \langle a, b \rangle$, find $L = A'$ using **[BCFLS]**, and use the hypothesised SL$(2, q)$-oracle to test whether $L \cong$ SL$(2, q)$. If this isomorphism holds, we also obtain a field $\mathbb{F} = \mathrm{GF}(q)$, an isomorphism $\Psi_L\colon \mathrm{SL}(2,q) \to L$ and a generating set $\mathcal{S}_L^*$ for L.

As in **[KS]**, Lemma 3.8, with probability $\geq \frac{1}{2}$ we have $A \cong \mathrm{SL}(2,q) \circ \langle a \rangle$; finding L uses a Monte Carlo algorithm that is made Las Vegas by the $\mathrm{SL}(2,q)$-oracle call. Finding a, A and L takes $O(\xi + \mu \log^2 q + \chi)$ time.

Elements z, $j(\gamma) \in A$. We know that A is a central product $A = L \circ \langle z \rangle$, and we have found L. We next construct $\langle z \rangle$.

Use our $\mathrm{SL}(2,q)$-oracle to find the two transvection groups T_1, T_2 in L normalised by a, as well as $h \in L$ of order $q-1$ normalising T_1 and T_2. Note that a has order dividing $(q-1)/(2,q-1)$, while h induces an element of order $(q-1)/(2,q-1)$ on T_1. Use the discrete log oracle inside L to find an integer n such that $0 \leq n < q-1$ and h^n and a induce the same scalar on T_1, and set $z = ah^{-n}$. Then $A = L \circ \langle z \rangle$, where z is found in $O(\mu e \log q + \chi)$ time.

Also find a transvection $j(\gamma) \in L$ such that $T_1^{j(\gamma)} = T_2$.

Finding Q and $Q(\alpha)$. Find $Q(\alpha) = \langle T_1, T_1^\tau \rangle$ and $Q = \langle T_1, T_1^{\tau^{j(\gamma)^{-1}}} \rangle$ in $O(\mu \log q)$ time. Since τ fixes the eigenspaces of a (on the vector space V we have not yet constructed and hence cannot use for algorithmic purposes), up to duality these are the groups of order q^2 consisting of all transvections having the same centre or axis as T_1, respectively.

The transvection groups $T_1, \ldots, T_6$. We already have transvection groups T_1 and T_2. Rename T_2 as T_4. In $O(\mu e)$ time find $T_2 = [Q(\alpha), z]$, $T_3 = T_2^{j(\gamma)}$, $T_6 = [Q, z]$ and $T_5 = T_6^{j(\gamma)}$.

In order to understand the behavior of these six groups on V, let T_1 and T_4 have centres α and γ, respectively, and let β denote the point of intersection of the axes of T_1 and T_4. Then $T_1 = T_{\alpha,\langle\alpha,\beta\rangle}$ "is" the group of transvections with centre α and axis $\langle \alpha, \beta \rangle$, $T_2 = T_{\alpha,\langle\alpha,\gamma\rangle}$, $T_3 = T_{\gamma,\langle\alpha,\gamma\rangle}$, $T_4 = T_{\gamma,\langle\beta,\gamma\rangle}$, $T_5 = T_{\beta,\langle\beta,\gamma\rangle}$ and $T_6 = T_{\beta,\langle\alpha,\beta\rangle}$. We have $Q = T_1 \oplus T_6$ and $Q(\alpha) = T_1 \oplus T_2$.

Redefine $L = \langle T_2, T_5 \rangle \cong \mathrm{SL}(2,q)$. This normalises Q and $Q(\gamma) = Q(\alpha)^{j(\gamma)}$.

Decompositions of Q and $Q(\alpha)$; linear algebra. We can omit **[KS]**, Corollary 3.12, since we already have L. We may assume that the oracle for L returns Ψ_L mapping $\left\{\left(\begin{smallmatrix} 1 & 0 \\ \rho^k & 1 \end{smallmatrix}\right) \mid 0 \leq k < e\right\}$ and $\left\{\left(\begin{smallmatrix} 1 & \rho^k \\ 0 & 1 \end{smallmatrix}\right) \mid 0 \leq k < e\right\}$ into T_2 and T_5 respectively. Let $c = \Psi_L\left(\begin{smallmatrix} 0 & 1 \\ -1 & 0 \end{smallmatrix}\right)$, and proceed as in Section 2.7.

Transitivity of Q and $Q(\alpha)$. We require a version of Lemma 2.1 for the case $d = 3$. In the context of this paper, we view the following algorithm as deterministic: the only use of randomness is that which is occurs in any practical algorithm to recognise $\mathrm{SL}(2,q)$.

Lemma 3.1. *There is a deterministic $O(\chi + \mu \log q)$ time algorithm to solve problems* (i) *and* (ii) *of Lemma* 2.1 *when $d = 3$.*

Proof. We redefine $h = \Psi_L(\mathrm{diag}(\rho^{-1}, \rho))$ and view it as acting on the transvection groups T_2 and T_5 as the scalars ρ^{-2} and ρ^2, respectively. It follows (by considering an appropriate isomorphism $\Psi\colon \mathrm{SL}(3, q) \to G$ extending Ψ_L) that we may view h as acting on T_1 and T_6 as the scalars ρ^{-1} and ρ, respectively.

In (i) we are given $Q(\alpha)^x$ and $Q(\alpha)^y$, each generated by two transvection groups (e.g., T_1^x and T_2^x) at least one of which is opposite T_1. Let $A_1 < Q(\alpha)^x$ and $A_2 < Q(\alpha)^y$ be transvection groups opposite T_1. For $i = 1, 2$, use the SL$(2, q)$-oracle for $\langle T_1, A_i\rangle$ to find an element h_i of order $q-1$ normalising T_1 and A_i. Each h_i induces an automorphism of order $(q-1)/(2, q-1)$ on T_1; use the discrete log oracle inside $\langle T_1, A_1\rangle$ to arrange for these to be the same automorphism as that induced by h^{-2} (which we know corresponds to the scalar ρ^2).

If q is even or $q \equiv 3 \pmod 4$, then ρ is the only square root of ρ^2 in $\mathbb{F}^*$ of order $q-1$, and hence h_1 and h_2 induce automorphisms on U/T_1 corresponding to ρ. If $q \equiv 1 \pmod 4$ then the automorphisms of order $q-1$ that h_1 and h_2 induce on U/T_1 correspond to $\pm\rho$, and we now ensure that they both induce the same automorphism ρ. Fix $1 \neq t \in T_6$. For $i = 1, 2$, use the $\langle T_1, A_1\rangle$-oracle to test whether $b_i = t^{h_i}(t^h)^{-1} \in T_1$; if not then replace h_i by $h_i^{(q+1)/2}$ (since in this situation h_i induces $-\rho$ on U/T_1). Each h_i now induces ρ on U/T_1.

We now construct $u \in U = Q(\alpha)Q = T_2T_1T_6$ such that $h_1^u \equiv h_2 \pmod{T_1}$. Note that $h_1^{-k}h_2^k$ centralises both T_1 and U/T_1 for any integer k, and hence lies in U. If q is odd set $u = (h_1^{(q-1)/2}h_2^{(q-1)/2})^{(p+1)/2}$; if q is even set $u = (h_1^{-1}h_2)^{h_1^j}$ for the integer j such that $0 \leq j \leq q-1$ and $\rho^j(1-\rho) = 1$, found using the discrete log oracle.

We claim that $h_1^u \equiv h_2 \pmod{T_1}$. Since $U\langle h_1\rangle/T_1 = U\langle h_2\rangle/T_1$, there is some $\tilde{u} \in U$ such that $h_1^{\tilde{u}} \equiv h_2 \pmod{T_1}$. We will compute using actions on the group $U/T_1 \cong \mathbb{F}^2$. Suppose that the element $\tilde{u}$ we seek acts on $\mathbb{F}^2$ via $v \mapsto v + c$ for some $c \in \mathbb{F}^2$. Since h_1 acts via $v \mapsto \rho v$ we have $h_2 = h_1^{\tilde{u}}\colon v \mapsto \rho v + (1-\rho)c$. If q is even then $(h_1^{-1}h_2)^{h_1^j}\colon v \mapsto v + c$, so that $u = (h_1^{-1}h_2)^{h_1^j}$ behaves as claimed; the case q odd is similar.

Next, we decompose $u = v'v$ with $v' \in Q(\alpha) = T_2T_1$ and $v \in Q = T_6T_1$. Use our SL$(2, q)$-oracle to find an element $h_{36} \in \langle T_3, T_6\rangle$ of order $q-1$ normalising T_3 and T_6. This element induces a scalar of order $q-1$ on T_2; use the discrete log oracle for L to arrange for this scalar to be ρ^{-1}. It follows (again by considering an extension Ψ of Ψ_L) that h_{36} induces ρ^2 on T_6. Use the discrete log oracle to find integers m and n such that $0 \leq m, n \leq q-1$, $\rho^{-m}(\rho^{-1}-1) = 1$ and $\rho^n[\rho^{2m}(\rho^2-1)-1] = 1$. We claim that $v = (u^{-1}[u, h_{36}]^{h_{36}^m})^{h^n}$ satisfies the conditions stated above.

We know that there are elements $v' \in T_2T_1$ and $v \in T_6T_1$ such that $u = v'v$. Computing once again inside the 2-space $U/T_1 \cong T_2 \oplus T_6$ we find, mod T_1, first that $[u, h_{36}] \equiv (v'+v)^{h_{36}} - (v'+v) \equiv (\rho^{-1}-1)v' + (\rho^2-1)v$, then that $u^{-1}[u, h_{36}]^{h_{36}^m} \equiv v(\rho^{2m}(\rho^2-1)-1)$, and finally that $(u^{-1}[u, h_{36}]^{h_{36}^m})^{h^n} \equiv v$. Thus, our choice of v behaves as claimed.

Finally, observe that $\langle A_1^u, T_1\rangle = \langle A_2, T_1\rangle$ (since $\langle h_i, T_1\rangle$ is in just one conjugate of L for $i = 1, 2$), and hence $\langle A_2, T_1, A_1^v\rangle/O_p(\langle A_2, T_1, A_1^v\rangle) \cong \mathrm{SL}(2, q)$ (since v' fixes the line joining α and α^{xu} in the target module V for G). Use our $\mathrm{SL}(2, q)$-oracle for this quotient group to find $t \in T_1$ such that $A_1^{vt} \equiv A_2 \pmod{O_p(\langle A_2, T_1, A_1^v\rangle)}$, and output vt. All of this takes $O(\chi + \mu \log q)$ time. (N.B.—We will use a similar, but simpler, method below in Section 4.2.) □

Remark. In the proof above we used our $\mathrm{SL}(2, q)$-oracle inside the black box group $K/O_p(K)$ where $K = \langle A_2, T_1, A_1^v\rangle$. In order to do this, we need a membership test for $O_p(K)$ (so that in $K/O_p(K)$ we can decide whether or not a given string represents the identity, and hence $K/O_p(K)$ is a black box group). For any black box group K there is an efficient test for whether a given $k \in K$ lies in $O_p(K)$: test whether or not $\langle k^K\rangle$ is a p-group (see **[BCFLS]**).

End of proof. Now we have all of the subgroups needed in order to complete the case $d = 3$ by direct imitation of **[KS]**, 3.6.3. The timing calculations in **[KS]**, 3.6.3, 7.2.2, show that the time for Theorem 1.1(1) is $O(\xi + e\chi + \mu \log^2 q)$ without the verification of a presentation, and $O(\xi + e\chi + \mu \log^2 q + |\mathcal{S}|(\chi + \mu \log q))$ with such a verification. The times for Theorem 1.1(2a) and (2b) are both $O(\chi + \mu \log q)$.

Note that the use of randomness in Theorem 1.1(2b) when $d > 3$ is forced upon us by the Las Vegas Lemma 2.1. When $d = 3$, however, the analogous Lemma 3.1 above is deterministic, upgrading Theorem 1.1(2b) to a deterministic algorithm in this case. This also accounts for the apparent discrepancy between the timings for $d = 3$ and $d = 4$ in Theorem 1.1(1).

However, as already noted at the end of Section 1, the existing algorithms for recognition of $\mathrm{SL}(2, q)$ involve the use of randomness, so, in practice, probabilities will need to be treated more carefully for $\mathrm{SL}(3, q)$ and $\mathrm{PSL}(3, q)$.

4. Other classical groups

We will now briefly outline an algorithm for a version of Theorem 1.1 for symplectic groups $\mathrm{PSp}(2m, q)$ and $\mathrm{Sp}(2m, q)$ when $2m \geq 4$ and $q > 9$; and then comment on the remaining classical groups.

4.1. $\mathrm{PSp}(2m, q)$, $2m \geq 6$. The group $G \cong \mathrm{PSp}(2m, q)$ or $\mathrm{Sp}(2m, q)$, $2m \geq 6$, is actually somewhat simpler to handle than the case considered in Section 2. In **[KS]** the only places a factor of q arises in timings are of five types:

(i) testing for isomorphism to $\mathrm{Sp}(4, q)$, $\Omega^+(4, q)$ or $\Omega^+(6, q)$ and finding such an isomorphism;

(ii) testing whether a given subgroup consisting of transvections is a full transvection group of order q;

(iii) performing a test using all conjugates a^b, $b \in B$, for a short group B and an element a;

(iv) listing a short root group of order q in order to perform linear algebra within that group or to perform a discrete log call; or

(v) listing a transvection group of order q in order to perform linear algebra within that group.

Type (i). These occur in **[KS]**, 5.2.1 and 5.2.2. For Sp$(4, q)$ see Section 4.2; for $\Omega^+(4, q)$ use **[KS]**, 3.6.2, together with our SL$(2, q)$-oracle; and for $\Omega^+(6, q)$ use Section 2 since P$\Omega^+(6, q) \cong$ PSL$(4, q)$.

Type (ii). This occurs in **[KS]**, 5.3.1. If T denotes the given group, let $g \in G$ and use our SL$(2, q)$-oracle to test whether $\langle T, T^g \rangle \cong$ SL$(2, q)$; if so use the resulting isomorphism to test whether $|T| = q$. This isomorphism occurs with high probability, but $\Theta(\log m)$ choices of g are needed to make the probability of success $> 1 - \frac{1}{8m^2}$.

Type (iii). This occurs in **[KS]**, 5.3.2. The context is as follows: at that point there is a group Q that (probably) is $O_p(N_G(T))$ for a transvection group T. The goal is a procedure which, when given conjugates T_1,T_2 of T, neither of which commutes with T, produces the unique element $u \in Q$ conjugating T_1 to T_2. We assume that T is not in $\langle T_1, T_2 \rangle$ (as otherwise we can use our oracle). Then we obtain u as follows: for a random $v \in Q$ use Section 4.2 to test whether $K = \langle T, T_1, T_2, T_1^v \rangle \cong$ Sp$(4, q)$; and if so, to find $u \in O_p(N_K(T)) \le O_p(N_G(T))$ conjugating T_1 to T_2 (see Section 4.2, **Type (ii)**). Once again $\Theta(\log m)$ choices of g are needed to make the probability of success $> 1 - \frac{1}{8m^2}$.

Type (iv). This occurs in **[KS]**, 5.4.2 and 5.4.3, and needs to be handled for just one subgroup A: all others that need to be considered are given as explicit conjugates of one of them. This subgroup A may be assumed to lie in a subgroup Sp$(4, q)$ that has already been constructed, in which case Section 4.2 can be applied. Our oracle also handles the discrete log call appearing in **[KS]**, 5.4.2.

Type (v). This occurs in **[KS]**, 5.4.4, and is handled by our SL$(2, q)$-oracle.

Proposition 5.18 in [KS]. This contains an algorithm for finding an SLP to any given element $g \in G$, and uses some of the previous procedures whose timings had factors of q.

End of proof. We can now complete the algorithm by direct imitation of **[KS]**, Sections 5 and 7.2.2.

4.2. PSp(4, q). When $G \cong$ Sp$(4, q)$ or PSp$(4, q)$, the places a factor of q appears in the timing in **[KS]**, 5.6.1, are as follows:

(i) testing isomorphism with SL$(2, q)$ and finding such an isomorphism;

(ii) performing a test using all conjugates a^b, $b \in B$, for a short group B and a known element a;

(iii) listing a short root group of order q in order to perform linear algebra within that group or to perform a discrete log call; or

(iv) listing a transvection group of order q in order to perform linear algebra within that group.

Type (i). This situation, in which our $\mathrm{SL}(2,q)$-oracle can be used, occurs when finding a subgroup $A \circ B$ isomorphic to the central product $\mathrm{SL}(2,q) \circ \mathrm{SL}(2,q)$, as well as when finding $Q = O_p(N_G(T))$ for a transvection group $T < A$.

Type (ii). The context is as in Section 4.1**(iii)**: we need a procedure which, when given conjugates T_1,T_2 of T, neither of which commutes with T, produces the unique element $v \in Q$ conjugating T_1 to T_2. For this purpose, for $i = 1, 2$ apply our oracle to $\langle T, T_i\rangle$ in order to obtain elements h_i that normalise T and T_i, where $|h_i| = 2$ if q is odd and $|h_i| = q-1$ if q is even. We will construct $u \in Q$ such that $h_1^u \equiv h_2 \pmod T$, and hence $\langle T, T_1\rangle^u = \langle T, T_2\rangle$, in which case our $\mathrm{SL}(2,q)$-oracle can be used to find $t \in T < Q$ such that $T_1^{ut} = T_2$; then we output $v = ut$. That the element u constructed below behaves as required follows from a straightforward calculation (cf. Section 3).

If q is odd then $u = (h_1h_2)^{(p+1)/2}$ works. If q is even then h_1 and h_2 induce generators of the same cyclic automorphism group of order $q-1$ on T; use our discrete log oracle to make them induce the same automorphism ρ, and also to find j such that $\rho^j(1-\rho) = 1$. Then $u = (h_1^{-1}h_2)^{h_1^j}$ works.

Type (iii). As noted above we have already constructed a subgroup $B < C_G(T)$ and an isomorphism $\Psi_B\colon \mathrm{SL}(2,q) \to B$. Use Ψ_B to find an element $h \in B$ of order $q-1$, as well as the two transvection groups T_1, T_2 of B normalised by h. Then h acts on the abelian groups $Q_i = [Q, T_i]T$ of order q^2; these contain the short root groups we need to handle without listing. Note that $Q/T = (Q_1/T) \times (Q_2/T)$.

We may assume that $\Psi_B\begin{pmatrix} \rho & 0 \\ 0 & \rho^{-1}\end{pmatrix} = h$. In view of the standard matrix representation of $\mathrm{Sp}(4,q)$ we may assume that h induces ρ on Q_1/T. That is, we *define* the action of $\langle\rho\rangle$ on Q_1/T by: $\rho^j(u_1T) = u_1^{h^j}T$ for all $u_1 \in Q_1$ and $0 \le j < q-1$; and similarly we define $\rho^j(u_2T) = u_2^{h^{-j}}T$ for $u_2 \in Q_2$. Thus, we can now perform discrete log calls within Q_i/T.

We use two different methods to find the Q_i/T–coordinates of a given uT in Q/T depending upon the parity of q. When q is odd, this is achieved by constructing an alternating form on the 2-space Q/T using commutators in Q. When q is even, Q is abelian, so we must use a different method. In this case, however, short root groups are $\mathrm{Aut}(G)$–conjugate to transvection groups and we can use the $\mathrm{SL}(2,q)$–oracle to perform linear algebra inside short root subgroups of Q_i.

Type (iv). As in Section 4.1**(v)** this is handled by our $\mathrm{SL}(2,q)$-oracle.

End of proof. We can now complete the algorithm by direct imitation of **[KS]**, 5.6.1, 7.2.2.

Alternative ending. Once we have obtained the exact structure of QB (recall that $B \cong \mathrm{SL}(2,q)$ centralises T), it is not difficult to proceed directly to a short presentation for G, as in **[KM]**.

4.3. Orthogonal and unitary groups. The algorithms in **[KS]** for orthogonal groups involve factors of q in their timings in more ways than in the other classical groups. Nevertheless, there is reasonable hope that these factors can be replaced by calls to our SL$(2, q)$-oracle.

On the other hand, the timing in **[KS]** for PSU(d, q) involves $q^{3/2}$, starting with the 3-dimensional groups. However, it seems likely that it will be possible to prove an analogue of Theorem 1.1 in this case, assuming the availability of a PSU$(3, q)$-oracle. Moreover, we hope that hypothesising such an oracle will be justified for PSU$(3, q)$, given as a group of 3×3 matrices, by a variation on ideas in **[CLG]**.

References

[BCFLS] L. Babai, G. Cooperman, L. Finkelstein, E. M. Luks and Á. Seress, Fast Monte Carlo algorithms for permutation groups, J. Comput. System Sci. 50 (1995) 296–308.

[Br] S. Bratus, Recognition of finite black box groups, Ph.D. Thesis, Northeastern Univ. 1999.

[CLG] M. Conder and C. R. Leedham-Green, Fast recognition of classical groups over large fields, in: Groups and Computation III (W. M. Kantor and Á. Seress, eds.), Ohio State Univ. Math. Res. Inst. Publ. 8, Walter de Gruyter, Berlin–New York, 2001, 113–121.

[KL] W. M. Kantor and R. A. Liebler, The rank 3 permutation representations of the finite classical groups, Trans. Amer. Math. Soc. 271 (1982) 1–71.

[KM] W. M. Kantor and K. Magaard, Black box exceptional groups of Lie type, in preparation.

[KS] W. M. Kantor and Á. Seress, Black box classical groups, Mem. Amer. Math. Soc. 149, Nr. 708, Amer. Math. Soc., Providence, RI, 2001.

Department of Mathematics
University of Oregon
Eugene, OR 97403, U.S.A.
brooks@geometry.uoregon.edu
kantor@math.uoregon.edu

Fast recognition of classical groups over large fields

Marston Conder and Charles R. Leedham-Green

Abstract. A central issue in computing with matrix groups over finite fields is the explicit recognition of groups isomorphic to classical groups. Algorithms that deal more generally with black-box groups are available, but these have complexity that is polynomial in the size of the field (and the rank of the group). As classical groups have faithful representations over the natural field (such as the natural representation) that are of degree which is independent of the field size, more efficient algorithms are needed to deal with these cases. For special linear and symplectic groups the problem can be reduced to finding an efficient explicit recognition algorithm for $\mathrm{SL}(2, q)$ and $\mathrm{Sp}(4, q)$ in their natural representations. In this paper we describe such an algorithm for $\mathrm{SL}(2, q)$ in time that is polynomial in $\log q$ given a discrete logarithm oracle, and exhibit its efficacy when implemented in MAGMA. Also we give a crucial ingredient of a similar algorithm for $\mathrm{Sp}(4, q)$, namely the production of transvections necessary for generating subgroups isomorphic to $\mathrm{SL}(2, q)$.

1991 Mathematics Subject Classification: primary 20-04; secondary 15A33, 20G40.

1. Introduction

An explicit recognition algorithm for a group H takes as input a group G on a generating set X, and determines whether or not G is isomorphic to H, and then if so, computes the image of X under such an isomorphism, and enables computation of the image under the inverse isomorphism of any given element of H in G. This usually involves computing the image in G of a canonical set of generators of H, there being an algorithm to express a given element of H by a straight line program from this set of generators (see **[14]**). In the most general algorithms, G is a black-box group. We are concerned with the case when G is given as a matrix group over a finite field.

An implicit recognition algorithm simply proves that the groups G and H are isomorphic, but without constructing the isomorphisms.

Both explicit and implicit (otherwise known as 'constructive' and 'non-constructive') algorithms usually work by searching for random elements of G satisfying certain properties. If G is isomorphic to H then a known proportion of elements of G will satisfy this condition. If such an element is not found after a suitable

Groups and Computation III
Ohio State Univ. Math. Res. Inst. Publ. 8

number of tries the algorithm will return 'FALSE', but only with a certain degree of confidence, which can be increased, converging exponentially fast to certainty by repeated searching. Thus these algorithms are one-sided Monte Carlo algorithms **[1]**.

Implicit recognition of $SL(d, q)$ in its natural representation by Neumann and Praeger **[17]** was subsequently made more efficient and generalised to arbitrary classical groups in their natural representations by Praeger and Niemeyer **[18, 19]** and more informative by Celler and Leedham-Green **[8]**. The random generation of classical groups is addressed in **[12]**. The construction of random group elements is undergoing considerable analysis and review. The basic technique, which is adequate for our purposes, is described in **[10]**.

Algorithms for explicit recognition of classical groups in their natural representations have also been developed by Celler and Leedham-Green **[7]** for $SL(d, q)$ and by Celler in his doctoral thesis **[6]** for $Sp(2d, q)$. These algorithms have polynomial complexity in d (which is good) and/but also in q (which is bad, for fields of large order). Algorithms for explicit black-box recognition of such groups have been developed by Cooperman, Finkelstein and Linton **[11]**, Bratus **[3]** and Kantor and Seress **[13]**. Kantor and Brooksbank **[5]** have pointed out that the explicit black box recognition of $SL(d, q)$ and $Sp(2d, q)$ can be made polynomial in d and $\log q$ given an oracle to recognise $SL(2, q)$; see also **[4]** for orthogonal groups in their natural representations.

As we are concerned with matrix groups rather than with black box groups, and as we wish to have algorithms that are polynomial in the size of the input, it suffices for us to consider the explicit recognition of $SL(2, q)$ where the input consists of matrices over $GF(q)$. A further reduction is possible to the natural representation. This will be described and justified in a later paper. The difficulty then reduces to finding a transvection (that is, an element of order p, where p is the characteristic of $GF(q)$), expressed using a straight line program from the given generating set for $SL(2, q)$. Since the proportion of elements of $SL(2, q)$ of order p is about $1/q$, a simple search has complexity $O(q)$.

In this paper we present an algorithm for the explicit recognition of $SL(2, q)$ and a crucial ingredient of a similar algorithm for $Sp(4, q)$. These algorithms are polynomial in $\log q$, given a discrete logarithm oracle and a prime factorisation of $q - 1$. We could reduce the $Sp(4, q)$ case to $SL(2, q)$ using the approach taken in **[5]**, but we like our solution.

The discrete logarithm algorithm works by setting up certain tables for the finite field in question. When these tables have been set up, computing discrete logarithms over this field becomes cheap. Thus the number of calls to the discrete logarithm is unimportant; our algorithm for $SL(2, q)$ contains two such calls. Factorising $q - 1$ into prime factors is implicit in the current discrete logarithm algorithm **[20]**, and hence is not an additional burden. For $Sp(4, q)$ we need the discrete logarithm over $GF(q^4)$ in even characteristic.

2. Explicit recognition of SL$(2, q)$

Given a set X of 2×2 matrices of determinant 1 over GF(q), where q is large, we seek to verify that $\langle X \rangle = \mathrm{SL}(2, q)$ by producing canonical generators for SL$(2, q)$ as words in the elements of X. As mentioned before, the hard part is to find a transvection. This will be achieved by finding two non-commuting elements of $\langle X \rangle$ with a common eigenspace (by given straight line programs from X).

Step 1: Find by random search in $\langle X \rangle$ an element A of order $q - 1$.

Orders are easily computed given the factorisation of $q-1$; see **[9]**. Approximately half the elements of SL$(2, q)$ have order dividing $q - 1$, and all such elements except $\pm I$ lie in conjugacy classes of size $q(q + 1)$, the conjugacy class being determined by the eigenvalues $a^{\pm i}$ for some fixed primitive element $a \in \mathrm{GF}(q)$, and some i. To obtain order $q - 1$ exactly we need $\gcd(i, q - 1) = 1$. If $\alpha = \phi(q-1)/(q-1)$ is the proportion of integers in the range $1 \le i \le q-1$ that are coprime to $q - 1$, then $\alpha > 1/\log\log q$ (see **[16**, §II.8**]**), and so we expect to make at most $2 \log\log q$ random selections to find a suitable A.

We now look for another element of $\langle X \rangle$ that shares an eigenspace with A.

Step 2: Compute eigenvectors u and v of A; let a and a^{-1} be the corresponding eigenvalues.

Step 3: Take a conjugate B of A by a random element of $\langle X \rangle$.

We assume that A and B do not share an eigenspace. If they do share an eigenspace and do not commute, this stage of the computation is completed; but as the probability that they share an eigenspace is negligible, if they do share an eigenspace we discard this B and try again.

Step 4: Take another random element C of $\langle X \rangle$ and find an i such that $B^i C$ fixes $\langle u \rangle$, if such an i exists.

This is the heart of the matter. Note that i exists if and only if $\langle u \rangle C^{-1}$ lies in the orbit of $\langle u \rangle$ under $\langle B \rangle$. When q is odd, since B has order $q - 1$ this orbit consists of approximately half the 1-dimensional subspaces of the natural module; on the other hand, in even characteristic the orbit has length $q - 1$ and consists of every 1-dimensional subspace other than the eigenspaces of B. Hence (in both cases) a suitable C can easily be found if it can be recognised.

To determine whether a given C is suitable, and if so then to find i, we first diagonalise B as we need to consider powers of B. So now

$$B = \begin{pmatrix} a & 0 \\ 0 & a^{-1} \end{pmatrix},$$

and we proceed as follows:

(a) With respect to the new basis, the subspace $\langle u \rangle$ contains a vector $(1, d)$; and d is easily computed.

(b) If $C = (c_{ij})$ with respect to the new basis then the condition that $B^i C$ fixes $\langle u \rangle$ becomes

$$(a^i c_{11} + da^{-i} c_{21},\ a^i c_{12} + da^{-i} c_{22}) \in \langle (1,\ d) \rangle$$

which is equivalent to

$$a^{2i} c_{12} + dc_{22} = da^{2i} c_{11} + d^2 c_{21}$$

or on rearrangement

$$a^{2i} = (d^2 c_{21} - dc_{22})/(c_{12} - dc_{11}).$$

(c) This requires that the RHS should be a square, which is the significant part of the condition that C be suitable.

(d) If this condition is satisfied then to find i (and hence $B^i C$) we are left with a discrete log problem, which will have a solution since a is primitive.

Step 5: Let $S = [A, B^i C]$.

As A and $B^i C$ lie in $\mathrm{SL}(2, q)$ and have a common eigenvector u, their commutator S is a transvection fixing u.

If S is the identity then the random matrix C can be written as $B^{-i} A^s$ for some s, and in this unlikely event, try again with a different C until S is a non-identity element (with eigenspace spanned by u).

Step 6: Use the above technique to find a random element $D \in \langle X \rangle$ and j such that v is an eigenvector for $B^j D$, and $T = [A, B^j D]$ is not the identity.

So now T is a non-trivial transvection fixing v.

Step 7: Now write A, S and T with respect to the ordered basis $(u,\ v)$.

These consist of a diagonal matrix whose diagonal entries are primitive, a lower unitriangular matrix, and an upper unitriangular matrix. It is now easy to write any element of $\mathrm{SL}(2, q)$ using a straight line program from $\{A, S, T\}$, and we are done.

3. Computation times

Using an implementation of MAGMA [2] on a Toshiba 490CT laptop, this procedure takes at most 7 seconds to handle examples in which $q = 2^e$ for $e \leq 50$, with the exception of 2^{31} and 2^{49}, for which the discrete logarithm fails due to $q - 1$ having too large a prime divisor for the laptop's memory. Similar results are achievable for odd characteristic. The exceptions (and further examples with q up to 2^{100}) are easily handled on a larger machine.

4. Explicit recognition of Sp(4, q)

We next turn to the question of making the explicit recognition of Sp($2d, q$) in its natural representation polynomial in d and $\log q$. A crucial ingredient for this the production of a transvection in the case $d = 2$. In this section we provide an algorithm for this purpose. (A different one is given in [5], but ours appears to be faster since it makes use of linear algebra, a tool not available in the black box setting of [5].) In odd characteristic we reduce the problem to recognition of SL($2, q$) in polynomial time, while in characteristic 2 we use the discrete logarithm over GF(q^4) to perform this reduction. The method is sketched below, without proofs. A preliminary version has been encoded in MAGMA.

Approximately three quarters of the elements of Sp($4, q$) have order dividing q^2-1, with most of the rest having order dividing q^2+1. The characteristic polynomial of an element of order dividing q^2-1 splits over GF(q^2), while the elements of order dividing q^2+1 (but not dividing the characteristic p) are precisely those whose characteristic polynomial does not split over GF(q^2). So let us call a p'-element of Sp($4, q$) *split* if it has order dividing q^2-1, and *non-split* otherwise.

It is easy to see that the eigenvalues of a non-split element are a, a^{-1}, a^q, a^{-q} where $a \in$ GF(q^4) has order dividing q^2+1. On the other hand, the eigenvalues of a split element are either of the form a, a^{-1}, a^q, a^{-q} where $a \in$ GF(q^2) has order dividing q^2-1, or of the form a, a^{-1}, b, b^{-1} where a is a non-zero element of GF(q^2) of order dividing $q+1$ or a non-zero element of GF(q), and similarly for b. The eigenspaces of a non-scalar split element of the latter form are complementary anisotropic spaces, and for large q, between 30 and 50 percent of the elements of Sp($4, q$) are of this form.

The elements excluded by the above dichotomy into split and non-split elements are the elements of order a multiple of the characteristic. The fact that these elements are scarce is the heart of the problem of finding transvections.

4.1. Odd characteristic

Step 1: Find a split element P of even order with eigenvalues a, a^{-1}, b, b^{-1} such that a and b have distinct orders. Some power of this element P will then be an involution A with eigenvalues $1, 1, -1, -1$.

Step 2: Take a random conjugate $B = R^{-1}AR$ of A, and continue doing this until one is found for which AB has odd order greater than 1.

Now let the order of AB be m. Since $\langle A, B\rangle$ is dihedral, the element $S = (AB)^{(m-1)/2}$ conjugates B back to A, so that RS centralises A and hence RS preserves eigenspaces of A. So P (from Step 1) and RS both preserve these eigenspaces. The subgroup of Sp($4, q$) preserving these spaces is the direct product of two copies of SL($2, q$). In general, $\langle P, RS\rangle$ will be the direct product of two copies of SL($2, q$), so we now assume that this is the case.

Step 3: Since about half the elements of SL(2, q) have order dividing $q+1$ and most of the rest have order dividing $q-1$, it is easy, by powering up, to construct (using straight line programs) elements of $\langle X \rangle$ that act as the identity on one eigenspace of P, and generate SL(2, q) on the other. The problem is now reduced to finding a transvection in SL(2, q).

4.2. Even characteristic

In characteristic 2 we find ourselves in the frustrating situation that almost all elements of Sp(4, q) are of odd order, and almost all pairs of elements of Sp(4, q) generate the whole of Sp(4, q); however we can proceed as follows:

Step 1: Find a non-scalar split element P with eigenvalues not all of the same order, using a straight line program from X. Let U be an eigenspace of P, centralised by a non-identity power A of P.

Step 2: Translate this U by a random element R of $\langle X \rangle$, and continue until one is found for which $W = UR$ is a complement for U in $V = \mathrm{GF}(q)^4$.

Step 3: Find a non-split element Q of order q^2+1, with eigenvalues $\alpha, \alpha^q, \alpha^{-1}, \alpha^{-q}$ (permuted among themselves by the automorphism $x \mapsto x^q$).

We aim to find j such that $U \cap WQ^j = U \cap URQ^j$ is a 1-dimensional subspace preserved by $H = \langle A, A^{RQ^j} \rangle$. In general H will have two eigenspaces of dimension 1 and one of dimension 2. The action on the 2-dimensional space will in general be as SL(2, q), producing the required reduction.

Step 4: Diagonalise Q, and write bases for U and W with respect to the new basis for V over $\mathrm{GF}(q^2)$.

If these bases are $\{u_1, u_2\}$ and $\{w_1, w_2\}$, then the condition $U \cap WQ^i \neq \{0\}$ is equivalent to the vanishing of a determinant in the entries of the u_i's and the entries of the w_i's multiplied by powers of α, giving a polynomial equation in $\beta = \alpha^j$ of the form

$$c_0\beta^{q-1} + c_1\beta^{-1-q} + c_2\beta^{-q+1} + c_3\beta^{1+q} = c_4.$$

Applying the field automorphism $x \mapsto x^q$ to this equation produces a system of four linear equations

$$\begin{array}{lllll}
c_0\beta^{q-1} & + c_1\beta^{-1-q} & + c_2\beta^{-q+1} & + c_3\beta^{1+q} & = c_4 \\
c_3^q\beta^{q-1} & + c_0^q\beta^{-1-q} & + c_1^q\beta^{-q+1} & + c_2^q\beta^{1+q} & = c_4^q \\
c_2^{q^2}\beta^{q-1} & + c_3^{q^2}\beta^{-1-q} & + c_0^{q^2}\beta^{-q+1} & + c_1^{q^2}\beta^{1+q} & = c_4^{q^2} \\
c_1^{q^3}\beta^{q-1} & + c_2^{q^3}\beta^{-1-q} & + c_3^{q^3}\beta^{-q+1} & + c_0^{q^3}\beta^{1+q} & = c_4^{q^3}
\end{array}$$

which all turn out to be equivalent to their sum

$$d_0\beta^{q-1} + d_1\beta^{-1-q} + d_2\beta^{-q+1} + d_3\beta^{1+q} = d_4 \qquad (*)$$

the coefficients d_0, d_1, d_2, d_3 of which are permuted among themselves by the automorphism $x \mapsto x^q$, and with constant term $d_4 = c_4 + c_4^q + c_4^{q^2} + c_4^{q^3} \in \mathrm{GF}(q)$.

Step 5: Find $a \in \mathrm{GF}(q^2)$ for which $a + a^q = d_4$.

If θ is a primitive element of $\mathrm{GF}(q^2)$ and $b = \theta + \theta^q$, then we can take $a = \theta d_4/b$ as $b^q = b$ and $d_4^q = d_4$.

Next note that if $\gamma = \beta^{q-1}$, then the equation $(*)$ may be rewritten

$$d_0\gamma + d_0^q\gamma^q + d_2\gamma^{-1} + d_2^q\gamma^{-q} = d_4 = a + a^q$$

(as $d_1 = d_0^q$ and $d_3 = d_2^q$ and $\gamma^q = \beta^{-1-q}$ and so on), and this rearranges to

$$(d_0\gamma + d_2\gamma^{-1} + a)^q = d_0\gamma + d_2\gamma^{-1} + a,$$

from which it follows that $d_0\gamma + d_2\gamma^{-1} + a$ is in $\mathrm{GF}(q)$.

Step 6: Find $c \in \mathrm{GF}(q)$ for which the quadratic equation $d_0x + d_2x^{-1} + a = c$ has a solution $x = \gamma$ in $\mathrm{GF}(q^4)$.

This can be achieved easily by taking successive powers of a primitive element of $\mathrm{GF}(q)$ as candidates for c and solving the corresponding quadratic equation. Roots of quadratic polynomials over finite fields of characteristic 2 may be found using the von zur Gathen–Kaltofen–Shoup algorithm **[21]** for example (as built into MAGMA), with complexity that is linear in $\log q$.

Step 7: Check that the order of γ divides $q^2 + 1$, and then solve $\alpha^{j(q-1)} = \gamma$ for j using the discrete logarithm in $\mathrm{GF}(q^4)$.

This produces j such that $U \cap WQ^j = U \cap URQ^j$ is a 1-dimensional subspace preserved by $\langle A, A^{RQ^j}\rangle$, as required.

References

[1] L. Babai, Randomization in group algorithms: conceptual questions, in: Groups and Computation II (L. Finkelstein and W. M. Kantor, eds.), DIMACS Ser. Discrete Math. Theoret. Comput. Sci. 28, Amer. Math. Soc., Providence, RI, 1997, 1–16.

[2] W. Bosma, J. Cannon and C. Playoust, The MAGMA Algebra System I: The User Language, J. Symbolic Comput. **24** (1997), 235–265.

[3] Sergey Bratus, Recognition of Finite Black Box Groups, PhD thesis, Department of Mathematics, Northeastern University, Boston, 1999.

[4] P. A. Brooksbank, A constructive recognition algorithm for the matrix group $\Omega(d, q)$, in: Groups and Computation III (W. M. Kantor and Á. Seress, eds.), Ohio State Univ. Math. Res. Inst. Publ. 8, Walter de Gruyter, Berlin–New York, 2001, 79–93.

[5] P. A. Brooksbank and W. M. Kantor, On constructive recognition of a black box $\mathrm{PSL}(d, q)$, in: Groups and Computation III (W. M. Kantor and Á. Seress, eds.), Ohio State Univ. Math. Res. Inst. Publ. 8, Walter de Gruyter, Berlin–New York, 2001, 95–111.

[6] Frank Celler, Konstruktive Erkennungsalgorithmen klassischer Gruppen in GAP, Dissertation, RWTH Aachen, Germany, 1997.

[7] F. Celler & C. R. Leedham-Green, A constructive recognition algorithm for the special linear group, in: The atlas of finite groups: ten years on (Birmingham, 1995), London Math. Soc. Lecture Note Ser. 249, Cambridge University Press, 1998, 11–26.

[8] F. Celler & C. R. Leedham-Green, A non-constructive recognition algorithm for the special linear and other classical groups, in: Groups and Computation II (L. Finkelstein and W. M. Kantor, eds.), DIMACS Ser. Discrete Math. Theoret. Comput. Sci. 28, Amer. Math. Soc., Providence, RI, 1997, 61–67.

[9] F. Celler & C. R. Leedham-Green, Calculating the order of an invertible matrix, in: Groups and Computation II (L. Finkelstein and W. M. Kantor, eds.), DIMACS Ser. Discrete Math. Theoret. Comput. Sci. 28, Amer. Math. Soc., Providence, RI, 1997, 55–60.

[10] F. Celler, C. R. Leedham-Green, S. H. Murray, A. C. Niemeyer & E. A. O'Brien, Generating random elements of a finite group, Comm. Algebra 23 (1995), 4931–4948.

[11] G. Cooperman, L. Finkelstein and S. Linton, Constructive recognition of a black box group isomorphic to $\mathrm{GL}(n, 2)$, in: Groups and Computation II (L. Finkelstein and W. M. Kantor, eds.), DIMACS Ser. Discrete Math. Theoret. Comput. Sci.28, Amer. Math. Soc., Providence, RI, 1997, 85–101.

[12] R. M. Guralnick, W. M. Kantor & J. Saxl, The probability of generating a classical group, Comm. Algebra 22 (1994), 1395–1402.

[13] W. M. Kantor and Á. Seress, Black box classical groups, Mem. Amer. Math. Soc. 149, Nr. 708, Amer. Math. Soc., Providence, RI, 2001.

[14] C. R. Leedham-Green, The computational matrix group project, in: Groups and Computation III (W. M. Kantor and Á. Seress, eds.), Ohio State Univ. Math. Res. Inst. Publ. 8, Walter de Gruyter, Berlin–New York, 2001, 229–247.

[15] C. R. Leedham-Green & E. A. O'Brien, Recognising tensor products of matrix groups, Internat. J. Algebra Comput. 7 (1997), 541–559.

[16] D. S. Mitrinović, J. Sándor and B. Crstici, Handbook of Number Theory, Mathematics and Its Applications 351, Kluwer Academic Publishers, 1996.

[17] P. M. Neumann & C. E. Praeger, A recognition algorithm for special linear groups, Proc. London Math. Soc. 65 (1992), 555–603.

[18] A. C. Niemeyer & C. E. Praeger, A recognition algorithm for classical groups over finite fields, Proc. London Math. Soc. 77 (1998), 117–169.

[19] A. C. Niemeyer & C. E. Praeger, Implementing a recognition algorithm for classical groups, in: Groups and Computation II (L. Finkelstein and W. M. Kantor, eds.), DIMACS

Ser. Discrete Math. Theoret. Comput. Sci. 28, Amer. Math. Soc., Providence, RI, 1997, 273–296.

[20] S. C. Pohlig & M. E. Hellman, An improved algorithm for computing logarithms over GF(p) and its cryptographic significance, IEEE Trans. Inform. Theory 24 (1978), 106–110.

[21] V. Shoup, A new polynomial factorization algorithm and its implementation. J. Symbolic Comput. 20 (1995), 363–397.

Department of Mathematics
University of Auckland
Private Bag 92019, Auckland, New Zealand
conder@math.auckland.ac.nz

School of Mathematical Sciences
Queen Mary College
University of London
Miles End Road, London E1 4NS, U.K.
crlg@maths.qmw.ac.uk

Parallel GAP: Mature interactive parallel computing

*Gene Cooperman**

Abstract. Parallel GAP (ParGAP) is now a mature system to support parallel applications. No prior experience with parallel programming is needed in order to quickly develop new parallel applications. This article provides a description of some of the features of ParGAP that can greatly ease the challenge of writing a parallel program. As is true with standard GAP, ParGAP can be run in a completely interactive manner. This simple interactive usage is illustrated by a short program that implements "parallel streaming", running distinct algorithms on the same problem, but under different processes. One then returns the first answer to be produced, and interrupts the other processes. This usage of ParGAP can be useful even on a single computer.

1991 Mathematics Subject Classification: primary 68N99, secondary 20-01, 20-04.

1. Introduction

The acceptance of ParGAP (Parallel GAP) as a GAP share package marks ParGAP's transition into a mature package to support parallel computation. The goals of ParGAP are to support *cheap* parallel computing suitable, for example, for a network of workstations or Beowulf cluster. It employs a simple programmer's model of parallelism that requires minimal learning time.

An earlier version is reported on, under the name of GAP/MPI **[Co97]**. Since then, the defining concepts of ParGAP have been both simplified and extended. Restrictions in the earlier version have also been considerably reduced. For example, the earlier GAP/MPI has had a tendency to crash or hang following a CTRL-C (interrupt) input by the user.

In an implementation on a network employing 10 Mb/s Ethernet, one often observes a message latency time of 1/100 of a second, and in the author's experience, the latency can sometimes approach 1/10 of a second under realistic conditions. In a fast Ethernet (100 Mb/s), one does *not* observe times that are ten times faster, since many of the delays occur in the workstation itself, as a message is copied from one buffer to another and then sent out as multiple packets.

*Supported in part by NSF Grant CCR-9732330.

Hence, a key issue for ParGAP is latency tolerance. The ParGAP model of parallelism is the same as that for TOP-C **[Co96]** (Task-Oriented Parallel C). Hence, one can also combine TOP-C with GAP's ability to link a standalone C program, and create a C binary that is callable from GAP and that will execute a parallel program in C. In general, the features of the ParGAP/TOP-C parallel model include:

- checkpointing;
- natural load balancing;
- latency tolerance;
- overlap of computation and communication;
- heterogeneous parallelism (multiple CPU and O/S architectures);
- robustness (when slave processes die, after interrupt, etc.); and
- small implementation (1260 lines of GAP code, and 757 lines of C code).

Experimental versions of this package have at various times included the following additional capabilities. (The citations describe the original implementations of these features in C.)

- dynamic attachment of new processors **[CG98]**;
- Meta-computing on the Web (attachment of remote processes) **[CG98]**; and
- an SMP version **[CH97]**.

The TOP-C model of parallelism has been used in several computations that were the largest of their kind at the time. These include

- construction of a permutation representation for Lyons's group acting on 9,606,125 points given a representation in GL(111,5) **[CFTY97]**;
- coset enumeration of Lyons's group on a 4-processor SMP architecture (8,835,156 cosets) **[CH97]**;
- condensation of Thompson's group to dimension 5,693 given a representation in GL(112,2) **[CHLM97]**; and
- construction of a permutation representation for J_4 on 173,067,389 points given a representation in GL(1333,11) **[CLMW99]**.

The rest of this paper summarizes some of the novel features of ParGAP. For the full details of ParGAP, it is recommended to read the ParGAP manual, which contains approximately 50 pages, including many further examples. The entire ParGAP distribution, including the manual, can be downloaded from

`http://www-gap.dcs.st-and.ac.uk/~gap/Share/pargapmpi.html`

The distribution is a compressed file of approximately 256 Kilobytes.

Section 2 covers trivial parallelism (the simplest and most common form of parallel application). This is followed in Section 3 by a description of how to use ParGAP interactively, and Section 4 illustrates these principles with a short implementation of parallel streaming. Streaming refers to simultaneously running different algorithms for the same problem in different "streams" or processes, and accepting the first answer that returns. The remaining processes for the other algorithms are then terminated. ParGAP allows those processes to reside on a single CPU or on different CPUs. Hence, in the case of streaming, ParGAP is potentially useful even if one has only a single computer available, since multiple streams can still reside locally in separate processes.

For more sophisticated (non-trivial) parallel applications, the TOP-C model of parallelism is recommended, and it is described in Section 5. Finally, the advantages of ParGAP over a lower level parallel standard, such as MPI (Message Passing Interface) **[GLS94,MPI94]**, are briefly reviewed in Section 6.

2. Trivial Parallelism

In parallel computation, perhaps 80% of the applications fall under the heading of "trivial parallelism". These are situations in which one must compute many unrelated cases. For example, perhaps 200 different cases must be computed and results returned for each case and each case has no interaction with any other one. In the absence of shared data, it would be common to start 20 GAP jobs on 20 distinct workstations in a student computer laboratory. If there are 200 cases, then one writes 20 different "batch jobs", each batch job handling 10 distinct cases.

ParGAP provides strong support for this common situation. Conceptually, one is always talking to ParGAP on a master process (the local workstation), and the master process talks to each of various slave processes. In its simplest form, one merely generates a list of inputs for the cases, and writes a suitable function to provide the case information. Effectively, trivial parallelism can be expressed by a parallel version of GAP's `List()` function, which is called `ParList()` in ParGAP.

The following is an example of a ParGAP session that uses the `ParList()` function. Observe the use of `BroadcastMsg()` to ensure that the function `AnalyzePrimitivePermGroupsOfOrder()` is known to the slaves.

```
gap> AnalyzePrimitivePermGroupsOfOrder := function(ord)
>      #returns a list of data records of all primitive
>      #permutation groups of order ord that are represented
>      #as permutation groups in GAP's primitive group database
>
>      local PrimitiveGroupsOfOrder, AnalyzePrimitivePermGroup;
```

```
>
>      PrimitiveGroupsOfOrder :=
>        ord -> List( [1..NrPrimitiveGroups(ord)],
>                     i -> PrimitiveGroup(ord,i) );
>
>      AnalyzePrimitivePermGroup :=
>        grp -> rec( order := Size(grp),
>                    degree := NrMovedPoints(grp),
>                    baseSize := Size(BaseOfGroup(grp)) );
>
>      return List( Filtered( PrimitiveGroupsOfOrder(ord),
>                             IsPermGroup ),
>                   AnalyzePrimitivePermGroup );
>    end;;
gap> #Define AnalyzePrimitivePermGroupsOfOrder on slaves
gap> BroadcastMsg(
>      PrintToString("AnalyzePrimitivePermGroupsOfOrder := ",
>                    AnalyzePrimitivePermGroupsOfOrder ) );
gap> ParList( [2..256], AnalyzePrimitivePermGroupsOfOrder );
[ [ rec( order := 2, degree := 2, baseSize := 1 ) ],
  [ rec( order := 3, degree := 3, baseSize := 1 ),
      rec( order := 6, degree := 3, baseSize := 2 ),
        .
        .
    <GAP output truncated>
```

As one expects, the answer is computed in parallel in a time that decreases linearly with the number of processors. Hence a Beowulf cluster of 16 processors (1 ParGAP master and 15 ParGAP slave) will return this information approximately 15 times as fast as a single processor, greatly facilitating interactive exploration of the properties of various groups.

If each single case can be executed quickly, then the execution time for `ParList()` can be dominated by the network overhead imposed by message passing between the master and slaves. In this case, an optional third parameter to `ParList()` (to be implemented in the next version of ParGAP, ParGAP-1.1.0) will specify the number of cases to be submitted to a slave in a single message (modulo how many cases are left, of course). As an example, suppose we are interested in finding those of the first 100,000 integers that have many distinct prime factors, but to reduce the network overhead we wish to batch 1000 cases at a time to the slaves. Then with ParGAP-1.1.0, one will be able to execute the following.

```
gap> ParList([1..100000],
>            i -> Length(PrimePowersInt(i))/2, 1000);
```

3. Using ParGAP interactively

Having seen an example where ParGAP can be useful, one next needs to know what is involved in setting up and using ParGAP. ParGAP runs only on UNIX, due to its dependence on an MPI subset (included in the distribution) that calls UNIX sockets. It should be possible to port ParGAP to Windows, either through the use of Cygwin (UNIX emulation on Windows) or through a native Windows MPI implementation. Similarly, the code has been written in a "vanilla" manner, so that in principle, it should be easy to port to another interactive language, such as MAGMA or LISP. An older version was in fact written in LISP **[Co95]**, although the current version is not implemented in LISP. In the following, recall that ParGAP employs a master-slave architecture, with the local process being the "master" process, and all others being "slaves".

ParGAP is installed in the same way as other GAP share packages. After extracting the ParGAP files, one invokes `configure` and `make`. One next writes a file with a list of slave processors to attach, and an indication of where ParGAP is located on that processor. In ParGAP, this file is called `procgroup`, and must reside in the current directory. An example is:

```
# This is a sample ParGAP procgroup file.
# Lines starting with a # are comments.
# Blank lines are also treated as comments.

# The following line generates the MASTER process
local 0

# Each of the following 'localhost' lines generates
# a SLAVE process on the same machine as the MASTER
localhost 1 /proj/sparc/gap4r1/pkg/pargap/bin/pargap.sh
localhost 1 /proj/sparc/gap4r1/pkg/pargap/bin/pargap.sh

# The following line generates a SLAVE process on a remote
# machine: 'iron.ccs.neu.edu' (called via rsh or ssh or ...)
iron.ccs.neu.edu 1 /proj/alpha/gap4r1/pkg/pargap/bin/pargap.sh
```

Next, instead of invoking gap, one then invokes the supplied shell script, `pargapmpi.sh`, (to be renamed `pargap.sh` in ParGAP-1.1.0). Observe that unlike other GAP packages, there is no need to call `RequirePackage()` function from within a GAP session. As usual, master and slave processes read a `.gaprc` file in the home directory, if present. ParGAP looks for a file named `procgroup` file in the current directory, but can be directed to use a different file through the `-p4pg` switch. The following example assumes that `pargap` is a symbolic link or alias for the full pathname of `pargap.sh`. Hence, if one always uses ParGAP with the slave processes specified in myprocgroup.big, one may wish to set up an alias or else a symbolic link to a script containing the equivalent of:

```
pargapmpi.sh -p4pg PATH/myprocgroup.big
```

Observe also that heterogeneous computing is supported, in that each slave machine may be a different dialect of UNIX running under different hardware.

Within ParGAP, the standard message-passing commands (send, receive, probe, etc.) are available. The slaves are numbered consecutively starting at 1. If no slave parameter is supplied for SendMsg() or SendRecvMsg(), then the default slave is 1. Most other commands take a default slave parameter, MPI_ANY_SOURCE (meaning "ANY_SLAVE"). PingSlave() sends a "null message", and waits for an acknowledgement that the slave is alive.

The command SendMsg(string,i) sends the string message to slave i. Slave i then evaluates the string as a GAP command, and returns an answer. If a command has no answer, (such as "x:=3;"), then the string "<no_return_val>" is returned. If multiple commands are sent, such as "x:=3; y:=4; x+y;", then only the last return value is returned. Note that the final semicolon, normally required by GAP, is optional in a command string for a message.

Most of the other commands are clear from the context. Every SendMsg() to a slave must be balanced by a RecvMsg(), except where the commands FlushAllMsgs() and ParReset() are used to flush any pending messages from a slave. A SendRecvMsg(string,i) command is equivalent to the consecutive commands SendMsg(string,i) and RecvMsg(i). The command BroadcastMsg() sends its string argument to each slave, and there is no reply message. ParEval() is like BroadcastMsg(), but it also evaluates on the master process. If one is unsure whether there is a pending message, one can invoke ProbeMsgNonBlocking() or ProbeMsg(); if no slave parameter is provided, the default is MPI_ANY_SOURCE. The command ParRead() is a parallel version of GAP's Read() command; it is useful for ensuring that the slaves and master share a similar environment. Other examples of commands that are analogous to GAP's built-in commands include: ParReread, ParEval, ParInstallGlobalFunction and ParInstallGlobalFunction.

We now illustrate the usage of the commands mentioned above with a sample ParGAP session, for which a procgroup file has defined two slaves.

```
gap> PingSlave(1); #a 'true' response indicates Slave 1 is alive
true
gap> # Print() on slave appears on standard output
gap> # i.e. after the master's prompt.
gap> SendMsg( "Print(3+4)" );
gap> 7
gap> # A <return> was input above to get a fresh prompt.
gap> #
```

```
gap> # To get special characters (including newline: '\n')
gap> # into a string, escape them with a '\'.
gap> SendMsg( "Print(3+4,\"\\n\")" );
gap> 7

gap> # Again, a <return> was input above after the '7'and '\n'
gap> # were printed to get a fresh prompt.
gap> #
gap> # Each SendMsg() is normally balanced by a RecvMsg().
gap> SendMsg( "3+4", 2);
gap> RecvMsg( 2 );
7
gap> # The following is equivalent to the two previous commands.
gap> SendRecvMsg( "3+4", 2);
7
gap> # Flush any messages that are pending. The response is
gap> # the number of messages flushed. (Above, the two
gap> # SendMsg("Print...") (to the default slave: 1) did not
gap> # have corresponding RecvMsg() command.
gap> FlushAllMsgs();
2
gap> # As with Print() the result of Exec() appears on standard
gap> # output. Print() and Exec() are each 'no-value' functions,
gap> # and so the result of a RecvMsg() in these cases
gap> # is "<no_return_val>".
gap> SendRecvMsg( "Exec(\"pwd\")" );
/home/gene
"<no_return_val>"
gap> # Put default slave into an infinite loop.
gap> SendMsg("while true do od");
gap> # Default slave can't execute the next command until it's
gap> # finished with the previous command.
gap> SendMsg("Print(\"WAKE UP\\n\")");
gap> # Check to see if a message is waiting to be collected but
gap> # return immediately (i.e. don't get blocked by waiting for
gap> # a message to appear). A 'false' response indicates the
gap> # infinite loop hasn't terminated and produced a value yet!
gap> ProbeMsgNonBlocking();
false
gap> # Send an interrupt to each slave, slave 1 will see the
gap> # following command and print 'WAKE UP', and then all
gap> # pending messages are flushed.
gap> ParReset();
```

```
... resetting ...
WAKE UP
0
gap> # The return value, 0, from ParReset() indicates there
gap> # were 0 pending messages flushed, confirming correctness
gap> # of ProbeMsgNonBlocking() when it returned "false"
gap> SendRecvMsg( "a:=45; 3+4", 1 );
7
gap> # Note "a" is defined on slave 1, not slave 2.
gap> SendMsg( "a", 2 ); # Slave prints error, output on master
gap> Variable: 'a' must have a value

gap> # <return> entered above to get fresh prompt.
gap> RecvMsg( 2 ); # "No Value" msg from last SendMsg() command
"<no_return_val>"
gap> SendMsg( "a", 1 ); # 'a' is known on slave 1
gap> RecvMsg( 1 );
45
gap> myfnc := function() return 42; end;;
gap> # Use PrintToString() to define myfnc on slave processes
gap> BroadcastMsg( PrintToString( "myfnc := ", myfnc ) );
gap> SendRecvMsg( "myfnc()", 1 );
42
gap> FlushAllMsgs(); # There are no messages pending.
0
gap> # Execute analogue of GAP's List() in parallel on slaves.
gap> squares := ParList( [1..100], x->x^2 );
[ 1, 4, 9, 16, 25, 36, 49, 64, 81, 100, 121, 144, 169, 196,
  225, 256, 289, 324, 361, 400, 441, 484, 529, 576, 625, 676,
  729, 784, 841, 900, 961, 1024, 1089, 1156, 1225, 1296, 1369,
  1444, 1521, 1600, 1681, 1764, 1849, 1936, 2025, 2116, 2209,
  2304, 2401, 2500, 2601, 2704, 2809, 2916, 3025, 3136, 3249,
  3364, 3481, 3600, 3721, 3844, 3969, 4096, 4225, 4356, 4489,
  4624, 4761, 4900, 5041, 5184, 5329, 5476, 5625, 5776, 5929,
  6084, 6241, 6400, 6561, 6724, 6889, 7056, 7225, 7396, 7569,
  7744, 7921, 8100, 8281, 8464, 8649, 8836, 9025, 9216, 9409,
  9604, 9801, 10000 ]
gap> # Ensure problem environment read into master and slaves.
gap> ParRead( "/home/gene/myprogram.g");
```

4. Streaming

Next, we demonstrate a simple implementation of "streaming" as an example of how easy it is to use the ParGAP tools to provide new functionality. To the best of this author's knowledge, the term "streaming" originated with Charles Leedham-Green at the conference on which this proceedings is based, where he made a general request for streaming functionality to be implemented in some software system.

The idea of streaming is that one may have two algorithms for solving a problem. One would like to solve the problem simultaneously on two processors, each using a distinct algorithm. Whichever processor finds the solution first should report back to the master process. The master process should then interrupt the other slave, so as to make it available for further computation. This is useful when one has a heuristic available that may finish early with the correct answer, or may continue for a very long time. The heuristic can then be "streamed" alongside the standard algorithm, in full confidence that the standard algorithm will provide a reasonable upper bound on the time to compute an answer.

One way to provide "streaming" functionality is by the implementation of a function we describe below that is inspired by GAP's `First()` function. The GAP function `First()` takes a list and boolean function as arguments, and returns the first element of the list for which the boolean function returns true. In contrast, we define a corresponding ParGAP function, `ParFirstResult()`, which takes a list and a (general) function as arguments. Messages are sent to the slaves causing the ith slave to evaluate the function on the ith list element. The value returned is the value obtained from whichever slave finishes first. Note that in consistency with the goal of streaming, the function signals an error if asked to evaluate a list with more entries than the number of slaves.

```
ParFirstResult := function( list, fnc )
  local i, result;
  if Length(list) > TOPCnumSlaves then
    Error("too few slaves");
  fi;
  for i in [1..Length(list)] do
    SendMsg( PrintToString(
                   "fnc :=", fnc, "; fnc(", list[i], ");"),
             i );
  od;
  result := RecvMsg(); # default is MPI_ANY_SOURCE
  ParReset(); # Interrupt all other slaves
  return result;
end;
```

We now demonstrate the use of `ParFirstResult()` in "streaming". Assume that `ParFirstResult()` is defined on the master process and that we have two slaves.

```
StreamingFactInt := function(i, x)
  local alg;
  alg :=
   [ x -> [ "MPQS ALGORITHM",  FactorsMPQS(Factorial(x)+1) ],
     x -> [ "CFRAC ALGORITHM", FactorsCFRAC(Factorial(x)+1) ]
   ];
  return alg[i](x);
end;
```

Both `StreamingFactInt(1, x)` and `StreamingFactInt(2, x)` factor an integer, but one uses multiple polynomial quadratic sieve algorithm (MPQS), and the other uses the continued fraction algorithm (CFRAC). We demonstrate the "streaming" of these algorithms in determining the factorization of $35! + 1$. It is important to ensure that the "FactInt" package has been loaded on the slaves. Currently, one must add `RequirePackage("FactInt");` to one's `.gaprc` file before beginning our `ParGAP` session, in order to ensure this. This will be fixed in a future version. For more information, see `?FactInt` within GAP on the FactInt package **[Ko99]**.

```
gap> # RequirePackage("FactInt"); in .gaprc to define on slaves.
gap> # Now, define StreamingFactInt() on each of the slaves.
gap> BroadcastMsg( PrintToString( "StreamingFactInt :=",
>                                 StreamingFactInt ) );
gap> ParFirstResult([1,2], i->StreamingFactInt(i, 35) );
... resetting ...
[ "CFRAC ALGORITHM",
  [ [ 137, 379, 17839, 340825649, 32731815563800396289317 ],
    [ ] ] ]
```

5. TOP-C model for non-trivial parallelism

There are many examples where trivial parallelism does not suffice. Typically, this happens either when there are global variables that must be "read" by each slave process, and possibly updated, or else when the input for the next slave process depends on the result that arrived from the last slave process. In this paper, we provide only the basics of the parallel model. The parallel model was described in more detail in **[Co97]**, and a still more detailed description is contained in the manual.

For non-trivial parallelism, ParGAP uses the TOP-C model **[Co96]**. ParGAP typically invokes the TOP-C model via a command such as

```
MasterSlave( GenerateTaskInput, DoTask, CheckTaskResult,
             UpdateSharedData );
```

The four routines, `GenerateTaskInput`, `DoTask`, `CheckTaskResult`, and `UpdateSharedData` are "callback" functions written by the application programmer, and the names of those callback functions are arbitrary. (In fact, we illustrate using the choice of callback function names from the upcoming ParGAP-1.1 manual. Although the ParGAP-1.0 manual uses slightly different names for its examples, the illustrations here run verbatim in either version.)

The only ParGAP command above is `MasterSlave()`. A task is an arbitrary routine (here called `DoTask()`) executed on a slave process, that takes input from the master process and returns its output to the master process. The diagram below gives some idea of the flow of control as a task is processed.

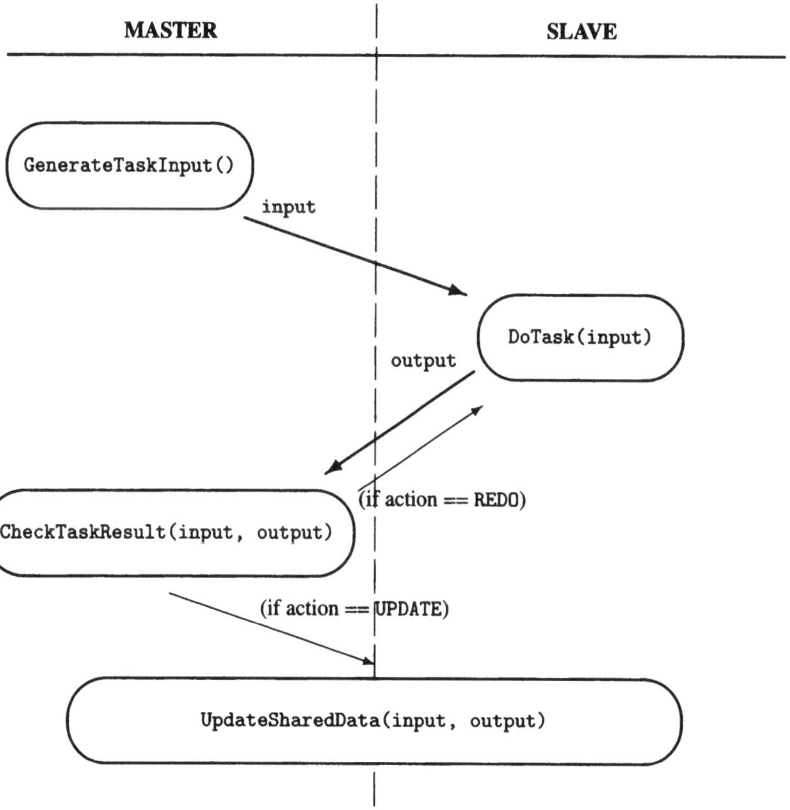

TOP-C Programmer's Model
(Life Cycle of a Task)

The diagram above is meant to represent the three main abstractions of the TOP-C model: (1) the task, (2) the shared data, and (3) the action. The ***shared data*** consists of any globally shared data (which should be readable by all processes). The ***task*** is a procedure executed on a slave process that takes a ***task input***, and the shared data, and produces a ***task output***, which depends only on the task input and shared data. Finally, the task input and task output are sent to the master process, which must then decide upon an ***action***.

Typical actions are NO_ACTION (and the master process might save the task output in a private list of results), UPDATE_SHARED_DATA (send a message to all slaves to update the local copy of the globally shared data), and REDO (re-do the computation in case the shared data was changed by another slave process). (In ParGAP-1.0.x, uses UPDATE_ENVIRONMENT. ParGAP-1.1.0 will offer the alternative UPDATE_SHARED_DATA. We standardize on the latter in this description.)

An example invocation of MasterSlave() is shown below, where we pass the four application functions as direct arguments of MasterSlave(). The routine below implements a simplified version of the ParList() function described in Section 2.

```
ParInstallTOPCGlobalFunction( "MyParList",
function( list, fnc )
  local result, iter;
  result := [];
  # Each invocation of GAP's iterator
  # returns next element of list.
  iter := Iterator(list);
  MasterSlave(
      # GenerateTaskInput():
      function() if IsDoneIterator(iter) then return NOTASK;
                 else return NextIterator(iter); fi; end,
      # DoTask():
      fnc,
      # CheckTaskResult():
      function(input,output) result[input] := output;
                             return NO_ACTION; end,
      # UpdateSharedData():
      Error # We never see action:  UPDATE_SHARED_DATA
      );
  return result;
end );
```

The function ParInstallTOPCGlobalFunction() installs MyParList on all ParGAP processes. It also defines the version of MyParList() on the master differently from on a slave, so that a call, MyParList([1..100], x->x^2);, on the master automatically causes MyParList([1..100], x->x^2); to be invoked on the slave with the same arguments. This is required for the TOP-C model. Note that the distinct function, ParInstallGlobalFunction(), exists for the equivalent of BroadcastMsg("InstallGlobalFunction(MyParList, function ... end)");. Both functions should be called only from the master (possibly from inside a Read() command).

The shared data consists only of the variable fnc, which is read by all processes, but in this case is never "updated". Note that one need not explicitly declare variables that are in the shared data. A TOP-C shared data variable is defined as a variable whose value is read by more than one process, and which is modified only through a call to the application routine UpdateSharedData(). If we would like to see the messages, as they are passed back and forth between master and slave, we can optionally set the variable ParTrace. We are then ready to execute our new parallel function.

```
gap> ParTrace := true;;
gap> MyParList([2..256], AnalyzePrimitivePermGroupsOfOrder);
```

The example above in fact requires only trivial parallelism. Hence the CheckTaskResult() parameter to MasterSlave() is especially simple. Since the action is never REDO, we achieve the maximum concurrency among slave processes, resulting in a speedup that is linear in the number of slaves.

In general, the parallel efficiency of an application, with its concommitant decisions about concurrency of slave tasks are typically determined by the actions chosen by CheckTaskResult(). The sample below is a standard ParGAP "idiom" that allows one to easily set up reasonable concurrency in a parallel application.

```
CheckTaskResult := function( taskInput, taskOutput )
  if taskOutput = fail then return NO_ACTION;
  elif not IsUpToDate() then return REDO_ACTION;
  else return UPDATE_ACTION;
  fi;
end;
```

The function IsUpToDate() is a boolean function provided by ParGAP that returns true if there has been no UPDATE_SHARED_DATA action since the time that the "corresponding" task input was generated by a call to the application routine GenerateTaskInput(). Otherwise, IsUpToDate() returns false. The "corresponding" task input is the task input associated with that task output which was most recently seen on the master process. In the context of the application routine CheckTaskResult(), it is the first argument to that routine. See the ParGAP manual for more sophisticated methods to provide high concurrency of the slave tasks, which go beyond the use of the IsUpToDate() utility.

6. Comparison of ParGAP/TOP-C with MPI (Message Passing Interface)

There already exists a standard for parallel programming using messages. It is MPI (Message Passing Interface) **[GLS94,MPI94]**. Hence, it is reasonable to ask how

ParGAP compares. ParGAP is in fact written on top of a subset MPI implementation that is included with the distribution. However, MPI was intended as a lower level standard. Hence, the MPI standard defines a binding to FORTRAN and to C, but not to any interactive language, such as GAP, Visual Basic, etc. This is due to the history of MPI, which was originally intended as a standard specification for message passing, that could then be implemented by each computer vendor, who might also provide specialized hardware assistance to improve the latency and throughput of messages.

Some ParGAP issues were not covered by the original MPI standard because they do not occur in compiled languages. Some examples are:

- a slave process executes error and enters break loop;
- a slave process enters infinite loop;
- a user types ^C (interrupt); and
- a runaway slave process occurs that doesn't recognize `quit;`.

See **[Co95]** for a discussion of these interactive issues in the context of an implementation of the TOP-C model for LISP. At a higher level, the MPI programmer is still required to carry out several error-prone tasks. The issues below assume some knowledge of MPI.

- Each send command must be matched with a corresponding receive command.
- One must avoid deadlock situations in which every process is blocked on an outstanding receive request, and so no process can reach the corresponding send command.
- One must avoid deadlock arising because a receive is executed on one process before a barrier (such as occurs in MPI's collective communication), and a send is executed on a second process after the barrier.

Finally, there are still more subtle MPI traps for the unwary application programmer, such as the example below. We illustrate the trap using C syntax, while writing "..." for arguments that do not matter in this example. A bug arises because MPI allows the two invocations of `MPI_ANY_SOURCE` to refer to two distinct source (slave) processes. Hence, the call to `MPI_Probe()` will fill in the status for one process, and a buffer will be allocated based on the pending message from that first process. However the call to `MPI_Recv()` employs `MPI_ANY_SOURCE` a second time, and can cause a different, larger message from a second slave process to be received and truncated to fit in the buffer provided.

```
MPI_Probe(MPI_ANY_SOURCE, ..., &status);
MPI_Get_size( &status, datatype, &count);
buf = malloc(count * 4);
MPI_Recv(buf, count, datatype, MPI_ANY_SOURCE, ...);
```

Acknowledgement. The author gratefully acknowledges the many valuable suggestions and careful reading of the reviewer. The detailed comments not only improved the paper, but also provided insights into how better to document future versions of the ParGAP distribution. I would also like to acknowledge the encouragement and feedback from many in the community.

References

[Co97] G. Cooperman, GAP/MPI: Facilitating Parallelism, in: Groups and Computation II (L. Finkelstein and W. M. Kantor, eds.), DIMACS Ser. Discrete Math. Theoret. Comput. Sci. 28, Amer. Math. Soc., Providence, RI, 1997, 69–84.

[Co96] G. Cooperman, TOP-C: A Task-Oriented Parallel C Interface, 5th International Symposium on High Performance Distributed Computing (HPDC-5), IEEE Press, 1996, 141–150.

[Co95] G. Cooperman, STAR/MPI: Binding a Parallel Library to Interactive Symbolic Algebra Systems, Proc. Internat. Symp. on Symbolic and Algebraic Computation (ISSAC '95), ACM Press, 126–132.

[CFTY97] G. Cooperman, L. Finkelstein, M. Tselman and B. York, Constructing Permutation Representations for Matrix Groups, J. Symbolic Comput. 24 (1997), 1–18.

[CG98] G. Cooperman and V. Grinberg, TOP-WEB: Task-Oriented Metacomputing on the WEB, Internat. J. Parallel and Distributed Systems and Networks 1 (1998), 184–192; a shorter version appears as: TOP-WEB: Task-Oriented Metacomputing on the Web, G. Cooperman and V. Grinberg, Proceedings of Ninth IASTED International Conference on Parallel and Distributed Computing and Systems (PDCS-97), IASTED/Acta Press, Anaheim, 1997, 279–286.

[CH97] G. Cooperman and G. Havas, Practical parallel coset enumeration, Proc. of Workshop on High Performance Computation and Gigabit Local Area Networks (G. Cooperman, G. Michler and H. Vinck, eds.), Lecture Notes in Control and Inform. Sci. 226, Springer-Verlag, 1997, 15–27.

[CHLM97] G. Cooperman, G. Hiss, K. Lux, and J. Müller, The Brauer tree of the principal 19-block of the sporadic simple Thompson group, J. Experiment. Math. 6 (4) (1997), 293–300.

[CLMW99] G. Cooperman, W. Lempken, G. Michler and M. Weller, A New Existence Proof of Janko's Simple Group J_4, Progr. Math. 173, Birkhäuser, 1999, 161–175.

[CT96] G. Cooperman and M. Tselman, New Sequential and Parallel Algorithms for Generating High Dimension Hecke Algebras using the Condensation Technique, Proc. of International Symposium on Symbolic and Algebraic Computation (ISSAC '96), ACM Press, New York, 1996, 155–160.

[GAP00] The GAP Group, GAP — Groups, Algorithms, and Programming, Version 4.2; Aachen, St Andrews, 2000 (`http://www-gap.dcs.st-and.ac.uk/~gap`).

[GLS94] W. Gropp, E. Lusk and A. Skjellum, Using MPI, MIT Press, 1994.

[Kn91] D. E. Knuth, Notes on Efficient Representation of Perm Groups, Combinatorica 11 (1991), preliminary version circulated since 1981, 57–68.

[Ko99] S. Kohl, FactInt, GAP share package, http://www-gap.dcs.st-and.ac.uk/~gap/Share/factint.html.

[MPI94] Message Passing Interface Forum (author), MPI: A Message-Passing Interface Standard, Internat. J. Supercomput. Appl. 8 Number 3/4 (1994).

[Si71] C. C. Sims, Computation with Permutation Groups, in: Proc. Second Symposium on Symbolic and Algebraic Manipulation (S. R. Petrick, ed.), ACM Press, New York, 1971, 23–28.

College of Computer Science
Northeastern University
Boston, MA 02115, U.S.A.
gene@ccs.neu.edu

Computing with infinite polycyclic groups

Bettina Eick

Abstract. We consider polycyclic groups given by power-conjugate presentations and describe practical algorithms for some group theoretic questions in such groups. First, we discuss the computation of complements to a normal subgroup. Then, based on that, we introduce a method to compute the conjugacy classes of finite subgroups and an algorithm to determine the torsion subgroup if it exists. Moreover, we outline methods to compute various types of subgroups of finite index.

2000 Mathematics Subject Classification: primary 20F16, 20-04; secondary 68W30

1. Introduction

Each polycyclic group can be represented by a consistent power-conjugate presentation (*pc presentation*). This is a presentation of the form

$$\begin{aligned} \langle g_1, \ldots, g_n \mid \ g_i^{g_j} &= g_{j+1}^{e(i,j,j+1)} \cdots g_n^{e(i,j,n)} \text{ for } 1 \le j < i \le n, \\ g_i^{g_j^{-1}} &= g_{j+1}^{f(i,j,j+1)} \cdots g_n^{f(i,j,n)} \text{ for } 1 \le j < i \le n, \\ g_i^{r_i} &= g_{i+1}^{\ell(i,i+1)} \cdots g_n^{\ell(i,n)} \text{ for } i \in I \,\rangle \end{aligned}$$

for some $I \subseteq \{1, \ldots, n\}$ and integers $r_i > 1$. See **[14]** for details.

A pc presentation reflects the polycyclic structure of the group which it defines and this feature can be used to design practical algorithms for polycyclic groups. For various other representations of polycyclic groups there are methods known to convert them to a pc presentation; for example, see **[10]**, **[7]** or **[12]**. In this paper we call a group defined by a pc presentation or a subgroup of such a group a *pc group*.

In finite pc groups practical methods for several group-theoretic concepts are known. However, in the infinite case much less has been done so far. It is the aim of this paper to describe some practical algorithms for arbitrary, possibly infinite pc groups. In particular, we are interested in the determination of certain finite subgroups as well as various subgroups of finite index.

The basis of all our algorithms is the determination of complement classes to a normal subgroup in a pc group. For finite pc groups there is a method for this described in **[2]**. Here we discuss its extension to the infinite case.

Groups and Computation III
Ohio State Univ. Math. Res. Inst. Publ. 8

Then we consider the computation of certain finite subgroups in a pc group. By **[9]**, a polycyclic group has only finitely many conjugacy classes of finite subgroups. We introduce a method to compute a set of representatives for these classes and their normalizers in a pc group. Similarly, we can determine the normalizer of a given finite subgroup.

Let $T(G)$ be the set of elements with finite order in a polycyclic group G. In general $T(G)$ need not be subgroup; it is for nilpotent groups. We describe an algorithm to test in a pc group G whether $T(G)$ is a subgroup and determine it in this case. This also yields a method to check whether G is torsion-free. Further, we outline a method to compute the largest finite normal subgroup $TN(G)$.

Next we consider subgroups of finite index in a pc group. Here we first describe a 'low index subgroups' algorithm for pc groups; that is, a method to compute all subgroups or all normal subgroups of a given finite index. This is an alternative to the algorithm introduced in **[6]** and we include a runtime comparison of the two methods.

As observed in **[13]**, p. 148, every maximal subgroup of a polycyclic group G has prime power index in G. We outline a first approach to determine all maximal subgroups of a pc group with p-power index for a given prime p.

Moreover, as described in **[13]**, p. 153, each polycyclic group has a normal subgroup of finite index which is poly-$\mathbb{Z}$; that is, which has a polycyclic series with infinite factors only. We show how to determine a characteristic subgroup of this type together with a free abelian normal series through the subgroup.

The paper is organized as follows.

Algorithm	Section
Conjugacy classes of complements	4
Conjugacy classes of finite subgroups	5.1
Normalizer of a finite subgroup	5.1
$T(G)$ and $TN(G)$	5.2
Conjugacy classes of subgroups of given index	6.1
Normal subgroups of given index	6.1
Conjugacy classes of maximal subgroups of p-power index	6.2
A normal poly-$\mathbb{Z}$ subgroup of finite index	6.3

The algorithms introduced here have been implemented in the computer algebra system GAP 4, see **[15]**, and they are available as part of the "Polycyc" package, see **[4]**. To show the practicability of our methods we include in Section 7 example runtimes of our main methods and comparisons to other available methods.

2. Preliminaries

An account of the basic concepts for computing in pc groups is given in **[14]**. This includes a discussion of various collection algorithms; e.g., methods to determine the normal form, $g_1^{e_1} \dots g_n^{e_n}$ where $0 \leq e_i < r_i$ for $i \in I$, of an element of a pc presented group G. Then there are algorithms outlined to compute polycyclic generating sequences for a subgroup of G; that is, a generating set which exhibits a polycyclic series of the subgroup. Moreover, methods to determine a pc presentation for a factor group and to compute with homomorphisms between pc groups are introduced.

Furthermore, in **[1]** a large number of group theoretic problems including most of the problems considered in this paper are shown to be decidable in polycyclic-by-finite groups. However, practical methods for some central problems such as the computation of normalizers, centralizers and intersections of arbitrary subgroups of infinite pc groups are not yet available.

3. Normal series in pc groups

For induction purposes, we will need a method to compute a normal series with elementary or free abelian factors (*efa series*) in a pc group. Clearly, a pc group may have several efa series and the choice of a "good" one will be important for the efficiency of our methods.

The sum of the ranks of the infinite factors in an efa series of G is an invariant called the *Hirsch length*, see p. 147 in **[13]**. However, if G is infinite, then the number of finite factors in an efa series of G is not bounded above. For many of our applications a short efa series will be useful. Further, it is often necessary to determine a characteristic efa series.

In some situations it might be more useful to work with an efa series whose infinite factors are collected at the end of the series; see Section 6.3 for an algorithm. In general, such an efa series has more finite factors than an efa series obtained by the methods of this section.

3.1. Efa series of abelian pc groups. First we recall well-known ideas for an abelian pc group A. Using a Smith normal form computation on the relators of A we can compute a direct decomposition of A as product of cyclic groups. Thus let $A = C \times T$ where C is free abelian and T is the torsion subgroup of A and, moreover, let t be the exponent of T. Then we can derive two characteristic series of A:

$$A \rhd T \rhd 1 \quad \text{and} \quad A \rhd C^t \rhd 1.$$

To refine either of the two series to an efa series we only need to refine the finite factor $T/1$ or A/C^t to elementary abelian factors. If we choose the ranks of the elementary abelian factors as large as possible, then the efa series refining the first characteristic series is shortest possible for A. The efa series refining the second characteristic series has the infinite factor at the end of the series.

3.2. Efa series of arbitrary pc groups. Let G be a pc group. Clearly, one approach to determine an efa series of G is to compute the derived series of G and refine the abelian factors of this series with one of the methods outlined in Section 3.1. This seems to be a suitable approach if we want to determine an efa series whose infinite factors are rather at the end of the series. See Section 6.3 for an application.

However, if the aim is to obtain a short efa series, then we use the following algorithm. Note that the efa series of G obtained by this method might have larger finite factors than necessary, but the number of finite factors in the series is usually minimal and, moreover, the method is quite efficient in general.

We determine an efa series $G = G_1 \rhd G_2 \rhd \cdots \rhd G_l \rhd G_{l+1} = 1$ of G by a top-down approach. Thus suppose that we computed G_i and let $G_i/G_i' = C/G_i' \times T/G_i'$ be the decomposition of G_i/G_i' as direct product of a free abelian and a finite abelian factor as in Section 3.1. To determine G_{i+1} from G_i we distinguish two cases.

- If G_i/G_i' is infinite, then G_i/T is a non-trivial free abelian group. In this case we choose $G_{i+1} = T$.
- If G_i/G_i' is finite, then we choose a prime p dividing the order of G_i/G_i' and we define G_{i+1} as the smallest subgroup of G_i whose factor G_i/G_{i+1} is an elementary abelian p-group.

If the given pc group G is nilpotent, then the following theorem shows that the efa series of G obtained by the above approach is shortest possible. For a proof see **[13]**, p. 132.

Theorem 3.1. *Let G be an infinite nilpotent group. Then G/G' is infinite.*

Thus if G is nilpotent, then we obtain an efa series of G where the first factors are free abelian and the remaining factors are all finite. In particular, the obtained efa series exhibits the torsion subgroup of the nilpotent group G.

4. Complements and supplements

In **[2]** a method described to compute complement classes to normal subgroups in finite pc groups. This algorithm can be extended to the infinite case with some simple modifications. However, in the infinite case infinitely many complement classes may exist and we need to observe that this case cannot arise in our applications of the method.

4.1. The first cohomology group. Let A be an abelian normal subgroup of the pc group U and denote $F = U/A$. The groups of 1-cocycles and 1-coboundaries, resp., are defined by

$$Z^1(F, A) = \{\delta : F \to A \mid \delta(fh) = \delta(f)^h \cdot \delta(h) \text{ for } f, h \in F\}$$
$$B^1(F, A) = \{\delta \in Z^1(F, A) \mid \delta(f) = a^{-1}a^f \text{ for } f \in F \text{ and some } a \in A\}.$$

If A has a complement K in U, then K can be written as $\{r(f) \mid f \in F\}$ and the group $Z^1(F, A)$ corresponds to the set of all complements via

$$\delta \mapsto K_\delta = \{r(f)\delta(f) \mid f \in F\}.$$

Moreover, when we restrict this correspondence to coset representatives of $H^1(F, A) = Z^1(F, A)/B^1(Z, A)$ then we obtain a set of representatives for conjugacy classes of complements.

To compute $Z^1(F, A)$ and $B^1(F, A)$ we use an embedding of these groups into a direct power of A. Let $\{f_1, \ldots, f_m\}$ be a generating set of F. Then we define $\psi : Z^1(F, A) \to A^m : \delta \mapsto (\delta(f_1), \ldots, \delta(f_m))$ and denote by Z and B the images of $Z^1(F, A)$ and $B^1(F, A)$ under ψ, resp. If A is elementary or free abelian, then we can consider A^m as vector space over $\mathbb{F}_p$ or as $\mathbb{Z}$-lattice, resp. The first case is considered in [2] and it shown there that bases of Z and B can be determined as kernel and image of certain linear mappings over $\mathbb{F}_p$. This setting can be translated directly to the second case that A is free abelian with the modification that we need to determine kernel and image of linear mappings over $\mathbb{Z}$. Moreover, a single complement K to A in U can be computed as solution of an inhomogeneous system of linear equations over $\mathbb{F}_p$ or $\mathbb{Z}$, resp., again as described in [2].

4.2. An action on $H^1(F, A)$. Let R be a pc group with normal subgroups U and A such that $A \leq U$ and A is abelian. As in Section 4.1 we can determine the U-classes of complements to A in U. Now we want to derive R-classes of complements and determine the normalizer of each complement in R.

Let $\delta \in Z^1(F, A)$ and let K_δ be the corresponding complement to A in U. Note that R acts naturally on $F = U/A$ via conjugation. For $x \in R$ we define

$$\delta_x(f) = \delta(f^{x^{-1}})^x \quad \text{and} \quad t_x(f) = r(f)^{-1} \cdot r(f^{x^{-1}})^x.$$

Then R acts on $Z^1(F, A)$ via $\delta^x(f) = \delta_x(f)t_x(f)$. This action is affine with linear part δ_x and translation t_x. Moreover, we obtain $(K_\delta)^x = K_{\delta^x}$ and thus this action of R corresponds to the conjugation action of R on the set of complements. Hence the orbit of U is $B^1(F, A)$ and therefore $B^1(F, A)$ is a block for the orbits of R on $Z^1(F, A)$. This yields an induced action of R on $H^1(F, A)$ and we summarize its properties in the following lemma.

Lemma 4.1. *Let R be a pc group with normal subgroups A and U such that $A \leq U$ and A abelian. Suppose that A has a complement K in U.*

a) *The orbits of R in $H^1(F, A)$ correspond one-to-one to the R-classes of complements to A in U.*

b) *Let $\delta \in Z^1(F, A)$ and define $S = \mathrm{Stab}_R(\delta\, B^1(F, A))$; i.e. for each $s \in S$ we have $\delta^s = \delta \cdot \gamma_s$ for some $\gamma_s \in B^1(F, A)$. Let γ_s be induced by taking commutators with $a_s \in A$. Then $N_R(K_\delta) = \{s a_s \mid s \in S\}$.*

Hence we can determine a set of representatives for conjugacy classes of complements to A in U and their normalizers in R whenever $H^1(F, A)$ is finite. The following well-known lemma proves this for the cases that we will need later.

Lemma 4.2. *Let U be a pc group with $A \lhd U$ abelian and $F = U/A$.*

a) *If A is finite, then $Z^1(F, A)$ is finite.*

b) *If F is finite, then $H^1(F, A)$ is finite.*

Note that the determination of complements which are normal in R can be derived easily from Lemma 4.1. In this case we need to determine the elements $\delta \in Z^1(F, A)$ with $\delta^x = \delta$ for all $x \in R$.

4.3. Determination of complement classes. Let G be a pc group and N a normal subgroup of G. To determine the classes of complements to N in G we first compute a G-invariant efa series $N = N_1 \rhd \cdots \rhd N_r \rhd N_{r+1} = 1$ of N as in Section 3. Then we use induction down this series; that is, we compute the complement classes to N/N_{i+1} in G/N_{i+1} using the complement classes to N/N_i in G/N_i. The inductive step is outlined in the following lemma. See [2] for a proof.

Lemma 4.3. *Let G be a pc group with normal subgroup N. Let $A \leq N$ be an abelian normal subgroup of G. Suppose that $U_1/A, \ldots, U_l/A$ are representatives for the complement classes to N/A in G/A. Furthermore, for $1 \leq j \leq l$ let $R_j/A = N_{G/A}(U_j/A)$.*

a) *For each complement K to N in G there exists a unique index j such that KA is conjugate to U_j.*

b) *Representatives for the G-conjugacy classes of complements K to N in G with KA conjugate to U_j can be computed as representatives of the R_j-conjugacy classes of complements to A in U_j.*

Note that the lemma can be extended to compute complement classes under the action of a group R which is normalizing G similarly as described in Section 4.2. Combining Lemma 4.2 and Lemma 4.3 we obtain the following remark.

Remark 4.4. Let G be a pc group and N a normal subgroup. If either N or G/N is finite, then there are only finitely many conjugacy classes of complements to N in G.

4.4. Computation of supplements. Let K be a supplement to the abelian normal subgroup A in a group U. Then $L = A \cap K$ is a normal subgroup of U and K/L is a complement to A/L in U/L. Thus to determine supplements in a pc group U we first determine the desired U-normal subgroups L of A and then apply the complement algorithm.

Let R be a group normalizing U and A. If we want to compute R-classes of supplements only, then we first determine orbits and stabilizers of the desired U-normal subgroups of A under the action of R and then compute complement classes under the action of the stabilizers.

We will use this supplement method in several variations later. Our applications include the following special cases.

- *A is an elementary abelian p-group.* In this case we can consider A as vector space over $\mathbb{F}_p$ and U acts as matrix group on A. Thus we determine the U-normal subgroups as invariant subspaces of A using the methods of **[5]**.

- *Supplements of given index l.* Note that $[U : K] = [A : L]$. Thus to restrict to supplements of index l, we only need to restrict the choice of U-normal subgroups suitably.

- *Maximal supplements.* A supplement K is a maximal subgroup of U, if and only if L is a maximal U-normal subgroup of A. The index $[A : L]$ is a prime power in this case. Hence $A^p \leq L$ for some prime p and we determine L/A^p as maximal U-normal subgroup of the elementary abelian factor A/A^p similar as in the first case.

5. Finite subgroups of polycyclic groups

If the set of all finite order elements $T(G)$ is a subgroup of G, then it is a characteristic finite subgroup of G and all finite subgroups are contained in $T(G)$. Thus in this case the finite subgroups of G can be computed using methods for finite pc groups. If $T(G)$ is not a subgroup, then there may be infinitely many finite subgroups, but they fall into finitely many conjugacy classes. In Section 5.1 we introduce an algorithm to compute these classes.

In Section 5.2 we describe a method to determine whether $T(G)$ forms a subgroup of G and to compute it in this case. Similarly, one can determine the maximal finite normal subgroup $TN(G)$ or check if G is torsion-free.

5.1. Computing the conjugacy classes of finite subgroups. Let G be a pc group and consider the last term A in an efa series of G. Suppose by induction that we know the conjugacy classes of finite subgroups of G/A. The following lemma can be proved similarly to Lemma 4.3.

Lemma 5.1. *Let G be a pc group with abelian normal subgroup A. Suppose that $U_1/A, \ldots, U_l/A$ are representatives for the conjugacy classes of finite subgroups of G/A. Furthermore, for $1 \leq j \leq l$ let $R_j/A = N_{G/A}(U_j/A)$.*

Then representatives for the conjugacy classes of finite subgroups of G can be obtained as representatives for the R_j-classes of finite subgroups K of U_j with $KA = U_j$ for $1 \leq i \leq j$.

Suppose that U/A is a finite subgroup of G/A with normalizer R/A. To determine the R-classes of finite subgroups K in U with $KA = U$ we distinguish two cases.

- A is elementary abelian.
  Then each subgroup of U is finite and hence we need to determine R-classes of supplements to A in U; see Section 4.4.

- A is free abelian.
  Then each finite subgroup K of U satisfies $A \cap K = 1$ and hence we need to compute the R-classes of complements to A in U; see Section 4.2. Note that U/A is finite and hence there are only finitely many such complement classes by Lemma 4.2.

5.2. Determination of $T(G)$ and $TN(G)$. We determine $T(G)$, if it exists, using induction up a polycyclic series of G. We assume that this series contains factors of infinite or prime order only. The following lemma yields the inductive step.

Lemma 5.2. *Let G be a polycyclic group and $N \lhd G$.*

a) $T(G) \cap N = T(N)$.

b) *If G/N is infinite cyclic, then $T(G) = T(N)$.*

c) *Suppose that G/N is cyclic of prime order and $T(N)$ is a group.*

   i) *If $N/T(N)$ has no complement in $G/T(N)$, then $T(G) = T(N)$.*

   ii) *If $N/T(N)$ has one complement $K/T(N)$ in $G/T(N)$, then $T(G) = K$.*

   iii) *Otherwise $T(G)$ is not a subgroup.*

Proof. a) Trivial.

b) Let $g \in G$ be of finite order. Then $gN = N \in G/N$, since G/N has no non-trivial finite subgroup. Therefore $g \in N$ and hence $g \in T(N)$.

c) By passing to $G/T(N)$ we may suppose $T(N) = 1$. Let $g \in G$ be of finite order. Since N is torsion-free, we obtain that $\langle g \rangle$ is a complement to N in G.

i) In this case no element of finite order exists in G and hence $T(G) = 1$.

ii) If there is just one complement K, then this complement is the unique maximal finite subgroup of G. Therefore, $T(G) = K$.

iii) If there are at least two complements K_1 and K_2, then $\langle K_1, K_2 \rangle$ contains an element from N and thus an infinite element. On the other hand, if $T(G)$ is a subgroup, then $K_1, K_2 \leq T(G)$ yielding a contradiction. □

Thus in the inductive step over a cyclic factor of prime order we have to compute the conjugacy classes of complements to $N/T(N)$ in $G/T(N)$. Since G/N is finite in this case, there are only finitely many complement classes by Lemma 4.4 and hence we can compute them using the method of Section 4.3. Note that G/N is a group with one generator and one relator and thus the complement computation will be very effective. Clearly, it will be an advantage for the efficiency to have a polycyclic series with few finite factors.

A similar approach is possible to compute the normal torsion subgroup $TN(G)$ as observed in the next lemma. We omit its proof which is similar to the proof of Lemma 5.2.

Lemma 5.3. *Let G be a polycyclic group and $N \lhd G$.*

a) $TN(G) \cap N = TN(N)$.

b) *If G/N is infinite cyclic, then $TN(G) = TN(N)$.*

c) *Suppose that G/N is cyclic of prime order. If there exists a unique complement $K/TN(N)$ to $N/TN(N)$ in $G/TN(N)$, then $TN(G) = K$. Otherwise $TN(G) = TN(N)$.*

In the inductive step of Lemma 5.3 c) we need to check if there exists a unique complement. If such a complement exists, then it is normal and hence $G/TN(N)$ is the direct product of $N/TN(N)$ with $K/TN(N)$. In particular, G/N centralizes $N/TN(N)$ in this case and this can be used as initial check.

Finally, the following remark can be used to check if a pc group G is torsion-free using the torsion subgroup algorithm as outlined in Lemma 5.2.

Remark 5.4. A group G is torsion-free, if and only if $T(G)$ is the trivial subgroup of G.

6. Subgroups of finite index in polycyclic groups

The subgroups of finite index play an important role in the theory of polycyclic groups. For example, a polycyclic group is residually finite and, moreover, each subgroup U

of a polycyclic group G can be written as the intersection of all subgroups of finite index in G containing U. Furthermore, each maximal subgroup of a polycyclic group has finite index.

Among the normal subgroups of finite index there are some which give a reflection of the structure of a polycyclic group. For example, there exists a normal finite-index subgroup N in a polycyclic group G which is poly-$\mathbb{Z}$ and, moreover, there exists a normal finite-index subgroup N in a polycyclic group G with N' nilpotent.

In the following sections we introduce algorithms to compute certain subgroups of finite index.

6.1. Computing (normal) subgroups of bounded index. Let $n \in \mathbb{N}$. We describe an algorithm to compute the conjugacy classes of subgroups of index dividing n using induction on an efa series of G. Let A be the last term in an efa series and suppose the subgroups of G/A of index dividing n are known. The following lemma yields the inductive step. It is proved similar to Lemma 4.3.

Lemma 6.1. *Let G be a pc group with abelian normal subgroup A. Suppose that $U_1/A, \ldots, U_l/A$ are representatives for the conjugacy classes of subgroups of index dividing n of G/A. Furthermore, for $1 \leq j \leq l$ let $R_j/A = N_{G/A}(U_j/A)$ and denote $l_j = [G : U_j]$.*

Then representatives for the conjugacy classes of subgroups of index dividing n of G can be obtained as representatives for the R_j-classes of subgroups K of index dividing n/l_j in U_j with $KA = U_j$ for $1 \leq i \leq j$.

Suppose that U/A is a subgroup of index l in G/A with normalizer R/A where $l \mid n$. To determine the R-classes of subgroups K of index dividing n/l in U with $KA = U$ we need to compute R-classes of supplements of index dividing n/l in U. This is described in Section 4.4.

This method can be modified readily to determine normal subgroups only. For this purpose we need to restrict the supplement algorithm of Section 4.4 to compute normal supplements only. This, in turn, relies mainly on the calculation of normal complements as briefly sketched in Section 4.2.

Moreover, the determination of normal low index subgroups could be used to compute normal subgroups of finite index with certain properties; e.g. a normal subgroup of finite index which is poly-$\mathbb{Z}$ or a normal subgroup of finite index which is nilpotent-by-abelian. However, for the first application we introduce a more efficient method below.

6.2. Determination of maximal subgroups. Let $G = G_1 \rhd \cdots \rhd G_l \rhd G_{l+1} = 1$ be an efa series of G. We consider each factor G_i/G_{i+1} separately and determine the maximal subgroups M of G containing G_{i+1} but not G_i. Thus M/G_{i+1} is a maximal subgroup of G/G_{i+1} supplementing G_i/G_{i+1}. Hence the maximal subgroup computation is as well an application of the supplement algorithm as outlined in Section 4.4.

Note that for finite pc groups there is a better method available, see [3]. This method has the advantage that it considers a special efa series instead of an arbitrary one. This special efa series is obtained using the lower nilpotent series of the given pc group. However, this series might not exist in an infinite pc group and therefore we cannot extended the method to our situation.

Next we consider the Frattini subgroup of a pc group G. In [1] it is shown that this subgroup can be computed in a pc group G, but an efficient algorithm is not yet available. Here we outline an "approximation" of the Frattini subgroup.

Lemma 6.2. *Let G be a polycyclic group and $M_1, \ldots, M_m$ a conjugacy class of maximal subgroups of p-power index in G supplementing an elementary abelian normal p-subgroup A. Define $C = C_G(A)$ and $K = C'C^p$. Then $K \leq \cap_{i=1}^m M_i$ and K has finite index in G.*

Proof. Consider $M = M_1$ and w.l.o.g. assume that $A \cap M = 1$. Then $C = C_M(A) \times A$, since M is a complement to A. Thus we obtain $K = C'C^p \leq C_M(A) \leq M$, since A is an elementary abelian p-group. □

Using Lemma 6.2, we can compute the core of a conjugacy class of maximal subgroups in G, even though there is no general and practical intersection algorithm known for infinite pc groups yet. As described in Lemma 6.2, we compute K and then determine $H/K = \mathrm{Core}_{G/K}(M_1/K)$ in the finite group G/K. Then we obtain H as the intersection of $M_1, \ldots, M_m$.

Note that the computation of the intersection $U \cap N$ for N normal under the action of U is efficiently possible as kernel of the natural homomorphism $U \to UN/N$. Thus we can determine the intersection of all maximal subgroups of p-power index $\phi_p(G)$ as the intersection of finitely many normal subgroups in G. Clearly, $\phi_p(G)$ is a characteristic subgroup of G and, moreover, it allows to describe the Frattini subgroup $\phi(G)$ as the intersection of infinitely many normal subgroups

$$\phi(G) = \bigcap_{p \in \mathbb{P}} \phi_p(G).$$

6.3. A poly-(free-abelian) normal subgroup of finite index. Here we outline an algorithm to compute a characteristic series $G > N = N_1 \rhd \cdots \rhd N_l \rhd 1$ of a pc group G such that $[G : N]$ is finite and each factor N_i/N_{i+1} is free abelian. Thus, in particular, we compute a characteristic subgroup N of G which is poly-$\mathbb{Z}$.

We start with an efa series as described in Section 3.1. In the following lemma we introduce a method to move a free abelian section downwards across an elementary abelian section. By an iterated application of this lemma we can successively move down all the free abelian factors of the given efa series.

Lemma 6.3. *Let B be a characteristic elementary abelian p-subgroup of a pc group A such that A/B is free abelian. Define $C = C_A(B)$ and let $K = C^q$, where $q = p$, if $p > 2$, and $q = 4$ otherwise.*

a) $K = \{c^q \mid c \in C\}$.

b) *K is a characteristic free abelian subgroup of A.*

c) *K has finite index in A; more precisely, $[A : K] = [A : C] \cdot p^t \cdot q^s$, where t and s are the ranks of B and A/B, resp.*

Proof. a) Let $a, b \in C$. Since $[b, a] \in B$ is centralized by C, we have

$$(ab)^q = a^q \cdot b^q \cdot [b, a]^{q(q-1)/2} = a^q \cdot b^q.$$

Thus the q-th powers in C form a subgroup and a) is proved.

b) Let $b \in B \cap K$. Then $b = c^q$ and c is an element of finite order. Since C/B is free abelian, we obtain $c \in B$. But B is elementary abelian and thus $b = 1$. Therefore $K \cap B = 1$ and $K \cong KB/B \leq C/B$ is free abelian. K is characteristic in A since C is characteristic.

c) Clearly, $[A : K] = [A : C][C : K]$ and for the second factor we observe $[C : K] = [C : KB][B : B \cap K] = q^s \cdot p^t$. □

Thus Lemma 6.3 yields a method to compute a characteristic free abelian subgroup of such a group A. We first have to compute the centralizer $C = C_A(B)$ of the finite elementary abelian group B. Then, if $C = \langle c_1, \ldots, c_s, B\rangle$ is given, we obtain K as $\langle c_1^q, \ldots, c_s^q\rangle$.

In Lemma 6.3 c) we observe that the index of K in A might be rather large. However, every normal free abelian subgroup L of A intersects trivially with B and centralizes B. Thus the index of L in A is at least $[A : C] \cdot p^t$. Hence the subgroup K as determined in Lemma 6.3 is at most an index q^s away from a solution with smallest index. Note that there exist groups A in which the subgroup K as determined in Lemma 6.3 is the largest free abelian characteristic subgroup of A.

7. Implementation and runtimes

Recently, we (joint work with W. Nickel) started to implement a basic setup for computing with infinite pc groups in GAP 4. This package includes a from-the-left collector for pc groups, methods to compute polycyclic generating sequences for subgroups and algorithms to work with homomorphisms and factor groups. This can be used now as a basis to implement higher level algorithms for pc groups.

7.1. Constructing examples of infinite pc groups. Infinite polycyclic groups are often represented as finitely presented groups or matrix groups on a computer. If an arbitrary finitely presented group is given, then we can compute a pc presentation for a polycyclic quotient using the methods in **[10, 11]** or **[7, 8]**. Both of these have become a valuable source for determining interesting infinite pc groups.

Another approach to compute pc presentations for infinite polycyclic groups is to start with a pc group G and a matrix representation for G over $\mathbb{Z}$ or $\mathbb{F}_p$ and compute extensions of G with the underlying module M. For this purpose we can compute $H^2(G, M)$ and then construct extensions with given cocycle. Note that $H^2(G, M)$ might be infinite and thus it might not be possible to determine explicit pc presentations for all extensions of G by M.

7.2. Examples. We use the following polycyclic groups to outline runtimes. We describe the groups briefly and list the relative orders of their pc presentations; that is, the indices of the underlying polycyclic series where we use a 0 to denote an infinite index. However, the presentations themselves are omitted here to save space. The groups will be available as example groups in the "Polycyc" package.

1. The pc groups G_1, G_2 and G_3 are nilpotent groups determined by W. Nickel using the NQ share package of GAP. Their relative orders are
   - [0, 0, 0, 0, 0, 0, 0, 0, 0, 0, 5, 4, 0],
   - [0, 0, 0, 0, 0, 0, 0, 2, 0, 2, 2, 0, 0, 2, 2, 0, 5] and
   - [0, 0, 0, 0, 0, 0, 0, 0, 0, 0, 5, 4, 0, 5, 5, 4, 0, 6, 4, 0, 2].

2. The pc groups G_4 and G_5 are extensions of a maximal finite subgroup of $\mathrm{GL}(6, \mathbb{Z})$ of isomorphism type $(C_2 \times S_4) \wr C_2$ with $\mathbb{Z}^6$. This maximal finite subgroup of $\mathrm{GL}(6, \mathbb{Z})$ can be obtained as 'ImfMatGroup(6, 7, 1)' in GAP and the extensions have been computed as described in Section 7.1. The first group is the split extension, the second extension is non-split. Both groups have relative orders [2, 2, 2, 2, 2, 3, 3, 2, 2, 2, 2, 0, 0, 0, 0, 0, 0].

3. The pc groups G_6 and G_7 are split extensions $\mathbb{Z}^n \rtimes U$ for $U \cong \mathbb{Z}^2$ and $n = 2$ or $n = 4$, resp. The generators of U act on $\mathbb{Z}^n$ as follows.

$$\text{Action for } G_6\text{:} \quad \begin{pmatrix} 0 & 1 \\ 1 & 1 \end{pmatrix} \quad \text{and} \quad \begin{pmatrix} -1 & 0 \\ 0 & -1 \end{pmatrix}$$

$$\text{Action for } G_7\text{:} \quad \begin{pmatrix} 0 & 1 & 0 & 0 \\ 0 & 0 & 1 & 0 \\ 0 & 0 & 0 & 1 \\ 1 & 1 & 0 & 0 \end{pmatrix} \quad \text{and} \quad \begin{pmatrix} 1 & 1 & 0 & -1 \\ -1 & 0 & 1 & 0 \\ 0 & -1 & 0 & 1 \\ 1 & 1 & -1 & 0 \end{pmatrix}$$

4. G_8 is 'SmallGroup(1920, 5000)' in the small groups catalogue. It is a finite group with relative orders [2, 2, 2, 2, 2, 2, 2, 3, 5].

7.3. Runtimes. All runtimes listed below have been obtained on an 450 Mhz Pentium II computer running under Linux. They are given in seconds or as min:sec if necessary.

In Table 1 we outline runtimes of the methods to compute $T(G)$, $TN(G)$ and the conjugacy classes of finite subgroups of G as described in Section 5. We include the orders of $T(G)$ and $TN(G)$ and the number of finite subgroup classes in Table 1. The groups G_4 and G_5 have no torsion subgroup. The groups G_6 and G_7 are omitted from Table 1, since they obviously have no finite subgroups. The group G_8 is finite and thus we can also use the GAP 4 method for conjugacy classes of subgroups to determine the classes of finite subgroups of G_8; GAP 4 needs 17.04 sec for this computation.

	$T(G)$		$TN(G)$		Classes of finite subgroups	
	order	time	order	time	number	time
G_1	$2 \cdot 5$	0.04	$2 \cdot 5$	0.03	4	0.16
G_2	$2 \cdot 5$	0.06	$2 \cdot 5$	0.06	4	0.29
G_3	$2^3 \cdot 3 \cdot 5^3$	0.31	$2^3 \cdot 3 \cdot 5^3$	0.27	384	54.64
G_4	-	0.02	1	0.46	27508	41:23.03
G_5	-	0.02	1	0.74	2998	21:09.01
G_8	1920	0	1920	0	244	7.05

Table 1: Algorithms for various finite subgroups

In Table 2 we compare two methods to compute the normal subgroups of index n for a given n. With algo1 we denote the algorithm described in Section 6.1 and the algorithm outlined in [6] is called algo2. Note that we used a C implementation of algo2 to obtain runtimes, while algo1 is implemented in the GAP language. The GAP language is interpreted and thus slower than C. In the cases denoted by ? algo2 did not come back after a few hours of runtime. As a general remark we noticed that algo2 is faster, if the pc presentation of the input group has few generators, and it is slower otherwise.

	n	algo1	algo2	number of groups
G_1	5	0.47	23.66	6
G_2	10	28.14	?	18
G_3	10	1:16.22	?	18
G_4	10	1.97	12.85	0
G_5	10	1.80	12.85	0
G_6	60	10.65	0.48	178
G_7	60	16.78	4:04.41	168
G_8	60	1.37	0.10	1

Table 2: The normal subgroups of index n

Finally, we briefly consider the computation of maximal subgroups of p-power index. It took about 5.20 sec to compute the maximal subgroups of 2-power index of G_1 which was the largest runtime encountered for this application among the given groups. The groups G_6 - G_8 typically needed 0.05 sec to determine the maximal subgroups of p-power index for a given prime p.

Acknowledgements. I thank W. Nickel for various discussions concerning the implementation of these algorithms in the 'Polycyc' package and M. F. Newman for comments on the paper.

References

[1] G. Baumslag, F. B. Cannonito, D. J. S. Robinson, D. Segal, The algorithmic theory of polycyclic-by-finite groups, J. Algebra 142 (1991), 118–149.

[2] F. Celler, J. Neubüser, C. R. B. Wright, Some remarks on the computation of complements and normalizers in finite soluble groups, Acta Applic. Math. 21 (1990), 57–76.

[3] B. Eick, Special presentations of finite soluble groups and computing Frattini and prefrattini subgroups, in: Groups and Computation II (L. Finkelstein and W. Kantor, eds.), DIMACS Ser. Discrete Math. Theoret. Comput. Sci. 28, Amer. Math. Soc., Providence, RI, 1997, 101–112.

[4] B. Eick, W. Nickel, Algorithms for polycyclic groups, GAP package, see [15].

[5] D. F. Holt, C. R. Leedham-Green, E. A. O'Brien, S. Rees, Smash, Matrix Groups and G-Modules, GAP share package, see [15].

[6] E. H. Lo, Enumerating finite index subgroups of polycyclic groups, unpublished report.

[7] E. H. Lo, A polycyclic quotient algorithm, PhD thesis, Rutgers University, 1996.

[8] E. H. Lo, The PCQA share package of GAP, see [15].

[9] A. I. Mal'cev, On certain classes of infinite soluble groups, Mat. Sb. 28 (1951), 567–588.

[10] W. Nickel, Computing Nilpotent Quotients of Finitely Presented Groups, in: Geometric and Computational Perspectives on Infinite Groups (G. Baumslag et al., eds.), DIMACS Ser. Discrete Math. Theoret. Comput. Sci. 25, Amer. Math. Soc., Providence, RI, 1996, 175–191.

[11] W. Nickel, The NQ share package of GAP, see [15].

[12] G. Ostheimer, Algorithms for polycyclic-by-finite matrix groups, in: Groups and Computation II (L. Finkelstein and W. Kantor, eds.), DIMACS Ser. Discrete Math. Theoret. Comput. Sci. 28, Amer. Math. Soc., Providence, RI, 1997, 297–308.

[13] D. J. S. Robinson, A course in the theory of groups, 2nd. ed., Grad. Texts in Math. 80, Springer-Verlag, 1996.

[14] C. C. Sims, Computation with finitely presented groups, Encyclopedia Math. Appl. 48, Cambridge University Press, 1994.

[15] The GAP Group, GAP — Groups, Algorithms, and Programming, Version 4.2; Aachen, St Andrews, 2000 (http://www-gap.dcs.st-and.ac.uk/~gap).

Fachbereich 17
Universität Kassel
Heinrich-Plett-Str. 40, 34132 Kassel, Germany
eick@mathematik.uni-kassel.de

Computing the maximal subgroups of a permutation group I

Bettina Eick and Alexander Hulpke

Abstract. We introduce a new algorithm to compute up to conjugacy the maximal subgroups of a finite permutation group. Or method uses a "hybrid group" approach; that is, we first compute a large solvable normal subgroup of the given permutation group and then use this to split the computation in various parts.

1991 Mathematics Subject Classification: primary 20B40, 20-04, 20E28; secondary 20B15, 68Q40

1. Introduction

Apart from being interesting themselves, the maximal subgroups of a group have many applications in computational group theory: They provide a set of proper subgroups which can be used for inductive calculations; for example, to determine the character table of a group. Moreover, iterative application can be used to investigate parts of the subgroups lattice without the excessive resource requirements of computing the full lattice. Furthermore, algorithms to compute the Galois group of a polynomial proceed by descending from the symmetric group via a chain of iterated maximal subgroups, see **[Sta73, Hul99b]**.

In this paper, we present a new approach towards the computation of the conjugacy classes of maximal subgroups of a finite permutation group. For this purpose we use a "hybrid group" method. This type of approach to computations in permutation groups has been used recently for other purposes such as conjugacy classes **[CS97, Hul]**, normal subgroups **[Hul98, CS]** or the automorphism group **[Hol00]**.

For finite solvable groups there exists an algorithm to compute the maximal subgroups using a special pc presentation, see **[Eic93, Eic97, EW]**. Our approach incorporates a generalization of this method. Vice versa, if the group in question is finite solvable, then our algorithm reduces to this known method.

For finite groups in general the commonly used approach so far to obtain the maximal subgroup has been via a computation of the full subgroup lattice, see **[Neu60, Hul99a, CCH]**. Our approach in contrast avoids the time and space consuming calculation of all subgroups and thus is not only more efficient but also capable of dealing with much larger groups.

Groups and Computation III
Ohio State Univ. Math. Res. Inst. Publ. 8

2. Outline

Let G be a finite group and S its *solvable radical*; that is, its largest solvable normal subgroup. Then the factor G/S does not contain any solvable normal subgroups; i.e. it is *Fitting free*, and thus the socle $\mathrm{Soc}(G/S)$ is the direct product of nonabelian simple groups.

If G is a permutation group on d points, then we can compute S and a faithful permutation representation of G/S on at most d points **[LS97, Hol97]**. Similarly, given G/S as a permutation group we can compute $L/S = \mathrm{Soc}(G/S)$ and a faithful permutation representation of G/L **[CS**, Corollary 2.3**]**.

We use these algorithms to determine separately the maximal subgroups of G in the following three families:

I) Maximal subgroups that do not contain S. This is done using a generalization of the solvable group method, see Section 3.

II) Maximal subgroups that do contain S but do not contain L. These groups are preimages of the maximal subgroups of G/S that do not contain $\mathrm{Soc}(G/S)$, see Section 4.

III) Maximal subgroups that do not contain L; i.e. preimages of the maximal subgroups of G/L. Since L is a non-trivial subgroup of G and we have a permutation representation for G/L we obtain these subgroups by a recursive call of the algorithm.

Our algorithm for the first step is very efficient in general. Moreover, the groups G/L as considered in step III) are often solvable and very small. Hence the last step is usually easy for our methods. Therefore, we are left with step II) as the main time-consuming part of the algorithm.

3. The solvable radical

We first consider case I) of Section 2 and thus we want to compute the maximal subgroups M of G which do not contain the solvable radical S of G.

Let $S = S_1 \rhd S_2 \rhd \ldots \rhd S_l \rhd S_{l+1} = \{1\}$ the lower nilpotent series of S; that is, S_{i+1} is defined as the minimal normal subgroup of S_i with nilpotent factor group. Furthermore let $S_i^*/S_{i+1} = \Phi(S_i/S_{i+1})$ the Frattini subgroup of the nilpotent factor S_i/S_{i+1}. Then the refined lower nilpotent series

$$S = S_1 > S_1^* \geq S_2 > \cdots > S_l > S_l^* \geq S_{l+1} = \{1\}$$

is a characteristic series of S and hence a normal series of G. Note that this series can be computed effectively by the methods described in **[Eic93]**. The factors S_i/S_i^*

are called *heads* of S and they are non-trivial direct products of elementary abelian groups.

In the following lemma we relate maximal subgroups of G with the refined lower nilpotent series and the heads of G. The proof of it for solvable groups in **[Eic97]** will carry over directly also to nonsolvable groups.

Lemma 3.1. *Let M be a maximal subgroup of G with $MS = G$.*

a) *M covers all but one of the factors of the refined lower nilpotent series of S. This non-covered factor is a head of S.*

b) *Suppose that M does not cover the head S_i/S_i^* for some i. Define $N = S_i \cap M$. Then N/S_i^* is a maximal G-normal subgroup of S_i/S_i^*.*

c) *M/N is a complement to S_i/N in G/N.*

Thus for each maximal subgroup M of G with $MS = G$ there exists a unique head S_i/S_i^* which is not covered by M. For each $i \in \{1, \dots, l\}$ we will separately compute the maximal subgroups which do not cover S_i/S_i^*.

For this purpose we proceed in two steps: First we compute all maximal G-normal subgroups N/S_i^* of S_i/S_i^*. Since S_i/S_i^* is a direct product of elementary abelian p-groups, we may consider each Sylow p-subgroup of S_i/S_i^* separately and compute its maximal G-normal subgroups. This, in turn, can be translated into a calculation of the maximal submodules of a finite dimensional $\mathbb{F}_pG$-module which can be obtained efficiently using the MeatAxe, see **[HLOR95]**.

Then in the second step we consider each of the computed subgroups N in turn and compute the conjugacy classes of complements to S_i/N in G/N. By Lemma 3.1 these are the conjugacy classes of maximal subgroups as desired. To compute the complements we distinguish again two cases. If $i > 1$ and hence $S_i < S$, we can use a very efficient method described in Section 3.2. However, if $i = 1$, this algorithm cannot be applied and we have to determine the complements by a more general and less efficient approach, see Section 3.1.

To simplify notation, we assume from now on that $S_i^* = 1$ and that S_i is an elementary abelian p-group.

3.1. The top head S_1/S_1^*. Recall that $S_1 = S$ and let N be a maximal G-normal subgroup of S as considered in Lemma 3.1.

We wish to compute the conjugacy classes of complements to S/N in G/N. For this purpose we need a finite presentation for G/S. A polynomial length presentation can be obtained efficiently by the methods described in **[BGK⁺97, KS99]**, since we have a permutation representation of G/S.

If the finite presentation is given, then the calculation can be performed as described in **[CNW90]**. Since S/N is an elementary abelian p-group, this can be reduced to solving linear equations over the finite field with p elements.

Still, the calculation of a presentation and subsequential complement calculation may be time-consuming. In general, there seems to be no better method available, but in the special case when S is central under the action of G we can avoid the computation of a presentation with the following method.

Lemma 3.2. *Let S be a central normal p-subgroup of the finite group G. Furthermore let $R = G'G^p$. Then the maximal subgroups of G which do not contain S are preimages of those maximal subgroups of the elementary abelian p-group G/R which do not contain SR/R.*

Proof. Any maximal subgroup M of G which does not contain S must be normal of index p in G, since S is central, and therefore we obtain $R \leq M$. Thus we can translate the problem to G/R which completes the proof. □

To check the condition of Lemma 3.2, we consider the elementary abelian p-group G/R as a vector space V over the field with p-elements. Then SR/R corresponds to a subspace W of V and we need to find all those maximal subspaces of V which do not contain W.

3.2. The lower heads S_i/S_i^* with $i > 1$. In this case we can determine complements more efficiently than in the case of Section 3.1. In particular, we can always avoid the computation of a presentation of G/S_i^*. We use the following theorem on the head S_i/S_i^* and again we assume that $S_i^* = 1$ and S_i is an elementary abelian p-group. We use the following well-known lemma to prove the main theorem of this section.

Lemma 3.3. *Let G be finite group with normal elementary abelian p-subgroup A and nilpotent p'-subgroup T such that $TA \lhd G$ and $[A, T] = A$. Then $N_G(T)$ is a complement to A in G and all complements to A in G are conjugate.*

Proof. First we show that $N_G(T)$ is a complement to A in G. Let $g \in A \cap N_G(T)$. Then for each $t \in T$ we obtain $[g, t] = g^{-1} \cdot g^t \in A$ and $[g, t] = t^{-g} \cdot t \in T$. Hence $[g, t] = 1$ for all $t \in T$ and thus $g \in C_A(T)$. However, since $[A, T] = A$ and A and T are of coprime order, this yields $g = 1$. Therefore $A \cap N_G(T) = 1$.

Now let $g \in G$. Note that TA is a solvable normal subgroup of G with p-complement T. Thus $T^g \leq TA$ and T^g is another p-complement of TA. Hence $T^g = T^a$ for some $a \in A$. Therefore $ga^{-1} \in N_G(T)$ and we obtain that $AN_G(T) = G$. Hence $N_G(T)$ is a complement to A in G.

Finally let K be an arbitrary complement to A in G. Since $TA \lhd G$, we have that $|TA \cap K| = |TA/A|$ and hence $TA \cap K$ is another p-complement of TA. Thus $TA \cap K$ is conjugate to T. Therefore $K = N_G(TA \cap K)$ is conjugate to $N_G(T)$. □

Now we can apply the above lemma to show the following theorem.

Theorem 3.4. *Let $i \in \{2, \dots, l\}$ and suppose that N is a maximal G-normal subgroup of the elementary abelian p-group S_i. Then there exists exactly one conjugacy class of maximal subgroups M of G with $S_i \cap M = N$. A representative of this class can be computed as $N \cdot N_G(T_{i,p})$ where $T_{i,p}$ is a p-complement of S_{i-1}.*

Proof. Note that in **[EW]** a proof is given for the case that G is solvable. We give an alternative proof which shows the theorem in its more general version.

Let L be an arbitrary G-normal subgroup of S_i and consider $A = S_i/L$ and $T = T_{i,p}L/L$. Since A is the last term of the lower nilpotent series of G/L, we obtain that $[A, T] = A$ and hence we can apply Lemma 3.3. This yields that there exists exactly one conjugacy class of complements to S_i/L in G/L.

Thus, for $L = 1$ we obtain that $N_G(T_{i,p})$ is a complement to S_i in G.

Moreover, for $L = N$ a maximal G-normal subgroup of S_i the above argument shows that there is exactly one maximal subgroup M of G with $S_i \cap M = N$. However, $N \cdot N_G(T_{i,p})$ is a complement to S_i/N and thus it is a maximal subgroup of this type. □

Thus instead of computing complements by the standard approach we determine the normalizer $N_G(T_{i,p})$. Again, we do not use the a general method for this purpose, but we introduce a special approach. Note that the existence of a nilpotent p-complement $T_{i,p}$ is crucial for our special method. As there is no such p-complement in G/S_1^*, we cannot apply this method to the top head S/S_1^*, but only to all the lower heads.

3.3. Computing a normalizer of $T_{i,p}$. To simplify notation let $A = S_i$ and $T = T_{i,p}$. Since $N_G(T)$ is a complement to A in G, for every $g \in G$ there exists an element $a \in A$ with $ga \in N_G(T)$. We introduce a method to compute such an element a as solution of a linear equation over $\mathbb{F}_p$. This enables us to compute a generating set of $N_G(T)$ from a generating set of G.

Remark 3.5. The factor TA/A is the p-complement of the nilpotent group S_{i-1}/A. Since S_{i-1}/A is normal in G/A, we obtain that TA/A is normal in G/A, and hence TA is normal in G.

Let $t \in T \leq TA$ and $g \in G$. Then $[t, g] \in TA$ and thus $[t, g] = sb$ for some $s \in T$ and $b \in A$. Note that s and b are unique, since TA is the semi-direct product of the p-group A and the p'-group T.

If $ga \in N_G(T)$, then $[t, ga] \in T$. Since $[t, ga] \equiv [t, g] \bmod A$, we obtain that $[t, ga] = s$ in this case. On the other hand we have

$$\begin{aligned} [t, ga] &= t^{-1} \cdot a^{-1} \cdot g^{-1} \cdot t \cdot g \cdot a \\ &= (a^{-1})^t \cdot [t, g] \cdot a \\ &= (a^{-1})^t \cdot s \cdot b \cdot a \\ &= s \cdot (a^{-1})^{ts} \cdot b \cdot a. \end{aligned}$$

Combining the above formulas we obtain

$$[t, ga] = s \quad \text{if and only if} \quad a^{ts}a^{-1} = b.$$

Note that ts acts on A as $t[t, g] = t^g$ and hence the action of ts can be computed easily from t and g. Moreover, the factorisation $[t, g] = sb$ is easy to compute, since b is just the p-part of the commutator $[t, g]$.

Altogether we can compute the element a as solution of the equations $a^{t^g}a^{-1} = b$ where t runs over a generating set of T. Since A is an elementary abelian p-group, we can translate this set of equations into a system of inhomogeneous linear equations over $\mathbb{F}_p$.

4. Fitting-free groups

Now we turn our attention to the maximal subgroups in case II) of Section 2. Thus we want to determine the maximal subgroups of G which contain the solvable radical S, but do not contain L, where L/S is the socle of G/S. Recall that we may determine a permutation representation of G/S efficiently and hence, without loss of generality, we assume that $S = 1$ and G is a Fitting free group with socle L.

We first compute a G-chief series $L = N_0 > \cdots > \langle 1 \rangle$ of L, see **[KL90, CH97, Hul]**. We will parametrize the maximal subgroups according to the largest of the N_i they contain. We shall describe this process only for one index, to obtain all maximal subgroups one has to run in a loop over all N_i.

Let M be a maximal subgroup of G with $L \not\leq M$ and let j the smallest index such that $N_j \leq M$. To simplify the notation we assume now that $N_j = \langle 1 \rangle$ by translation to the factor G/N_j and we denote N_{j-1} by N. Note that we do not need to construct a faithful representation for this factor group in our algorithm, since all necessary calculations can be performed with representatives for the elements and full preimages of subgroups.

Thus we have reduced to the case that N is a minimal nonabelian normal subgroup of G and we want to determine the maximal subgroups M of G with $N \not\leq M$.

Let $C = C_G(N)$ and $K = \mathrm{Core}_G(M) = \bigcap_{g \in G} M^g \lhd G$. Then $K \leq M$ and therefore $N \not\leq K$. Moreover, $N \cap K = \langle 1 \rangle$, since N is a minimal normal subgroup, and thus the normal subgroup K centralizes N. Hence we obtain $K \leq C$ and we can summarize the situation in the picture below.

Remark 4.1. In this notation we obtain that G/K has a faithful primitive permutation representation on the cosets of M/K and M/K can be described as the point stabilizer of this permutation representation. Thus to determine all maximal subgroups of G not containing N we first compute all possible candidates for the normal subgroup K, then we calculate all faithful primitive permutation representations of G/K and finally we obtain the maximal subgroup M as point stabilizer of the permutation representation.

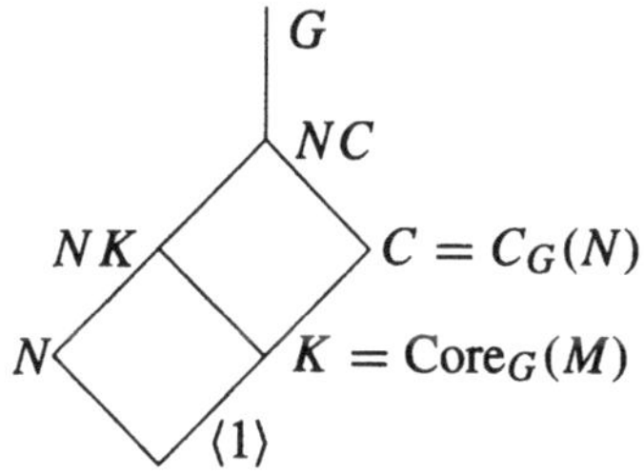

The structure of the socle of a primitive permutation group can be described in detail. We will use this to determine the structure of G/K more precisely.

Theorem 4.2. *Consider the notation as in the above picture. Then* $\mathrm{Soc}(G/K) = NC/K$. *Moreover, one of the following two cases holds for G.*

A) $K = C$ *and* $\mathrm{Soc}(G/K) = NK/K \cong N$ *is a minimal subgroup of* G/K. *(We say* G/K *is primitive of type A.)*

B) $K < C$ *and* $\mathrm{Soc}(G/K) = NK/K \times C/K \cong N \times N$ *is a direct product of two permutation isomorphic subgroups of* G/K. *(We say* G/K *is primitive of type B.)*

Proof. First we observe that $\mathrm{Soc}(G/K) = NC/K$. Clearly, NK/K is a minimal normal subgroup of G/K and hence $NK/K \leq \mathrm{Soc}(G/K)$.

Next we observe that $C/K \leq \mathrm{Soc}(G/K)$. This holds trivially if $C = K$. Thus we assume that $K < C$. By definition of K the factor G/K has a faithful primitive permutation representation. Its nontrivial normal subgroups NK/K and C/K centralize each other and both act regularly by **[DM96**, Theorem 4.2A]. Thus C/K is a minimal normal subgroup of G/K and the claim follows.

Now we show that $\mathrm{Soc}(G/K) \leq NC/K$. Let $A \lhd G$ such that A/K is a minimal normal subgroup of G/K. If $N \leq A$, then we obtain $A \leq NK \leq NC$. Otherwise $N \cap A = \langle 1 \rangle$ and thus $A \leq C \leq NC$. Thus we proved that $\mathrm{Soc}(G/K) = NC/K$.

By Theorem 4.3B of **[DM96]** the socle of a primitive permutation group is either a minimal normal subgroup or a direct product of two permutation isomorphic minimal normal subgroups. Applying this to our situation, we obtain directly that G/K must be of type A or B, proving the theorem. □

Applying Theorem 4.2 we now distinguish the two cases A) and B) on the centralizer C of N in G and its relationship to the core K of the desired maximal subgroup M of G.

To compute all maximal subgroups in type A which do not contain N we compute the action of G on N. The kernel of this action is C and in type A we have $C = K$. Following **[Hul**, section 3] the image Q of this action can be represented as a permutation group of small degree. We then must test whether Q has faithful primitive permutation representations of type A and compute point stabilizers of these representations to determine the desired maximal subgroups. See Section 5 for further details.

To determine the maximal subgroups in type B we first must obtain all possible kernels $K < C$. Again, we obtain C as the kernel of the action of G on N. Then we intersect the subgroups N_i of the chief series with C and remove duplicates. This yields a series $C = C_0 > C_1 > \cdots > \langle 1 \rangle$ of normal subgroups in C whose factors are G-chief factors.

For every possible K there is a minimal index k with $C_k \leq K$. In this case C_{k-1}/C_k is isomorphic to $C/K \cong N$ and hence C_{k-1}/C_k is nonabelian simple. Thus we obtain $K = C_{C/C_k}(C_{k-1}/C_k)$. In particular, each chief factor of C isomorphic to N can give rise only to at most one possible subgroup K.

To obtain all potential kernels K we loop over the factors C_{k-1}/C_k for all k. If $C_{k-1}/C_k \cong N$, we compute the simultaneous action φ of G on N and on C_{k-1}/C_k. Let Q be the image of this action. If the index of $\ker\varphi = C_{C/C_k}(C_{k-1}/C_k)$ in C is $|N|$, then the socle of Q is a direct product of two normal subgroups isomorphic to N and only in this case φ might give a primitive factor of type B.

If this is the case, we have to test whether Q has faithful primitive permutation representations of type B and to compute the maximal subgroups afforded by these representations. Again, see Section 5.

By the Jordan–Hölder Theorem primitive actions of type B can only arise if there is *another* chief factor N_{i-1}/N_i of G with $i < j$ which is isomorphic to N. This can be used as a quick initial test.

Note, that we do not need to get explicit isomorphisms of the chief factors, but only isomorphism types. However, the isomorphism type of a finite simple group is determined already by the size of the simple group except for $|\mathrm{PSL}_3(4)| = |A_8|$ and $\left|\mathrm{PSp}_{2n}(q)\right| = |\mathrm{P}\Omega_{2n+1}(q)|$, see **[Cam81]**; to distinguish between groups of same size one can either consider the size of p-element centralizers or use the approach of **[AB00]**.

5. Primitive permutation representations

By the reduction in the previous section we can assume that we have given a permutation group G and we need to determine the faithful primitive permutation representations of G. We know that the socle $H = \mathrm{Soc}(G)$ is the direct product of isomorphic simple groups; that is, $H = T_1 \times \cdots \times T_m$ with $T_i \cong T$ nonabelian simple. We say that H is *homogeneous* and the simple group T is called the *type* of H.

The O'Nan–Scott Theorem **[Sco80]** gives a structural description of primitive groups. We recall a version of it below in Theorem 5.1. A proof of the theorem can be found for example in **[LPS88]** or **[DM96**, chapter 4].

Our version of this theorem splits the primitive permutation groups in five types. In the first type the socle of the primitive group is abelian and hence we do not need to consider this type for the purpose of the computation of maximal subgroups. However, we include this type in the list for completeness. The types 2–5 on the other hand have non-abelian socles and we will have to consider these types.

Theorem 5.1 (O'Nan–Scott). *Let G be a group which acts primitively and faithfully on Ω with $|\Omega| = n$. Let $H = \mathrm{Soc}(G)$ and $\omega \in \Omega$. Then H is homogeneous of type T and exactly one of the following cases holds.*

1. *"Affine". T is abelian of order p, $n = p^m$ and $\mathrm{Stab}_G(\omega)$ is a complement to H which acts irreducibly on H.*

2. *"Almost simple". $m = 1$ and $H \lhd G \leq \mathrm{Aut}(H)$.*

3. *"Diagonal type". $m \geq 2$ and $n = |T|^{m-1}$. Further, G is a subgroup of $V = (T \wr S_m).\mathrm{Out}(T) \leq \mathrm{Aut}(T) \wr S_m$ in diagonal action and either*

   a) *$m = 2$ and G acts intransitively on $\{T_1, T_2\}$ or*
   b) *$m \geq 2$ and G acts primitively on $\{T_1, \ldots, T_m\}$.*

   In case a) *T_1 and T_2 both act regularly. Moreover, the point stabilizer V_ω of V is of the form $\mathrm{diag}(\mathrm{Aut}(T)^{\times m}).S_m \cong \mathrm{Aut}(T) \times S_m$ and thus $H_\omega = \mathrm{diag}(T^{\times m})$.*

4. *"Product type". $m = rs$ with $s > 1$. We have that $G \leq W = A \wr B$ and the wreath product acts in product action with A acting primitively, but not regularly, on d points and B acting transitively on s points. Thus $n = d^s$. The group A is primitive of either*

   a) *type* 3a *with socle T^2 (i.e. $r = 2$, $s < m$),*
   b) *type* 3b *with socle T^r (i.e. $r > 1$, $s < m$) or*
   c) *type* 2 *(i.e. $r = 1$, $s = m$).*

   We have that $W_\omega \cap A^s \cong A_1^{\times s}$ and $\mathrm{Soc}(G) = \mathrm{Soc}(W)$. Furthermore $W = A^{\times s}G$.

5. *"Twisted wreath type". H acts regularly and $n = |T|^m$. G_ω is isomorphic to a transitive subgroup of S_m. The normalizer $N_{G_\omega}(T_1)$ has a composition factor isomorphic to T. Thus, in particular, $m \geq k+1$ where k is the smallest degree of a permutation group which has T as a composition factor.*

Proof. The case distinction and closer description of the different cases is proved in **[DM96**, chapter 4]. It remains to show that G is of exactly one of the types listed and cannot be listed under two different categories. In type 1 the socle is abelian, in type 5 the socle is nonabelian but regular; type 2 is the only type with a simple nonabelian socle. To distinguish type 3 from type 4, we observe that in type 3 the point stabilizer in H is a diagonal and thus simple, while in type 4 this point stabilizer is a direct product of point stabilizers of constituent groups and thus is not simple. Cases 3a and

3b are distinguished on whether the socle is a minimal normal subgroup. The three subcases of type 4 are distinguished by the action of the point stabilizer in the socle on its orbits. □

Remark 5.2. The labelling of types in the literature is inconsistent. The following table gives translations of the labellings used.

Type	1	2	3a	3b	4a	4b	4c	5
[**DM96**, Section 4.8]	i	iii	iv	iv	v	v	v	ii
[**LPS88**]	I	II	IIIa	IIIa	IIIb	IIIb	IIIb	IIIc
[**Neu86**]	I	V	II	III	II	III	IV	IV

Note that for case 3a/b) we change the case distinction of [**DM96**, Theorem 4.5A] from degree 2/> 2 to intransitive/primitive.

Remark 5.3. Type A in Theorem 4.2 corresponds to primitivity of type 2, 3b, 4b, 4c or 5. Type B corresponds to type 3a or 4a.

We now use Theorem 5.1 to determine primitive permutation representations of our given group G. We will need to answer the following questions for each of the types 2–5 of Theorem 5.1:

i) What extra conditions on G are necessary and sufficient for G to have a faithful primitive action of a certain type.
ii) If G has such an primitive action, then classify all these actions up to conjugacy of the point stabilizers in G and determine their point stabilizers.

For type 2 these questions are answered easily: A group G has faithful primitive representations of type 2 if and only if its socle is nonabelian simple. The classification of finite simple groups [**Gor82**] can be used to identify the socle T of G and then we can employ pre-tabulated results as well as parametrizations for some of the series to obtain the maximal subgroups of G. See [**Sax95**] for a recent survey.

Another approach to obtain the primitive permutation representations of an almost simple group is by using the data base of table of marks together with generator words for subgroup representatives, see [**Mer98**]. Such a data base is available in GAP, see [**GAP00**]. (In particular, such a data base can comprise all almost simple groups, for which computation of the full lattice would be still feasible.) However, to use this precomputed data, algorithms for the constructive recognition of almost simple groups are required. Much work has been done in this area recently, see for example [**BP00, CFL97, KS**].

For types 3–5 the situation is more complex. Theorem 4.3 in [**Kov86**] can give a parametrization of maximal subgroups of type 3 and 4 though we have not studied the feasibility of this approach. Similarly, [**Bad93**] describes groups of type 5 in more detail. However, the computation of complements to a nonsolvable normal subgroup might become a problematic subtask in this case.

A more detailed examination of whether a group can be primitive in one of these cases will be the subject of a subsequent study.

6. Runtime requirements

The solvable part of the algorithm requires the solvable radical. This can be obtained efficiently using the methods in **[LS97, Hol97]**. Then we need to construct the lower nilpotent series and Frattini subgroups for the nilpotent factors involved. A practical algorithm for this purpose is outlined in **[Eic93]**. Furthermore, as one can consider these subgroups as O^{Σ}, for Σ the class of cyclic p-groups, **[KL90]** asserts that the calculation is possible in polynomial time.

For the computation of maximal subgroups we use the algorithms of **[LMR94]** and **[HLOR95]**. **[Rón90]** proves (for equivalent algorithms) that these calculations can be done in polynomial time.

The remaining steps for the computation of complements are linear algebra which is of polynomial time, provided the factor presentation is of polynomial length, which **[KS99]** assures.

An implementation of the solvable part of the algorithm shows that this method is highly efficient.

For the Fitting-free case we need to compute the action on chief factors. The required chief series, kernel, and intersections with normal subgroups again are in the polynomial time toolkit. Constructive recognition of the factor constituents **[KS]** then permits to write down a representation for the action of the chief factor(s).

We have not yet examined the determination of possible faithful primitive actions. However even for the almost simple case it seems impossible to give a general complexity, as we have to rely on pretabulated data (to get the maximal subgroups of almost simple groups) and we might only be able to bound the runtime by a function of the size of the involved simple groups and the width of the nonabelian composition factors.

7. Final comments

We have presented the algorithm for permutation groups. However – as with many other "hybrid" algorithms – the permutational group structure is only needed within subtasks such as the computation of the solvable radical or a chief series. Once all these subproblems are solvable for a group in another representation, our algorithm becomes available immediately.

The authors would like to thank the referee for helpful comments.

Part of the work was undertaken while the second author was at the University of St Andrews, supported by EPSRC Grant GL/L21013.

References

[AB00] Christine Altseimer and Alexandre V. Borovik, A non-deterministic algorithm for recognition of orthogonal and symplectic groups, in: Groups and Computation III (W. M. Kantor and Á. Seress, eds.), Ohio State Univ. Math. Res. Inst. Publ. 8, Walter de Gruyter, Berlin–New York, 2001, 1–20.

[Bad93] R. W. Baddeley, Primitive permutation groups with a regular nonabelian normal subgroup, Proc. London Math. Soc. (3) 67 (1993), no. 3, 547–595.

[BGK+97] L. Babai, A. J. Goodman, W. M. Kantor, E. M. Luks, and P. P. Pálfy, Short presentations for finite groups, J. Algebra 194 (1997), 97–112.

[BP00] Sergey Bratus and Igor Pak, Fast constructive recognition of a black box group isomorphic to S_n or A_n using Goldbach's conjecture, J. Symbolic Comput. 29 (2000), 33–57

[Cam81] Peter J. Cameron, Finite permutation groups and finite simple groups, Bull. London Math. Soc. 13 (1981), 1–22.

[CCH] John Cannon, Bruce Cox, and Derek Holt, Computing the subgroup lattice of a permutation group, J. Symbolic Comput., to appear.

[CFL97] Gene Cooperman, Larry Finkelstein, and Steve Linton, Constructive recognition of a black box group isomorphic to $GL(n, 2)$, in Finkelstein and Kantor **[FK97]**, pp. 85–100.

[CH97] John Cannon and Derek Holt, Computing chief series, composition series and socles in large permutation groups, J. Symbolic Comput. 24 (1997), 285–301.

[CNW90] Frank Celler, Joachim Neubüser, and Charles R. B. Wright, Some remarks on the computation of complements and normalizers in soluble groups, Acta Appl. Math. 21 (1990), 57–76.

[CS] John Cannon and Bernd Souvignier, On the computation of normal subgroups in permutation groups, Internat. J. Algebra Comput., to appear.

[CS97] John Cannon and Bernd Souvignier, On the computation of conjugacy classes in permutation groups, Proceedings of the 1997 International Symposium on Symbolic and Algebraic Computation (Wolfgang Küchlin, ed.), The Association for Computing Machinery, ACM Press, 1997, 392–399.

[DM96] John D. Dixon and Brian Mortimer, Permutation groups, Grad. Texts in Math. 163, Springer-Verlag, 1996.

[Eic93] Bettina Eick, Spezielle PAG-Systeme im Computeralgebrasystem GAP, Diplomarbeit, RWTH Aachen, 1993.

[Eic97] Bettina Eick, Special presentations for finite soluble groups and computing (pre-)Frattini subgroups, in: Finkelstein and Kantor [FK97], 101–112.

[EW] Bettina Eick and Charles Wright, Computing formation-theoretic subgroups and certain complements in finite solvable groups, in preparation.

[FK97] Larry Finkelstein and William M. Kantor (eds.), Groups and Computation II, DIMACS Ser. Discrete Math. Theoret. Comput. Sci. 28, Amer. Math. Soc., Providence, RI, 1997.

[GAP00] The GAP Group, GAP — Groups, Algorithms, and Programming, Version 4.2; Aachen, St Andrews, 2000 (`http://www-gap.dcs.st-and.ac.uk/~gap`).

[Gor82] Daniel Gorenstein, Finite simple groups, Plenum Press, 1982.

[HLOR95] Derek F. Holt, Charles R. Leedham-Green, Eamonn A. O'Brien, and Sarah Rees, Smash – matrix groups and G-modules, 1995, A GAP share package.

[Hol00] Derek F. Holt, Computing the automorphism group of a finite group, in: Groups and Computation III (W. M. Kantor and Á. Seress, eds.), Ohio State Univ. Math. Res. Inst. Publ. 8, Walter de Gruyter, Berlin–New York, 2001, 201–208.

[Hol97] Derek F. Holt, Representing quotients of permutation groups, Quart. J. Math. Oxford Ser. (2) 48 (1997), no. 191, 347–350.

[Hul98] Alexander Hulpke, Computing normal subgroups, Proceedings of the 1998 International Symposium on Symbolic and Algebraic Computation (Oliver Gloor, ed.), The Association for Computing Machinery, ACM Press, 1998, 194–198.

[Hul99a] Alexander Hulpke, Computing subgroups invariant under a set of automorphisms, J. Symbolic Comput. 27 (1999), 415–427, (ID jsco.1998.0260).

[Hul99b] Alexander Hulpke, Techniques for the computation of Galois groups, Algorithmic Algebra and Number Theory (B. H. Matzat, G.-M. Greuel and G. Hiss, eds.), Springer-Verlag, 1999, 65–77.

[Hul] Alexander Hulpke, Conjugacy classes in finite permutation groups via homomorphic images, Math. Comput. 69 (2000), no. 232, 1633–1651.

[KL90] William M. Kantor and Eugene M. Luks, Computing in quotient groups, in: Proceedings of the 22nd ACM Symposium on Theory of Computing, Baltimore, ACM Press, 1990, 524–563.

[Kov86] L. G. Kovács, Maximal subgroups in composite finite groups, J. Algebra 99 (1986), 114–131.

[KS] W. M. Kantor and Á. Seress, Black box classical groups, Mem. Amer. Math. Soc. 149, Nr. 708, Amer. Math. Soc., Providence, RI, 2001.

[KS99] William M. Kantor and Ákos Seress, Permutation group algorithms via black box recognition algorithms, Groups St Andrews 1997 in Bath (C. M. Campbell, E. F. Robertson, N. Ruskuc and G. C. Smith, eds.), London Math. Soc. Lecture Note Ser. 260/261, Cambridge University Press, 1999, 436–446.

[LMR94] Klaus Lux, Jürgen Müller, and Michael Ringe, Peakword Condensation and Submodule Lattices: An Application of the Meat-Axe, J. Symbolic Comput. 17 (1994), 529–544.

[LPS88] Martin W. Liebeck, Cheryl E. Praeger, and Jan Saxl, On the O'Nan-Scott theorem for finite primitive permutation groups, J. Austral. Math. Soc. Ser. A 44 (1988), 389–396.

[LS97] Eugene M. Luks and Ákos Seress, Computing the Fitting subgroup and solvable radical for small-base permutation groups in nearly linear time, in Finkelstein and Kantor [FK97], 169–181.

[Mer98] Thomas Merkwitz, Markentafeln endlicher Gruppen, Diplomarbeit, Lehrstuhl D für Mathematik, Rheinisch-Westfälische Technische Hochschule, Aachen, 1998.

[Neu60] Joachim Neubüser, Untersuchungen des Untergruppenverbandes endlicher Gruppen auf einer programmgesteuerten elektronischen Dualmaschine, Numer. Math. 2 (1960), 280–292.

[Neu86] Peter M. Neumann, Some algorithms for computing with finite permutation groups, Groups – St Andrews 1985 (Edmund F. Robertson and Colin M. Campbell, eds.), Cambridge University Press, 1986, 59–92.

[Rón90] Lajos Rónyai, Computing the structure of finite algebras, J. Symbolic Comput. 9 (1990), no. 3, 355–373.

[Sax95] Jan Saxl, Finite simple groups and permutation groups, Finite and locally finite groups (Istanbul, 1994), Kluwer Acad. Publ., Dordrecht, 1995, 97–110.

[Sco80] Leonard L. Scott, Representations in characteristic p, The Santa Cruz conference on finite groups (Providence, RI) (Bruce Cooperstein and Geoffrey Mason, eds.), Proc. Sympos. Pure Math. 37, Amer. Math. Soc., 1980, Corrigendum in [LPS88], 318–331.

[Sta73] Richard P. Stauduhar, The determination of Galois groups, Math. Comp. 27 (1973), 981–996.

Fachbereich 17
Universität Kassel
Heinrich Plett Str. 40, 34132 Kassel, Germany
`eick@mathematik.uni-kassel.de`

Department of Mathematics
The Ohio State University
231 W 18th Avenue, Columbus, OH 43210, U.S.A.
`ahulpke@math.ohio-state.edu`

On p-singular elements in Chevalley groups in characteristic p

Robert M. Guralnick and Frank Lübeck*

Abstract. We give upper bounds for the proportion of p-singular elements in finite Chevalley groups in characteristic p.

1991 Mathematics Subject Classification: primary 20G40; secondary 20D60.

1. Introduction

Let $\boldsymbol{G}$ be a reductive algebraic group over an algebraic closure of a prime field of characteristic p, defined over a finite field F_q with q elements. Let σ denote a (possibly twisted) Frobenius endomorphism of $\boldsymbol{G}$ with field of definition F_q. We denote $G = \boldsymbol{G}(q)$ the corresponding finite group of F_q-rational points of $\boldsymbol{G}$, i.e., the fixed points of σ. Since we can take $\boldsymbol{G}$ to be simply connected or adjoint (or anything in between) and since (as we see below) a center will not change our computations, the finite groups we are considering are those with $J(q) \leq G \leq \mathrm{InnDiag}\, J(q)$ where $J(q)$ is a simple finite group of Lie type defined over F_q.

Kantor and Seress **[KS]** have asked for an upper bound on the proportion of p-singular elements (i.e., those elements whose order is divisible by p) in G. As far as we know, Kantor was the first to conjecture that a bound of c/q was likely. The motivation came from a specific need within computational group theory. See also **[NP]** where estimates for some of the classical groups are given (not necessarily simple).

Spaltenstein observed that one gets asymptotic results (at least for a fixed characteristic) by counting the number of components of codimension 1 in the hypersurface which is the complement of regular semisimple elements. Asymptotic results can also be obtained from results in **[FJK98, FJ93, FJ93b, FJ94]** which give formulae for the number of regular semisimple classes.

The issue of a lower bound was addressed in **[IKS95]**. See also **[NP]**.

Let S denote the set of p-singular elements in G (i.e., the unipotent part is nontrivial). Let R denote the set of regular semisimple elements in $\boldsymbol{G}$ – i.e., the set of elements whose connected centralizer is a (necessarily maximal) torus or, equivalently, those

*Partially supported by an NSF grant.

Groups and Computation III
Ohio State Univ. Math. Res. Inst. Publ. 8

elements which do not commute with any nontrivial unipotent element. Clearly, R and S do not intersect. It is well known that R is an open subset of $\boldsymbol{G}$ and so R', the complement of R contains the closure of S. In fact R' is the closure of S: R' consists of the elements which commute with a nontrivial unipotent element – this implies that any element x in R' commutes with a root subgroup U (see Theorem 4.1) – then xU is an irreducible variety and with at most one exception consists of p-singular elements; in particular, x is in the closure of S.

So we define $s(G)$ to be the proportion of p-singular elements in G and $r(G)$ to be the proportion of regular semisimple elements in G. Let $r'(G) = 1 - r(G)$. So $r'(G)$ is the proportion of elements which commute with an element of order p. Clearly, $s(G) < r'(G)$ and it is $r'(G)$ that we will obtain estimates for.

So if c is the number of components of R' of codimension 1, it follows from the Lang–Weil estimates for the number of points on an absolutely irreducible variety that

$$r'(G) = |R'(q)|/|\boldsymbol{G}(q)| \le c/q + O(q^{-3/2}).$$

As mentioned above, this approach was suggested by Spaltenstein. The weakness of this approach is that one does not obtain effective bounds, only asymptotic ones. In particular, the nature of the error term conceivably depends on the characteristic of the underlying field. In this note, we obtain effective bounds as follows.

Theorem 1.1. *If $\boldsymbol{G}$ is simple then $s(\boldsymbol{G}(q)) < r'(\boldsymbol{G}(q)) < 3/(q-1) + 2/(q-1)^2$.*

As noted above, this theorem applies to all groups G with $J(q) \le G \le \mathrm{InnDiag}(J(q))$, where $J(q)$ is a simple finite group of Lie type defined over the field of q elements and $\mathrm{InnDiag}(J(q))$ is the full group generated by inner and diagonal automorphisms (see **[Car72]** for definitions). Note if we take $\boldsymbol{G}$ to be simply connected, then $G = \boldsymbol{G}(q)$ is almost always quasisimple. Since the center is a p'-group, there is no harm in passing to the quotient in computing either $r(q)$ or $s(q)$. Similarly, if we take $\boldsymbol{G}$ to be the adjoint group, then G will be the full group of inner and diagonal automorphisms of $J(q)$.

In the main body of the paper, we give more precise bounds for each type of group which are also asymptotically correct for growing q. One can modify our methods to obtain asymptotic lower bounds with the correct main term; see **[NP]**. In that paper, they also obtain results for fixed q as the rank tends to infinity.

We handle the classical groups first. The method we use is to show that any p-singular element commutes with a conjugate of one of a family of unipotent subgroups (for algebraic groups, this family can be taken to be the root subgroups). For the classical groups, we estimate how many elements are in the union of the set of conjugates of these subgroups. For exceptional groups, one can compute exactly the number of p'-elements of $\boldsymbol{G}(q)$ and $|R'(q)|$.

In Section 4, we discuss asymptotic results for the variety of p-singular elements and results for simple algebraic groups. In the final section, we make some brief remarks about the general case of almost simple groups.

2. Classical groups

In the case of GL_n or SL_n, the subset R is the collection of elements with distinct eigenvalues – a condition that can be expressed by saying that the discriminant of the characteristic polynomial is nonzero. There are similar descriptions for the other classical groups.

Let $J = G(q)$ denote a simple classical group (i.e., linear, unitary, symplectic or orthogonal) over F_q with $q = p^a$. Let $J \leq G \leq A$, where A is the full group of inner and diagonal automorphisms. We view J, G and A all acting on its natural module V (since whatever center is allowed has order prime to p, this has no effect on the proportion of p-singular elements or of elements which commute with a nontrivial p-element). So the groups we are considering for J are $\mathrm{PSL}_n(q), n \geq 2$, $\mathrm{PSp}_n(q)$, $n \geq 4$ and even, $\mathrm{PSU}_n(q)$, $n \geq 3$ or $\mathrm{O}_n^{\pm}(q)$, $n \geq 7$ (where the latter group is the commutator subgroup of the orthogonal group). The natural module is defined over F_q except in the unitary case where it is defined over F_{q^2}.

If $x \in G$, we will let s_x and u_x denote the semisimple and unipotent parts of x (so x is the commuting product of s_x and u_x and these element are the p' and p-parts of x).

Our strategy is as follows. We first identify semisimple elements in R' in terms of their module structure on the natural module. We then choose a unipotent subgroup U which is contained in the center of a Sylow subgroup of the centralizer of this semisimple element. It follows that any element of R' with this given semisimple part commutes with a conjugate of U. Fortunately, the choice of U (up to conjugacy) will not depend very much on the semisimple element, but only on the homogeneous components of the semisimple element. It follows that R' is the union of the conjugates of the centralizers of these families of unipotent subgroups and this is the estimate we use.

In the case of algebraic groups (and not just the classical groups), it turns out that we need only consider U to be a root subgroup (so in case of one root length, only one subgroup and in general at most two).

In this section we will repeatedly use the following easy lemmas.

Lemma 2.1. *Let G be a finite group and U be a subgroup of G with $t = |N_G(U) : U|$. Then $|\bigcup_{g \in G} U^g| < |G|/t$.*

Proof. There are $|G|/|N_G(U)| = |G|/(t|U|)$ different subgroups U^g, all of order $|U|$. □

Lemma 2.2. *Let $q > 1$ a real number. Then*

$$\sum_{e=1}^{\infty} 1/(q^e - 1) < 1/(q-1) + 1/(q-1)^2.$$

Proof. This follows from

$$\sum_{e=2}^{\infty} 1/(q^e - 1) < \sum_{e=2}^{\infty} 1/(q^e - q) = 1/(q(q-1)) + (1/q) \sum_{e=2}^{\infty} 1/(q^e - 1). \qquad \square$$

Recall that $r'(G) = |R'(q)|/|G(q)|$.

In the remaining part of this section, we prove the following:

Theorem 2.3. *If G is almost simple with socle J of classical type and G is contained in the group of inner-diagonal automorphisms of J, then $r'(G)$ is bounded above as given in the following table:*

J	$\|G : J\|$	*Upper bound for* $r'(G)$
$\mathrm{PSL}_2(q)$	1	$(2, q-1)/(q-1)$
$\mathrm{PSL}_2(q)$	2	$1/(q-1)$
$\mathrm{PSL}_n(q)$, $n \geq 3$		$1/(q-1) + 2/(q-1)^2$
$\mathrm{PSU}_n(q)$, $n > 2$		$1/(q-1) + 4/(q-1)^2$
$\mathrm{Sp}_n(q)$, $n = 2k > 2$, *q even*		$2/(q-1) + 1/(q-1)^2$
$\mathrm{PSp}_n(q)$, $n = 2k > 2$, *q odd*	1	$3/(q-1) + 1/(q-1)^2$
$\mathrm{PSp}_n(q)$, $n = 2k > 2$, *q odd*	2	$2/(q-1) + 1/(q-1)^2$
$\mathrm{O}_{2n}^{\pm}(q)$, $n \geq 4$		$1/(q-1) + 2/(q-1)^2$
$\mathrm{O}_{2n+1}(q)$, $n \geq 3$, *q odd*		$2/(q-1) + 2/(q-1)^2$

Proof. Let $x = su$ be the Jordan decomposition of $x \in R'$. Since s is not regular, it follows that s has a homogeneous component W on V which has multiplicity $m > 1$. Define e by $\mathrm{End}_{F_q}(W) = M_m(F_{q^e})$.

Now $C_G(s)$ leaves W (and each homogeneous component) invariant. In particular, u does. If U is a central subgroup of a Sylow p-subgroup of $C_G(s)$, then we see that the union of all G-conjugates of $C_G(U)$ contains su. Thus, choosing one U for each s and taking all conjugates covers all of R'. We will see that in each case we can choose only a small number of U and that we can bound the union of all these conjugates.

Case 1. $J = \mathrm{PSL}_2(q)$.

If $G = \mathrm{SL}_2(q)$, we can get an exact formula. R' is precisely the union of elements of the form $\pm u$ with u unipotent. Thus, $r'(G) = (2, q-1)q/(q^2 - 1) < (2, q-1)/(q-1)$.

If q is odd and G/J has order 2, then R' does not change but $|G| = 2|J|$.

Case 2. $J = \mathrm{PSL}_n(q)$, $n > 2$

In this case, we choose $U = U_e$ to be a root subgroup of the centralizer acting on W (note this centralizer is $\mathrm{GL}(m, q^e) \cap G$). Note that this conjugacy class does not depend on m, but only on e.

Thus R' is the union of the centralizers of the F_{q^e}-root subgroups. If $e < n/2$, then $|N_G(U)/C_G(U)| \geq (q^e - 1)$ (because all elements in the root subgroup are conjugate to a fixed one via some toral elements which normalize U – just consider the group over the extension field). If $e = n/2$, then similarly we see that $|N_G(U)/C_G(U)| \geq (q^e - 1)/(2, q - 1)$. Thus, by 2.1 and $U \subseteq C_G(U)$ we have $r'(G) < (2, q - 1)/(q^{n/2} - 1) + \sum_{e=1}^{n/2-1} 1/(q^e - 1)$ if n is even and $r'(G) < \sum_{e=1}^{(n-1)/2} 1/(q^e - 1)$ if n is odd.

This yields easily that

$$r'(G) < 1/(q-1) + 1/(q-1)^2,$$

if q is even or $n \neq 4$. In the remaining case,

$$r'(G) < 1/(q-1) + 2/(q-1)^2.$$

Case 3. $J = \mathrm{PSU}_n(q), n > 2$.

The first possibility is that W is nondegenerate. Then $C_G(s)$ (acting on W) is $U_m(q^{e/2})$ and we choose U to be the center of a root subgroup, so U consists of transvections (over the larger field) and has order $q^{e/2}$. Note that e is even, the class of U is independent of m and that $N_G(U)/C_G(U)$ has order at least $q^{e/2} - 1$ unless q is odd, n is even and $e = n$. In that case $|N_G(U)/C_G(U)| \geq (q^{e/2} - 1)/2$. Note that $1 \leq e/2 \leq n/2$.

The other possibility is that W is totally singular with $m \geq 2$. In this case, $C_G(s)'$ acts as $\mathrm{SL}_m(q^e)$ on W – so $|U| = q^e$ and $|N_G(U)/C_G(U)| \geq (q^e - 1)$ or $(q^e - 1)/2$ if q is odd, n is even and $e = n$.

Summing over all possible e, we obtain the estimates

$$r'(G) < 1/(q-1) + 1/(q-1)^2 + 1/(q^2-1) + 1/(q^2-1)^2$$

for n odd or q even, and

$$r'(G) < 1/(q-1)+1/(q-1)^2+1/(q^2-1)+1/(q^2-1)^2+1/(q^n-1)+1/(q^{n/2}-1)$$

for n even and q odd.

In all cases, we obtain the inequality:

$$r'(G) < 1/(q-1) + 4/(q-1)^2.$$

Case 4. $J = \mathrm{PSp}_n(q), n = 2k > 2$.

There are various possibilities for the parameters in this case.

The first possibility is that W is nondegenerate and $C_G(s)$ leaves invariant an F_{q^e}-alternating form on W. Note in this case $s = \pm 1$ on W. So $e = 1$. In this case, we take U to be a long root subgroup of $\mathrm{Sp}(m, q)$ which is a long root subgroup of G. So $|N_J(U)/C_J(U)| \geq (q-1)/(2, q-1)$. If $G \neq J$, then similarly we see that $|N_G(U)/C_G(U)| \geq (q - 1)$.

The second possibility is that W is totally singular. Then W is paired with another totally singular homogeneous component, W^* and $C_G(s)$ acts as $\mathrm{GL}(m, q^e)$ on W (and via the dual on W^*). In this case, we take U to be a root subgroup of $\mathrm{GL}(m, q^e)$. Again, we note that the conjugacy class of U is independent of m. Note that $N_G(U)/C_G(U)$ has order at least $q^e - 1$.

The final possibility is that W is nondegenerate but that $C_G(s)$ does not preserve an alternating form over F_{q^e} on W. In this case, $C_G(s)$ acts on W as a unitary group over F_{q^e} and we take U to be the center of a Sylow p-subgroup of order q^e consisting of unitary transvections (over the bigger field). In particular, e is even. Again, this class is independent of m. Reducing to the case $m = 2$, we see that this subgroup is actually contained in $U_2(q^e)$ and that it is a short root subgroup of $\mathrm{Sp}_4(q^e)$, a class of subgroups already allowed from the previous paragraph.

So as above, we obtain the estimates:

$$r'(G) < 3/(q-1) + 1/(q-1)^2$$

for q odd and $G = J$, and

$$r'(G) < 2/(q-1) + 1/(q-1)^2$$

for q even or G/J of order 2.

Case 5. Orthogonal Groups, $n \geq 7$.

We may assume that q is odd if n is odd (because otherwise the orthogonal group is the symplectic group, a case already dealt with).

We first note some elementary facts. If x is semisimple and a is an eigenvalue of x, let V_a denote the corresponding eigenspace (over the algebraic closure). Then V_a and V_b are orthogonal unless $ab = 1$. If $ab = 1$ and $a^2 \neq 1$, then $V_a \oplus V_b$ is nonsingular. In particular, a and a^{-1} have the same multiplicity.

These cases are all quite similar to the symplectic case. The possible components of centralizers of semisimple elements that we have to deal with correspond to the totally singular case where we get components of the form $\mathrm{GL}(m, q^e)$ and the nondegenerate case where the components are either of the form $\mathrm{O}^{\pm}(m, q)$ or $U(m, q^{e/2})$.

In the first case, the class of U depends only on e and $|N_G(U)/C_G(U)| \geq q^e - 1$ except that possibly if e is maximal, then only $|N_G(U)/C_G(U)| \geq (q^e-1)/(2, q-1)$. In the last case, we see that the conjugacy class of U depends only on e and not on m. Indeed, we see that we obtain the same conjugacy classes of U as in the totally singular case (these are 'long root subgroups' over $F_{q^{e/2}}$). The usual argument shows that this gives a contribution bounded by $1/(q-1) + 2/(q-1)^2$.

If the centralizer is an orthogonal group, then $e = 1$ because the semisimple element must be ± 1 on W. If it acts as -1, then $m \geq 4$ is even (it is even because the determinant must be 1 and as noted above all eigenvalues are paired with their inverses and $m \geq 4$ because otherwise there is no unipotent element centralizing the semisimple element on that component). If $m > 4$, then the center of a Sylow p-subgroup of $C_G(s)$ is a long root subgroup and if $m = 4$, a long root subgroup of G embeds in O_4^+ and so is contained in the center of a Sylow p-subgroup of $C_G(s)$. So this class of subgroups is already accounted for.

If it acts as 1, then $n-m$ is even (because of the pairing of eigenvalues noted above). Thus, if n is even, we obtain nothing new. If n is odd (and so q is odd), we pick up the conjugacy class U of short root subgroups. In this case, $|N_G(U)/C_G(U)| = (q-1)$.

So we have shown that if $J = \mathrm{O}_{2n}^{\pm}(q)$, $n \geq 4$,

$$r'(G) < 1/(q-1) + 2/(q-1)^2,$$

and that if $J = \mathrm{O}_{2n+1}^{+}(q)$, $n \geq 3$,

$$r'(G) < 2/(q-1) + 2/(q-1)^2.$$

This completes the proof of 2.3. □

3. Exceptional groups

In principal we could handle the exceptional groups in a similar way as the classical ones. But it would become much more complicated to distinguish the possible types of centralizers of semisimple elements.

Instead we use that for exceptional groups we can compute exactly the number of semisimple and regular semisimple elements in $\boldsymbol{G}(q)$. We did this with the help of computer programs written by the second named author. They get as input a complete root datum which determines the algebraic group $\boldsymbol{G}$ up to isomorphism and a twisting type of a Frobenius morphism of $\boldsymbol{G}$. The numbers we are interested in are then computed completely automatically. They can be given in the form of polynomials in q, after distinguishing a finite number of congruence classes for q.

Example. The number of semisimple elements of the exceptional groups of type $G_2(q)$ is given by:

c	number of semisimple elements if $q \equiv c \bmod 6$
1	$q^{14} - 2q^{13} + q^{12} + 3q^{11} - q^{10} - q^9 - q^8 + 2q^7 - 2q^6 - 3q^5 + q^4 + q^3 + 1$
2	$q^{14} - 2q^{13} + q^{11} + q^9 - q^8 + 2q^7 - q^6 - q^5 - q^3 + 1$
3	$q^{14} - 2q^{13} + q^{12} + 2q^{11} - q^{10} - q^8 + 2q^7 - 2q^6 - 2q^5 + q^4 + 1$
4	$q^{14} - 2q^{13} + 3q^{11} - q^9 - q^8 + 2q^7 - q^6 - 3q^5 + q^3 + 1$
5	$q^{14} - 2q^{13} + q^{12} + q^{11} - q^{10} + q^9 - q^8 + 2q^7 - 2q^6 - q^5 + q^4 - q^3 + 1$

Details of the programs and their mathematical background are described in **[Lüb]**. It is planned to publish a version of these programs for general use within the computer algebra package CHEVIE **[GHL+96]**.

We remark that these results in the case of simply connected $\boldsymbol{G}$ could in principal also be obtained from data which appear in the literature. This would involve simple but extremely lengthy and tedious computations by hand.

Of course the same computations could also be done for classical groups up to a certain rank (≤ 9, say) to improve the results of the last section.

Printing all the exact polynomials as in the example above would need a lot of space and is probably not very useful. They are available for interested readers from the authors. Instead we express for all types of simple exceptional groups the quotient $r'(q) = |R'(q)|/|G(q)|$ in the form

$$r'(q) = \frac{c}{q-1} + \frac{d}{(q-1)^2}$$

(similar to the estimating expressions in Section 2), taking for c the asymptotically correct value.

This leads to the following theorem.

Theorem 3.1. *Let G be simple and of exceptional type. The table below gives for each type constants c, x, y and $\tilde{d}$ such that for all q we have*

$$c/(q-1) + x/(q-1)^2 \leq r'(G(q)) \leq c/(q-1) + y/(q-1)^2$$

and such that

$$\tilde{d} = \lim_{q\to\infty} (r'(G(q)) - c/(q-1)) \cdot (q-1)^2.$$

In case of the Suzuki and Ree groups substitute q by q^2 in these expressions.

G	c	x	y	$\tilde{d}$
${}^2B_2(q^2)$, $q^2 = 2^{2m+1}$, $m \in \mathbb{N}$	1	−1	0	0
$G_2(q)$, *q even*	2	−3	−1	−3
$G_2(q)$, *q odd*	2	−4	−2	−4
${}^2G_2(q^2)$, $q^2 = 3^{2m+1}$, $m \in \mathbb{N}$	2	−1	0	−1
${}^3D_4(q)$	1	−1	0	−1
$F_4(q)$	2	−4	−1	−4
${}^2F_4(q^2)$, $q^2 = 2^{2m+1}$, $m \in \mathbb{N}$	2	−2	−1	−2
$E_6(q)$	1	−1	0	−1
${}^2E_6(q)$	1	−1	0	−1
$E_7(q)$	1	−1	0	−1
$E_8(q)$	1	−1	0	−1

Note that the estimates in the theorem are always the same for groups of simply connected and adjoint type (and hence they also hold for the corresponding finite simple groups) – although the exact numbers of semisimple elements do depend on the type of the group. Also, with the exception of $G_2(q)$, the estimates do not depend on the congruence classes of q. Finally, we remark that the lower bounds also estimate the complement of the set of all (not just regular) semisimple elements.

4. Algebraic groups

In this section, we show how our approach in Section 2 applies in the case of algebraic groups. As remarked in the introduction, these results will at least provide good asymptotic estimates for the finite groups and will also verify that the coefficient of $1/(q-1)$, as given in the table in Theorem 2.3, is correct. If we want the fixed points of the Frobenius endomorphism to be quasisimple, we must consider simply connected algebraic groups.

For this section, we refer to **[Car72]** and **[Bor91]** for general references about algebraic groups.

Let $\boldsymbol{G}$ be a reductive group over an algebraically closed field F as in the introduction. Recall that R' is the set of elements which commute with a unipotent element (i.e., R' is the complement of the set of regular semisimple elements in $\boldsymbol{G}$).

Theorem 4.1. *Let $x \in \boldsymbol{G}$, then $x \in R'$ if and only if x centralizes a root subgroup of $\boldsymbol{G}$.*

Proof. Let $x = s_x u_x$ the Jordan decomposition of x, let B a Borel subgroup of $\boldsymbol{G}$ containing x and T a maximal torus in B containing s_x. Let R^+ denote the set of positive roots with respect to $T \subset B$. Order the positive set of roots. Then $u_x = \prod_{R^+} u_\alpha(t_\alpha)$ where $t_\alpha \in F$ and the product is taken in the given ordering. This expression is unique. Since s_x centralizes u_x and normalizes each root subgroup, it follows that s_x centralizes each root subgroup U_α with $t_\alpha \neq 0$. Let X denote the subgroup generated by these root subgroups. Then X is a product of root subgroups and moreover its center contains a root subgroup. So s_x and u_x centralize this root subgroup as desired. □

Now assume that $\boldsymbol{G}$ is a simply connected algebraic group. Let $Z = Z_\alpha$ be a root subgroup of $\boldsymbol{G}$. There are 1 or 2 conjugacy classes of such subgroups (depending upon the number of root lengths). Let X be the variety of elements in $\boldsymbol{G}$ which commute with some conjugate of Z. Fix a maximal torus T in $N := N(Z)$.

First observe that X has codimension 1 in $\boldsymbol{G}$ (to see this, note that a generic element in the centralizer of a root subgroup U has centralizer SU where the connected component of S is a maximal torus in the centralizer of U – so S has codimension 1 in a maximal torus and the map from $\boldsymbol{G} \times S^0 U \to R'$ given by $(g, su) \mapsto (su)^g$ has generic fiber of dimension $1 + \mathrm{rank}(\boldsymbol{G})$).

First suppose that Z is a long root subgroup. Then $N := N(Z)$ contains a Borel subgroup and so is parabolic and in particular is connected. Moreover, $N = N'T$. Note that N' centralizes Z and so must be the connected component of $C(Z)$. Thus, the number of connected components of $C(Z)$ is precisely the number of connected components of $C := C_T(Z)$. The number of connected components of C is precisely the largest positive integer e relatively prime to p such that $\alpha = e\beta$ for some β in the weight lattice of $\boldsymbol{G}$. It is straightforward (using the description of the root and weight lattices in **[Car72]**) to see that $e = 1$ unless $e = 2$, $p \neq 2$ and one of the following holds:

$\boldsymbol{G}$ is of type A_1 or

$\boldsymbol{G}$ is of type C_l, $l \geq 2$.

Thus, the number of irreducible components in the variety X of elements which commute with a conjugate of Z is at most the number of components in C. So except in the cases above, X is an irreducible variety of codimension 1 in G.

In the two remaining cases one can check that the variety has 2 components. In fact, if X_1 is one component, then the other component is $\{A\} \times X_1$ where A is a generator of the center of G (and has order 2). Thus, both components have codimension 1.

Now assume that Z is a short root subgroup. Using estimates for the number of F_q-rational points in R' for exceptional groups, we see that the number of components of R' of codimension 1 in G is 1 for $G = G_2$ or F_4.

If $\boldsymbol{G} = B_\ell$ (or C_ℓ) with $p = 2$, we can argue as in the long root case (because short roots and long roots are interchanged when considering $\boldsymbol{G}$ as B_ℓ or C_ℓ). So $C(Z)$ is connected and so X is irreducible. Finally, assume that $p \neq 2$ and $G = C_\ell$ or B_ℓ. A straightforward computation shows that there are 2 connected components in $C(Z)$ and so X has at most 2 connected components. Indeed, we see that in fact there are two connected components for X. The one corresponding to the identity component of $C(Z)$ has codimension 1 in $\boldsymbol{G}$.

Suppose that $G = C_\ell, \ell \geq 2$. A short root subgroup Z has a 2-dimensional commutator space with the natural module which is a totally singular 2-space. Thus, the normalizer of Z must stabilize this subspace and so $N(Z) \leq P$, the corresponding parabolic subgroup which is the full stabilizer of the totally singular 2-space. Since the radical Q of P is Abelian and contains the short root subgroup, we see that $C(Z)$ contains Q. The Levi complement of P is $\mathrm{GL}(2) \times C_{\ell-2}$ where the second factor acts trivially on Q and so also on Q. The centralizer of Z in GL(2) is precisely the two dimensional orthogonal group which has connected a 1-dimensional torus of index 2. Thus, we see the two components and we see that any element not in the connected component centralizes a long root element and so these elements are already accounted for.

If $G = B_\ell, \ell \geq 2$, then again a short root subgroup Z has a 2-dimensional commutator space with the natural module. However, in this case, the radical W of these space is 1-dimensional and is invariant under $N(Z)$. Thus, $N(Z)$ is the contained in P, the parabolic subgroup stabilizing a 1-dimensional totally singular subspace. One sees that the two components of $C(Z)$ correspond to whether the eigenvalue on W is ± 1. This implies that any element in $C(Z)$ has a triple eigenvalue either ± 1. The component corresponding to the case of eigenvalue -1 must have -1 occurring with multiplicity at least 4 (because the multiplicity of -1 is even). Restricting to the nonsingular subspace of dimension at least 4 which is the -1 eigenspace of the semisimple part of an element in $C(Z)$ shows that any element in the non-identity component of $C(Z)$ already centralizes a long root element (as in the C_ℓ case).

So in all cases, we obtain a single component of codimension 1 of elements centralizing a short root subgroup which is not contained in the centralizer of a long root subgroup.

Putting these results together we get the following theorem.

Theorem 4.2. *Let* $\mathbf{G}$ *be simple simply connected. Then the number of components of* R' *of codimension* 1 *is as given in the following table.*

In particular the asymptotic results agree with the asymptotic estimates obtained in Sections 2 *and* 3 *(since we are dealing with simply connected groups, we must compare these estimates with the simple groups) – i.e., the first term in our estimates for upper bounds for* $r'(G)$ *cannot be improved.*

Group	*Number of root lengths*	*Number of components*
A_1	1	$(2, p-1)$
$A_l, l \geq 2$	1	1
$B_l, l \geq 3$	2	2
$C_l, l \geq 2$	2	$1 + (2, p-1)$
$D_l, l \geq 4$	1	1
$E_l, l = 6, 7, 8$	1	1
F_4	2	2
G_2	2	2

We also note the following result which we assume is known to the experts in the field.

Theorem 4.3. *If* $\mathbf{G}$ *is simple then the subset of nonregular semisimple elements has codimension* 3 *in* $\mathbf{G}$ *and the number of components of this variety is at most* 3.

Proof. Let Y denote the set of semisimple elements which are not regular. As in 4.1, we see that $Y = Y_1 \cup Y_2$ where Y_i are those semisimple elements which centralize a long root subgroup (for $i = 1$) and or a short root subgroup (for $i = 2$). Arguing as above we see that Y_1 is irreducible unless $p \neq 2$ and $\mathbf{G} = A_1$ or C_ℓ. In those cases, there are two components. A generic element (in either component in the latter cases) has centralizer TA_1 where T is a maximal torus and A_1 is the subgroup generated by a pair of long root subgroups. Computing the dimension of the generic fiber of the conjugation map shows that each component of Y_1 has codimension 3.

Similarly, we see that Y_2 is irreducible of codimension 3. This completes the proof. □

We remark that in the case of $\mathbf{G} = \mathrm{SL}$ or Sp, the closure of this variety is precisely the set of noncyclic matrices in each group.

5. Centers and outer automorphisms

In this section, we make some preliminary remarks about handling covering groups and other extensions of simple groups. We hope to return to this topic in a future article.

Fix a prime p. Let $S(G)$ denote the set of p-singular elements of G. Let $s(G) = |S(G)|/|G|$.

First consider the case that G has a center Z. If Z has order prime to p, then $s(G) = s(G/Z)$ (indeed, this is true for any normal subgroup with order prime to p). Furthermore, we note that:

Lemma 5.1. *Suppose that the center Z of G is a p-group. Then*

$$s(G) = s(G/Z)/|Z| + 1 - 1/|Z|.$$

Proof. Let $\pi : G \to G/Z$ be the natural map. Note that the inverse image of $S(G/Z)$ is contained in $S(G)$ and has cardinality $|Z||S(G/Z)$. Suppose that $x \in G$ is not p-singular in G/Z. If $z \in Z$, then $xz \in S(G)$ if and only if $z \neq 1$. Thus

$$|S(G)| = |Z||S(G/Z)| + |S'(G/Z)|(|Z| - 1) = |S(G/Z)| + |G/Z|(|Z| - 1).$$

Dividing by $|G|$ yields:

$$s(G) = s(G/Z)/|Z| + 1 - 1/|Z|.$$ □

Next we consider outer automorphisms. So let J be a finite simple group and G almost simple with socle J. We will consider the proportion of p-singular elements in a given coset xJ. So there is no harm in assuming that G/J is cyclic and generated by xJ. If p divides $|G/J|$, then every element in the coset xJ is p-singular.

So we shall assume that G/J is a p'-group. We restrict our attention to the case that J is a finite group of Lie type over the field F_q with q a power of p.

We have already handled the case of diagonal automorphisms. We do not answer the question completely but show that the answer can change in certain situations. Let p be a prime and let $\mathbf{G}$ be a simple algebraic group in characteristic p. Let q_0 be a power of p and e a positive integer with $q = q_0^e$. In a special case, we show that we can reduce the computation of the number of p-singular elements in a given coset to the same question for a group over the fixed field of a Frobenius map.

Theorem 5.2. *Let $G = \mathbf{G}(q)$. Assume that e is relatively prime to $|\mathbf{G}(q_0)|$. Let σ be the Frobenius automorphism of order e on $G(q)$. Set $H = \langle G, \sigma\rangle$ and $C = C_G(\sigma)$. Note that C is a group of the same type as $\mathbf{G}$ defined over the field of q_0 elements. The proportion of p-singular elements in the coset σG is equal to $s(C)$.*

Proof. Our hypotheses imply that $(e, |C|) = 1$ (since $C/\mathbf{G}(q_0)$ just involves diagonal automorphisms).

Let $y = \sigma g$ with $g \in G$. It is easy to see (**[GMS]** using (7.2) in **[GL83]**) that y lies in a unique conjugate of $D := \langle C, \sigma \rangle$.

Now assume that $y \in D$ (and so $g \in C$). Thus, y is p-singular if and only if g is. Thus, the number of p-singular elements in the coset σG is precisely $|S(C)||G : C|$ and so the proportion of p-singular elements in the coset is $s(C)$. □

Thus, since $s(C)$ is asymptotically of the form c/q_0 for $c = 1, 2$ or 3 (depending upon the type of $\boldsymbol{G}$), we see that the probability that a random element in the given coset is p-singular is much higher than for the simple group.

If e is not relatively prime to the order of $\boldsymbol{G}(q_0)$, the analysis becomes more difficult.

For graph automorphisms, the main case is when $\boldsymbol{G}$ has type A (for groups of type D, the involutory graph automorphism acts on the natural orthogonal module and an analysis as in Section 2 suffices). A modification of the methods of Section 2 show that the correct answer is still of the form c/q.

References

[Bor91] A. Borel, Linear algebraic groups, Grad. Texts in Math. 126, Springer-Verlag, New York, second edition, 1991.

[Car72] R. W. Carter, Simple Groups of Lie Type, Wiley-Interscience, London, 1972.

[FJ93a] P. Fleischmann and I. Janiszczak, The number of regular semisimple elements for Chevalley groups of classical type, J. Algebra 155 (1993), 482–528.

[FJ93b] P. Fleischmann and I. Janiszczak, The semisimple conjugacy classes of finite groups of Lie type E_6 and E_7, Comm. Algebra 21 (1) (1993), 93–161.

[FJ94] P. Fleischmann and I. Janiszczak, The semisimple conjugacy classes and the generic class numbers of Chevalley groups of type E_8, Comm. Algebra 22 (6) (1994), 2221–2303.

[FJK98] P. Fleischmann, I. Janiszczak, and R. Knörr, The number of regular semisimple classes of special linear and unitary groups, Linear Algebra Appl. 274 (1998), 17–26.

[GHL$^+$96] M. Geck, G. Hiss, F. Lübeck, G. Malle, and G. Pfeiffer, CHEVIE – A system for computing and processing generic character tables for finite groups of Lie type, Weyl groups and Hecke algebras, AAECC 7 (1996), 175–210.

[GL83] D. Gorenstein and R. Lyons, The local structure of finite groups of characteristic 2-type, Mem. Amer. Math. Soc. 276, Amer. Math. Soc., Providence, RI, 1983.

[GMS] R. Guralnick, P. Müller, and J. Saxl, The rational function analogue of a question of Schur and exceptionality of permutation representations, preprint.

[IKS95] I. M. Isaacs, W. M. Kantor, and N. Spaltenstein, On the probability that a group element is p-singular, J. Algebra 176 (1995), 139–181.

[KS] W. M. Kantor and Á. Seress, Prime power graphs for groups of Lie type, preprint.

[Lüb] F. Lübeck, Parameterization of semisimple conjugacy classes of finite groups of Lie type, in preparation.

[NP] P. Neumann and C. Praeger, Cyclic matrices in classical groups over finite fields, preprint.

Department of Mathematics
University of Southern California
1042 West 36th Place, DRB 155 Los Angeles, California 90089-1113, U.S.A.
guralnic@math.usc.edu

Lehrstuhl D für Mathematik
RWTH Aachen
Templergraben 64, D-52062 Aachen, Germany
Frank.Luebeck@Math.RWTH-Aachen.de

Experiments in coset enumeration

*George Havas and Colin Ramsay**

Abstract. Coset enumeration, based on the methods described by Todd and Coxeter, is one of the most important tools for investigating finitely presented groups. Such methods do not, in general, constitute an algorithm. Each problem has to be addressed individually, and determining whether or not an acceptable solution for a difficult problem can be found using given resources requires an empirical approach (i.e., experimentation). We discuss some of the ideas involved, and illustrate with examples which emphasize some of the possibilities.

1991 Mathematics Subject Classification: primary 20F05; secondary 20-04.

1. Introduction

Coset enumeration is long established as a technique for the investigation of finitely presented groups. It was used well before the days of electronic computers, apparently first by Moore **[14]**, and was later popularised by Todd and Coxeter **[17]**. The first computer implementation was that of Haselgrove in 1953. This, along with other early implementations, is described by Leech **[10]**. Detailed accounts of the techniques used in coset enumeration can be found in **[1, 6, 12, 15, 16]**, together with some experimental results.

Coset enumeration takes a finitely presented group and a finitely generated subgroup as input, and attempts to find the index of the subgroup in the whole group. In principle, it will succeed whenever this index is finite. However, the computational complexity of coset enumeration is not well-understood. Even using a fixed strategy, there is no computable bound on the number of cosets which need to be defined to obtain a complete coset table (in terms of the length of the input and a hypothetical index). Further, Sims **[16]** has proved that there is no polynomial bound, in terms of the maximum number of cosets, for the number of coset tables which can be derived by coset table operations such as those used in enumeration programmes.

Sims' result implies that the running time of coset enumeration procedures, in terms of the space available, may be very high. Pathological examples of, for example, the

*The work of both authors was supported by the Australian Research Council. We are grateful to The University of Queensland's High Performance Computing Unit for access to the Origin 2000 supercomputer.

trivial group exist which exhibit such behaviour. However, the main factor determining whether or not we can succeed in completing an enumeration is the amount of memory available; i.e., the number of cosets we can define. So the maximum and total number of cosets defined during an enumeration are often used as metrics, with smaller values being regarded as 'better'. Usually, the running time is proportional to the total number of cosets, and is not a limiting factor. There is generally a close correlation between maximum and total cosets, so we choose to concentrate on only one, total cosets, in this paper.

One contributing factor to our less than perfect understanding of coset enumeration is the freedom of choice of which coset to define next that we have during an enumeration. This freedom implies that the space of possible enumeration sequences, for a given coset enumeration problem, is extremely large. It is not clear how best to generate an enumeration which is space efficient, or from which an understandable formal proof can be extracted. In this paper we discuss how features of a modern implementation of coset enumeration can be used to explore the enumeration space and to gain insight into this challenge.

The coset enumeration implementation which we used was the one by Havas and Ramsay **[8]**, which is an enhanced version of the one described in **[6]**. The enumerations were performed on an SGI Origin 2000 computer which was equipped with enough memory to allow the definition of some hundreds of millions of cosets when necessary.

Our enumerator has a large number of parameters, with a wide choice of settings; we discuss the ones relevant to our examples. The most important are `CT` and `RT`, which indicate the relative weightings attached to defining new cosets using the coset table (traditionally called the Felsch strategy) and the relator tables (traditionally known as the HLT strategy). Each unit in `CT` indicates one definition, which is tested in all essentially different positions in all the relators until all consequent deductions and coincidences have been found and processed. Each unit in `RT` indicates one coset being scanned at all relators, with definitions made as necessary to close all scans. If both `CT` and `RT` are non-zero, then `RT` definitions and deductions are stacked, and tested at the start of the next `CT`-phase (so 'lookahead' is performed after each `RT` phase).

The `Mend` parameter controls a technique due to Mendelsohn **[13]**, wherein `RT`-style scanning is performed using all cyclic shifts of the relators; it can be set on or off. It is occasionally very beneficial in reducing maximum and total cosets, including enumerations for one of our examples.

As discussed in **[6]**, some number of the relators can be treated as subgroup generators, and scanned and closed at coset 1 (i.e., the subgroup) at the start of the enumeration. This is controlled by the `No` parameter; we will always use all or none of the relators, so we will say that `No` is on or off.

Also discussed in **[6]** is the idea of preferentially filling any gaps of length one noted when scanning relators, subject to a 'fill-factor' to ensure that some constant proportion of the coset table is always kept completed. Our programme implements this idea and allows gaps to be queued for later use (as in **[6]**), filled immediately, or filled immediately and the consequent deduction made immediately. We will say that `Fi` is on or off and, if on, explicitly state which fill mechanism we used. Unless stated otherwise, the default fill-factor will be used.

2. Can we do it?

Consider the group $W = \langle x, y \mid x^3y^4x^5y^7 = x^2y^3x^7y^8 = 1\rangle$. In 1996, D. L. Johnson [3] reported the failure of coset enumeration to answer a question of Malcolm Wicks in 1987; namely, is W the cyclic group of order 11, C_{11}? More recently, Havas, Holt, Kenne and Rees [7] have shown that $W \cong C_{11}$ by combining the Knuth–Bendix process (originally described in [9]) with the Todd–Coxeter process. They also note that the single coset of $\langle y\rangle$ (which generates W) and the eleven cosets of $\langle x\rangle$ (which is trivial) can be enumerated in, respectively, 22 million and 14 million cosets using coset enumeration alone.

These results leave unanswered the question of whether or not we can enumerate the cosets of $\langle 1\rangle$ in W directly. The number of cosets required for an enumeration tends to grow as the size of the subgroup shrinks. (Note that the result quoted above, where enumeration over $\langle x\rangle$ takes fewer cosets than over $\langle y\rangle$, is not typical.) Experience suggests that parameter combinations which work well for one subgroup often work well for other subgroups of the same group. This suggests that we should consider selecting parameters for $W/\langle 1\rangle$ on the basis of those that work well for $W/\langle x\rangle$ or $W/\langle y\rangle$. (For brevity here we denote the enumeration of the cosets of the group G over the subgroup H by G/H.)

To test this idea, we performed each of the enumerations $W/\langle x\rangle$ and $W/\langle y\rangle$ 121 times, varying the `CT` and `RT` parameters exponentially from 2^0, 2^1 to 2^{10}. This gives good coverage of the parameter space in a reasonably small number of examples, although, of course, we have no easy way of knowing whether or not we have missed any 'interesting' combinations. Both `Mend` and `No` were set on, `Fi` was set off, and the enumerator given enough memory for 50 million cosets (i.e., 800MB of memory). The total numbers of cosets defined (in thousands, to the nearest thousand) are given in Tables 1 and 2 (for brevity, omitting `RT` values 512 and 1024, which are not significantly different to numbers obtained for `RT` = 256); "o/f" means that the enumeration attempt overflowed the available space.

These results highlight two parameter combinations for further investigation; namely, the `CT:64; RT:2` enumeration for $W/\langle x\rangle$, and the `CT:512; RT:64` enumeration for $W/\langle y\rangle$.

For the enumeration $W/\langle 1\rangle$, we started by varying `CT` from 25 to 300 in steps of 25 and `RT` from 2 to 16 in steps of 2. The other parameters were as above, and the enumerator was given enough memory for 300 million cosets (i.e., 4800MB of memory). All 96 of these enumerations completed successfully; the results are plotted in Figure 1. More detailed investigations, varying `CT` and `RT` in steps of 1, revealed that

Table 1: Total cosets counts (thousands) for $W/\langle x\rangle$

	RT								
CT	1	2	4	8	16	32	64	128	256
1	14262	14225	14227	14229	14232	14234	14241	14246	14251
2	14671	14271	14237	14229	14232	14234	14241	14246	14250
4	15611	14681	14308	14240	14232	14234	14241	14246	14251
8	14318	15622	14713	14312	14233	14235	14241	14247	14251
16	11458	14441	15706	14703	14244	14235	14242	14247	14251
32	8072	10059	14379	15677	14681	14424	14416	14427	14432
64	11402	6864	7895	11945	15553	14939	14681	14682	14688
128	24958	13012	6925	7877	11992	16070	15435	15180	15171
256	35621	24934	12928	8879	8632	12519	17295	16631	16343
512	o/f	34681	35009	18365	12691	10645	15464	19101	19122
1024	o/f	o/f	40411	47665	30264	20811	18162	16159	15137

Table 2: Total cosets counts (thousands) for $W/\langle y\rangle$

	RT								
CT	1	2	4	8	16	32	64	128	256
1	30172	30093	30071	30063	30068	30072	30085	30098	30108
2	28214	30191	30099	30076	30068	30072	30085	30098	30108
4	26263	28222	30204	30105	30068	30072	30084	30097	30107
8	29179	26319	28234	30212	30068	30072	30084	30098	30107
16	30984	29330	26290	28372	30165	30074	30083	30097	30106
32	31022	31087	29207	26212	28310	30094	30084	30097	30107
64	29382	31176	34018	28926	26511	28271	30260	30260	30282
128	40451	29472	27616	35021	29543	26131	28075	30593	30581
256	o/f	36678	26261	22981	26492	28962	26836	27136	31218
512	o/f	o/f	31861	25873	21282	16871	14864	16216	27471
1024	o/f	o/f	o/f	37749	27671	24853	23903	21016	20072

the `CT:185; RT:4` enumeration is a local minimum, requiring 139 330 097 cosets. A similar search of the area around `CT:512; RT:64` revealed another local minimum at `CT:761; RT:59`, requiring 99 222 961 cosets.

So, in this case at least, we see that we can use an 'easy' enumeration to guide us as to how best to do a more difficult one. Although our tests were done on a machine with adequate memory, note that, if we had been working on a machine which allowed the definition of only, say, 150 million cosets, the technique discussed would have been crucial to success.

Despite what Figure 1 might suggest, in general the number of cosets required is not a 'smooth' function of the parameters. Figure 2 of the next section witnesses this on a much smaller scale. Additionally, despite being guided to good enumerations of $W/\langle 1\rangle$ by examining those for $W/\langle x\rangle$ and $W/\langle y\rangle$, the ratios of the number of cosets required for these three enumerations, for the same parameters, is very variable.

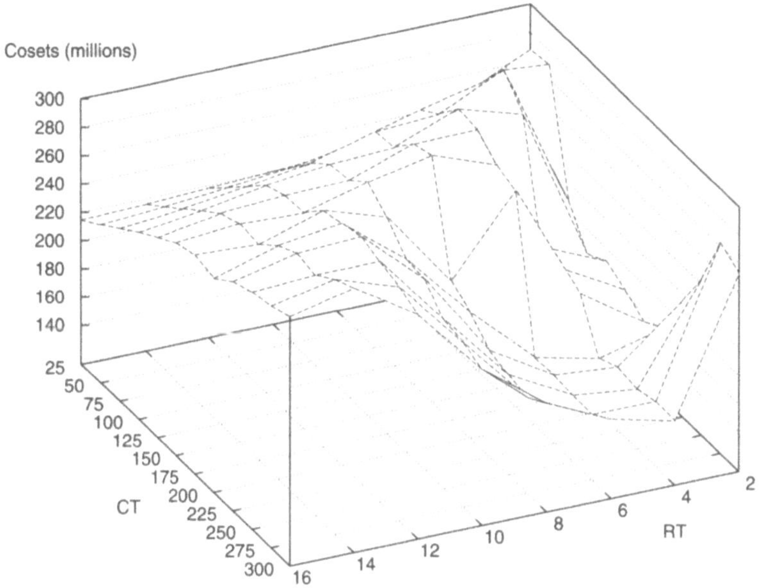

Figure 1: Total cosets counts for $W/\langle 1\rangle$

These points are illustrated in Table 3, where we list the total cosets required (both actual and normalised to the count for $W/\langle 1\rangle$) for the minimum parameters noted. We also include data for a parameter combination adjacent to one of the minima, and for a point in the 'flat' region where CT is low and RT high. Finally, it is important to note that, despite our success, we do not know whether or not we have, in fact, found the best possible parameter combination.

Table 3: Variations between the three enumerations

		$W/\langle y\rangle$		$W/\langle x\rangle$		$W/\langle 1\rangle$
CT	RT	Total	Normalised	Total	Normalised	Total
2	128	30098364	0.142	14246404	0.067	212011726
185	4	28331117	0.203	9096142	0.065	139330097
760	59	19357727	0.112	19305998	0.112	172843398
761	59	19379546	0.195	19283983	0.194	99222961

3. How well can we do it?

Consider the group $(8, 7 \mid 2, 3) = \langle a, b \mid a^8 = b^7 = (ab)^2 = (a^{-1}b)^3 = 1\rangle$. This has order 10752, so the octahedral subgroup $\langle a^2, a^{-1}b\rangle$ has index 448 in it. Leech

[11] notes that this enumeration is "far from straightforward", seemingly due to the difficulty of establishing that the given relations imply $(a^2b^4)^6 = 1$. This enumeration is one of those discussed by Cannon, Dimino, Havas and Watson [1] and by Havas [6]; the best enumeration listed there required the definition of 840 cosets.

The effect of permuting the order of the relators and of rotating the relators was discussed in [1], where it was shown that this can yield large variations in the number of cosets required in an enumeration. Our enumerator has the ability to generate and use all possible relator orderings, relator rotations, and relator inversions. For $(8, 7 \mid 2, 3)$, this amounts to a total of $(4!)(1.1.2.2)(2^4) = 1536$ possibilities. For each combination of `CT` and `RT` from 1 to 25 in steps of 1, we tested all of these possibilities and recorded the maximum and minimum values of the total number of cosets defined. `Mend` was set off, and `No` and `Fi` were set on, with fill definitions and their consequent deductions being made immediately. The results, for `CT` and `RT` up to 20, are plotted in Figure 2.

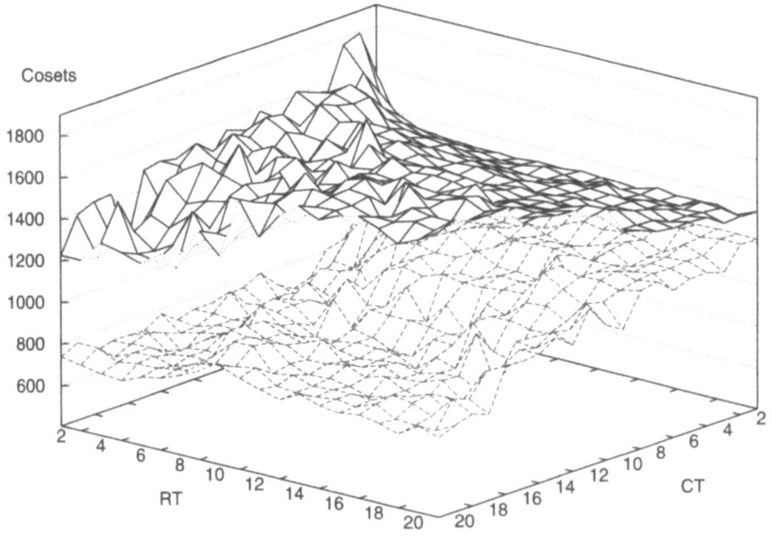

Figure 2: Minimum and maximum total cosets for $(8, 7 \mid 2, 3)/\langle a^2, a^{-1}b\rangle$

The parameters `CT:15; RT:3` yield the smallest minimum number of cosets, namely 662. This is an appreciable reduction on the 840 previously achieved, almost halving the number of redundant cosets. Interestingly, parameter combinations which yield short enumerations also seem to yield long enumerations as the presentation is varied. This suggests that, in the general case, an unfortunate choice of presentation can lead one badly astray when searching for the best enumeration. Note that `CT` low with `RT` high strategies, where most of the definitions are made in closing the relators, are conservative, in that, although they do not yield short enumerations, they do not yield overlong ones either. The minimum range noted for one parameter combination for the number of cosets was 1220–1313 for `CT:1; RT:19`, and the maximum was 796–1879 for `CT:19; RT:23`.

For the `CT:15; RT:3` parameters (which give the smallest minimum), the presentation as given needs 1004 cosets, while if the relators are sorted in length-increasing order only 965 cosets are needed. For these parameters the number of cosets needed for all 1536 equivalent presentations was also recorded, and the resulting distribution is shown in Figure 3. This distribution is highly skewed, with 67.3% of the runs taking 1000 or less cosets and 91.7% taking 1100 or less. Thus, randomly changing the initially given presentation into one of its equivalents has a better than 2 in 3 chance of yielding a superior enumeration.

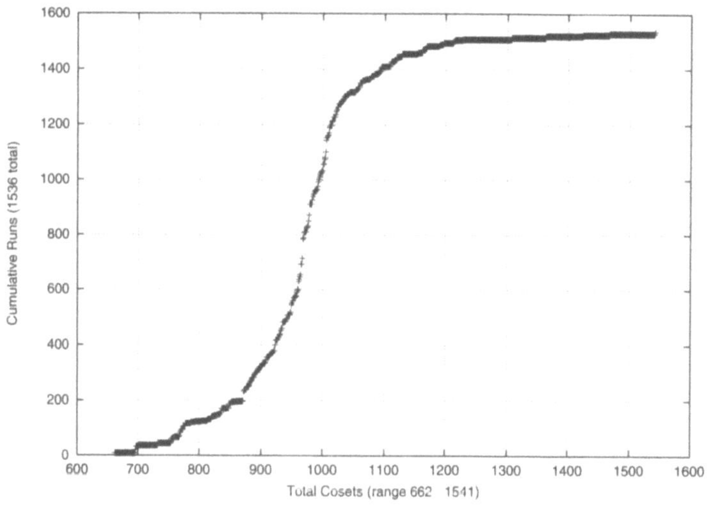

Figure 3: Distribution of total cosets for $(8,7 \mid 2,3)/\langle a^2, a^{-1}b\rangle$

4. Can we prove it?

Consider the Fibonacci group

$$F(2,7) = \langle a,b,c,d,e,f,g \mid ab=c,\ bc=d,\ cd=e,\ de=f,\ ef=g,\ fg=a,\ ga=b\rangle.$$

The first formal proof that $F(2,7) \cong C_{29}$, the cyclic group of order 29, was given by Havas **[5]**, and was derived from the workings of an index 1 enumeration over one of the generators. This enumeration used a total of 327 cosets, and the proof extended over several pages. It is reasonable to expect that the better the coset enumeration, the better any extracted proof. This focuses attention on trying to do even relatively easy enumerations, such as this, as well as possible. Leech **[12]** reports on a hand enumeration which took 129 cosets, and was then pruned to 55 cosets. (This was reduced to 53 by Edeson **[2]**. Interesting new work on pruning by Felsch, Hippe and Neubüser **[4]** includes the result that this enumeration can be done in 50 cosets.) Leech's enumeration yields a shorter (but no less impenetrable) proof, and Leech remarks that 'the initial hand number of 129 will serve as a target for "intelligent" computer algorithms'. The enumerator discussed by Havas **[6]** achieved a total of 169.

For $F(2,7)$, direct `RT`-style definitions do not seem beneficial, so our tests were done with the parameters `CT:1000; RT:0`. We also set `Mend` off and `No` on. Since all the relators of $F(2,7)$ have length three, gaps of length one will be common, so we set `Fi` on. However, the effects of `Fi` are very variable for $F(2,7)$, so we do not use the default value of the fill-factor. Instead, we regard the fill-factor as a useful means of exploring the space of possible enumerations; values in the range 10–30 were used.

Another useful source of variation, particularly effective with `CT`-style definitions, is the order of the columns (i.e., the generators) in the coset table. For our programme, this depends upon the order in which the generators are listed to the enumerator, and we have $7!$ possibilities for $F(2,7)$. Due to the symmetry of the presentation for $F(2,7)$, we can perform the enumeration over any generator; so there are 7 possibilities here.

Combining all these different possibilities with the $7!2^7 3^7$ different equivalent presentations (which can be obtained by permuting the order of relators, formally inverting relators and rotating relators) gives a very large number of enumerations; far too many to exhaustively test. To sample these, we select possible enumerations at random, and use the selected presentation as a starting point in a test of equivalent presentations. We allow each test to run for up to a few days before terminating it and starting another run. To maximise throughput, we give the enumerator only enough memory for, say, 150 cosets. Most enumerations quickly overflow and abort, allowing another to be started; this technique allowed approximately 1000 enumerations to be tested per second.

Over the course of 21 runs, a total of 3710×10^6 enumerations were performed. Of these, 5245 beat Leech's target, with 152 taking less than 120 cosets. Our best result, of 112 cosets, betters the target comfortably.

Of course, these enumerations can be manually pruned. In fact, we can improve the computer results slightly by specifying what the first few definitions are to be. (This is readily done by adding words which are formally trivial to the subgroup generators.) This 'encourages' the enumerator to process particular words; suitable words are chosen by examining the output of previous tests, and seeing which coincidence precipitated the final collapse. Using this technique, our best 'computer' result is 109 cosets.

5. Conclusions

The examples we have discussed indicate that, while we can only explore a very small proportion of the space of possible enumeration sequences, this exploration is worthwhile. It can enable us to complete an enumeration where we would otherwise fail, and it can yield dramatically shorter enumerations, perhaps leading to more insightful formal proofs. It is difficult to give general guidelines on how best to handle a particular enumeration, since cases seem to be highly individualistic. However, the techniques described here should certainly be considered when confronted with any new, difficult problem.

References

[1] John J. Cannon, Lucien A. Dimino, George Havas, and Jane M. Watson, Implementation and analysis of the Todd–Coxeter algorithm, Math. Comp. 27 (1973), 463–490.

[2] Margaret Edeson, Investigations in coset enumeration, Master's thesis, Canberra College of Advanced Education, 1989.

[3] Group-pub-forum e-mail list (`group-pub-forum@maths.bath.ac.uk`), Problem 11, 1996; see `http://www2.bath.ac.uk/~masgcs/problem/problem11.htm`.

[4] Volkmar Felsch, Ludger Hippe and Joachim Neubüser, Interactive Todd–Coxeter manual, available among the Share Packages at `http://www-gap.dcs.st-and.ac.uk/~gap`.

[5] George Havas, Computer aided determination of a Fibonacci group, Bull. Austral. Math. Soc. 15 (1976), 297–305.

[6] George Havas, Coset enumeration strategies, in: ISSAC'91 (Stephen M. Watt, ed.), Proceedings of the 1991 International Symposium on Symbolic and Algebraic Computation, ACM Press, 1991, 191–199.

[7] George Havas, Derek F. Holt, P. E. Kenne, and Sarah Rees, Some challenging group presentations, J. Austral. Math. Soc. Ser. A 67 (1999), 206–213.

[8] George Havas and Colin Ramsay, Coset enumeration: ACE version 3, 1999, available as `http://www.csee.uq.edu.au/~havas/ace3.tar.gz`.

[9] D. E. Knuth and P. B. Bendix, Simple word problems in elementary algebra, in: Computational Problems in Abstract Algebra, Pergamon Press, Oxford, 1970, 263–297.

[10] J. Leech, Coset enumeration on digital computers, Math. Proc. Cambridge Philos. Soc. 59 (1963), 257–267.

[11] John Leech, Computer proof of relations in groups, in: Topics in Group Theory and Computation (Michael P. J. Curran, ed.), Academic Press, 1977, 38–61.

[12] John Leech, Coset enumeration, in: Computational Group Theory (Michael D. Atkinson, ed.), Academic Press, 1984, 3–18.

[13] N. S. Mendelsohn, An algorithmic solution for a word problem in group theory, Canad. J. Math. 16 (1964), 509–516; Corrigendum: Ibid., 17 (1965), 505.

[14] E. H. Moore, Concerning the abstract groups of order $k!$ and $\frac{1}{2}k!$ holohedrically isomorphic with the symmetric and the alternating substitution-groups on k letters, Proc. London Math. Soc. (1) 28 (1897), 357–366.

[15] J. Neubüser, An elementary introduction to coset-table methods in computational group theory, In Groups – St. Andrews 1981, London Math. Soc. Lecture Note Ser. 71, Cambridge University Press, 1982, 1–45.

[16] Charles C. Sims, Computation with finitely presented groups, Cambridge University Press, 1994.

[17] J. A. Todd and H. S. M. Coxeter, A practical method for enumerating cosets of finite abstract groups, Proc. Edinburgh Math. Soc. 5 (1936), 26–34.

Centre for Discrete Mathematics and Computing
Department of Computer Science and Electrical Engineering
The University of Queensland
Queensland 4072, Australia
havas@csee.uq.edu.au
cram@csee.uq.edu.au

A presentation for the Thompson sporadic simple group

George Havas, Leonard H. Soicher and Robert A. Wilson

Abstract. We determine a presentation for the Thompson sporadic simple group *Th*. The proof of correctness of this presentation uses a coset enumeration of 143,127,000 cosets. In the process of our work, we determine presentations for ${}^3D_4(2)$, ${}^3D_4(2){:}3$, $G_2(3){:}2$, and $C_{Th}(2A)$ (of shape $2_+^{1+8}{\cdot}A_9$). We also provide, via the internet, matrices generating *Th* and satisfying our presentation.

1991 Mathematics Subject Classification: primary 20D08, 20F05, 20-04.

1. Introduction

Presentations for many sporadic simple groups and their automorphism groups have been published. For example, see **[PS97]** for M_{11}, M_{12}, M_{22}, M_{23}, M_{24}, J_1, J_2, *HS*, *McL*, *He*, *Suz*, *O'N*, Co_1, Co_2, Co_3, Fi_{22}, Fi_{23}, Fi_{24}; see **[HM69]** for J_3, **[Wei91]** for *Ru*, **[BN95]** for *HN*:2, **[JW96, HS99]** for *Ly*, **[SW88, Iva92]** for J_4, and **[Iva99]** for B and M. Indeed, until now, the Thompson group *Th* was the only sporadic simple group S for which there was no published presentation for S or $\mathrm{Aut}(S)$. In this paper, we apply computational group theory in various ways to determine a presentation for *Th* ($\cong$ Aut(*Th*)). In the process, we determine presentations for ${}^3D_4(2)$, ${}^3D_4(2){:}3$, $G_2(3){:}2$, and $C_{Th}(2A)$. We also provide, via the internet, matrices generating *Th* and satisfying our presentation.

Throughout, we use ATLAS notation **[ATLAS]** for conjugacy classes and group structures. The proof of correctness of the presentation for *Th* uses a coset enumeration of 143,127,000 cosets, first completed during the Conference.

2. The Thompson group

In the monster group M, the centralizer of an element in conjugacy class $3C$ is of shape $3 \times Th$, where *Th* is the Thompson (sporadic simple) group. The group *Th* has order $90,745,943,887,872,000 = 2^{15}.3^{10}.5^3.7^2.13.19.31$. The maximal subgroups of *Th* have been completely determined (see **[Wil88, Lin89, Lin91]**), the largest

Groups and Computation III
Ohio State Univ. Math. Res. Inst. Publ. 8

being ${}^3D_4(2){:}3$ of index 143,127,000. The next largest maximal subgroup is the Dempwolff group $2^5{\cdot}L_5(2)$, of index 283,599,225. The Thompson group contains just one conjugacy class of involutions, and the centralizer of an element in this class is a maximal subgroup of shape $2_+^{1+8}{\cdot}A_9$. There are exactly three conjugacy classes of elements of order 3 in *Th*, and the normalizer $N(3A)$ of the subgroup generated by an element in the smallest such class is a maximal subgroup of shape $(3 \times G_2(3)){:}2$.

3. The presentations

Theorem. *Consider the group T presented by generators a, b, c, d, e, s, t, u and relators* (1)–(6) *below:*

$$\begin{gathered} a^2,\ b^2,\ c^2,\ d^2,\ e^2,\ (ab)^3,\ (ae)^2,\ (bc)^3,\ (bd)^2, \\ (be)^2,\ a=(cd)^4,\ (ce)^2,\ (de)^3,\ (bcde)^8, \end{gathered} \tag{1}$$

$$s^7,\ [s,a],\ [s,b],\ [s,c],\ (sd)^2,\ [e,s]=e^{s^3}, \tag{2}$$

$$t^3,\ [t,a],\ [t,b],\ [t,c],\ [t,d],\ [t,e],\ s^t=s^2, \tag{3}$$

$$\begin{gathered} u^2=ac,\ [u,a],\ [u,c],\ [u,e],\ (ded^u)^2, \\ [u,(ac)^b]=e,\ [u^d,(ac)^b]=ue(ac)^b u^d ec, \end{gathered} \tag{4}$$

$$t^u=t^{-1}, \tag{5}$$

$$[e,u^{s^2}],\, ac=(us)^3=[u,s]^4,\ (du^{s^2})^4=acc^d c^{des^{-1}} c^{des^2}. \tag{6}$$

(Here, the equation $w_1 = w_2$ denotes the relator $w_1 w_2^{-1}$. The notation $w_1^{w_2}$ means $w_2^{-1}w_1w_2$, and $[w_1, w_2]$ means $w_1^{-1}w_2^{-1}w_1w_2$.)
Then the following hold:

1. $T \cong Th$.
2. *Generators a, b, c, d, e, s together with relators* (1), (2) *give a presentation for* ${}^3D_4(2)$.
3. *Generators a, b, c, d, e, s, t together with relators* (1), (2), (3) *give a presentation for* ${}^3D_4(2){:}3$.
4. *Generators a, b, c, d, e, u together with relators* (1), (4) *give a presentation for* $G_2(3){:}2$.
5. *Generators a, c, d, e, s, t, u together with the relators from* (1)–(6) *not involving b give a presentation for* $C_{Th}(2A)$.

Proof. Using the 248-dimensional $GF(2)$-matrix representation of *Th* available from the ATLAS of Finite Group Representations **[Wil98, W+99]**, we have constructed non-identity 248-dimensional $GF(2)$-matrices $\boldsymbol{a}, \boldsymbol{b}, \boldsymbol{c}, \boldsymbol{d}, \boldsymbol{e}, \boldsymbol{s}, \boldsymbol{t}, \boldsymbol{u} \in Th$ satisfying all the relators (1) – (6) above (in the obvious way). The interested reader can find these matrices in GAP-format (see **[GAP4]**) at

`http://www.mat.bham.ac.uk/atlas/gap/Th/`

and verify this. Later, we will see that these matrices generate *Th*. The determination of these matrices is described in Section 4.

Now the abstract generators a, b, c, d, e together with the relators (1) are easily seen to give a presentation for the group $U_3(3){:}2$ (this is essentially the same presentation as that for $U_3(3){:}2$ which appears in **[Soi92]**). Indeed, the reader is encouraged to (at least) check that the group so presented has order 12,096. A coset enumeration shows that $\langle a, b, c, d, e\rangle$ has index 17,472 in the group D presented by generators a, b, c, d, e, s and relators (1), (2). Thus D, as well as its matrix group image

$$\boldsymbol{D} := \langle \boldsymbol{a}, \boldsymbol{b}, \boldsymbol{c}, \boldsymbol{d}, \boldsymbol{e}, \boldsymbol{s}\rangle,$$

has order at most that of ${}^3D_4(2)$. Moreover, the subgroup $\boldsymbol{D}$ of *Th* contains $\langle \boldsymbol{a}, \boldsymbol{b}, \boldsymbol{c}, \boldsymbol{d}, \boldsymbol{e}\rangle \cong U_3(3){:}2$ (as opposed to a proper quotient of this group, in which case the relators would imply $\boldsymbol{a} = 1$, a contradiction), and also $\boldsymbol{D}$ contains $7^2 \cong \langle \boldsymbol{b}\boldsymbol{c}\boldsymbol{c}^{\boldsymbol{d}}, \boldsymbol{s}\rangle$. From the maximal subgroup structure of *Th* we conclude that the only possibility for $\boldsymbol{D}$ is ${}^3D_4(2)$. Since the order of D is bounded by that of ${}^3D_4(2)$ as well as having $\boldsymbol{D}$ as a homomorphic image, we conclude that $D \cong {}^3D_4(2)$, establishing Statement 2 of the theorem.

Given that $\boldsymbol{t}$ has order 3 and the relators (3), we conclude that t is an element of order 3 in T, normalizing the subgroup $\langle a, b, c, d, e, s\rangle \cong {}^3D_4(2)$ and centralizing its maximal subgroup $\langle a, b, c, d, e\rangle \cong U_3(3){:}2$. Since no inner element of ${}^3D_4(2)$ of order 3 has these properties, it follows that

$$\boldsymbol{H} := \langle \boldsymbol{D}, \boldsymbol{t}\rangle \cong {}^3D_4(2){:}3,$$

and Statement 3 follows.

Now $\boldsymbol{H}$ is a maximal subgroup of *Th*, and $\boldsymbol{u} \notin \boldsymbol{H}$ since, by relator (5), $\boldsymbol{u}$ must invert $\boldsymbol{t}$. Hence

$$Th = \langle \boldsymbol{a}, \boldsymbol{b}, \boldsymbol{c}, \boldsymbol{d}, \boldsymbol{e}, \boldsymbol{s}, \boldsymbol{t}, \boldsymbol{u}\rangle,$$

and we have that *Th* is a homomorphic image of the group T presented. Now the main Statement 1 of the theorem follows from the enumeration of 143,127,000 cosets of $H := \langle a, b, c, d, e, s, t\rangle$ in T. This enumeration is described in Section 5.

We now tackle Statement 4. The matrix $\boldsymbol{u}$ is in the normalizer of $\boldsymbol{t}$, with $\boldsymbol{u}$ inverting $\boldsymbol{t}$. Hence

$$\boldsymbol{G} := \langle \boldsymbol{a}, \boldsymbol{b}, \boldsymbol{c}, \boldsymbol{d}, \boldsymbol{e}, \boldsymbol{u}\rangle \leq N_{Th}(\boldsymbol{t}) \cong (3 \times G_2(3)){:}2,$$

and it is easy to see that $G \cong G_2(3){:}2$ (with the full normalizer of $\langle t\rangle$ being $\langle G, t\rangle$). A coset enumeration shows that $\langle a, b, c, d, e\rangle$ has index 702 in the group presented by generators a, b, c, d, e, u and relators (1), (4), establishing Statement 4.

Finally, we prove Statement 5. It is easily deduced from relators (1)–(3) that a commutes with c, d, e, s, and t. Then a coset enumeration shows that $\langle a, c, d, e, s, t\rangle$ has index 819 in $H = \langle a, b, c, d, e, s, t\rangle$, establishing that $\langle a, c, d, e, s, t\rangle = C_H(a) \cong 2_+^{1+8}{:}L_2(8){:}3 \cong \langle \boldsymbol{a}, \boldsymbol{c}, \boldsymbol{d}, \boldsymbol{e}, \boldsymbol{s}, \boldsymbol{t}\rangle$. Since $[\boldsymbol{u}, \boldsymbol{a}] = 1$ and $\boldsymbol{u} \notin \boldsymbol{H}$, we have that

$$\boldsymbol{C} := \langle \boldsymbol{a}, \boldsymbol{c}, \boldsymbol{d}, \boldsymbol{e}, \boldsymbol{s}, \boldsymbol{t}, \boldsymbol{u}\rangle = C_{Th}(\boldsymbol{a}).$$

Now let C be the group presented by generators a, c, d, e, s, t, u together with the relators from (1)–(6) not involving b. It is easy to show, using coset enumeration, that in C, $L := \langle d, e, s, t\rangle \cong L_2(8){:}3$, and then a coset enumeration of the 61,440 cosets of L in C completes the proof of 5. □

We now outline (roughly) how our presentation for *Th* was derived. We started by determining the presentation above for $H = \langle a, b, c, d, e, s, t\rangle \cong {}^3D_4(2){:}3$, starting with the known presentation for $U := \langle a, b, c, d, e\rangle \cong U_3(3){:}2$, and using the subgroup structure of ${}^3D_4(2){:}3$ given in **[ATLAS]** to determine the further generators and relators. In H, t is an outer element of order 3 centralized by U, and in *Th*, we can extend U to $G_2(3){:}2$ by an element u inverting t. Up to automorphisms of $G_2(3){:}2$ there is just one way to identify the elements a, b, c, d, e. Calculating in a permutation representation of $G_2(3){:}2$, we identified these elements, chose an outer element u centralizing a and satisfying other short relators, and then determined a presentation for $G_2(3){:}2$ on generators a, b, c, d, e, u. For this, the GAP4 function `PresentationViaCosetTable` was very helpful in determining the crucial relation $[u^d, (ac)^b] = ue(ac)^b u^d ec$, to make correct the subgroup $\langle a, c, (ac)^b, c^d, e, u, u^d\rangle \cong 2^3{\cdot}L_3(2).2$ in $G_2(3){:}2$. Defining relators for $C_{Th}(a) \cong 2_+^{1+8}{\cdot}A_9$ on generators a, c, d, e, s, t, u (for an appropriate choice of u) were more difficult to obtain, although starting with $\langle d, e, s, t\rangle \cong L_2(8){:}3$, we determined defining relators modulo the normal 2_+^{1+8} (which is the normal closure of $\langle c\rangle$ in $\langle a, c, d, e, s, t\rangle$). To exactly determine defining relators for $C_{Th}(a)$, we constructed the matrices generating *Th* and satisfying the relators (1)–(5) above, and then deduced the exact form of the relators (6).

Some of our relators for $T \cong Th$ are known to be redundant, but included for clarity or to aid coset enumeration. We also remark that, in T, $\langle a, c, (ac)^b, c^d, e, s, t, u, u^d\rangle$ is isomorphic to the Dempwolff group $2^5{\cdot}L_5(2)$.

4. Determining matrices in *Th* satisfying the presentation

Inside the Thompson group, a copy of ${}^3D_4(2){:}3$ can easily be found in a random search of subgroups generated by an involution and a $3C$-element. To find $3 \times U_3(3){:}2$, the centralizer of a $3C$-element in ${}^3D_4(2){:}3$, we first found two elements, of orders 21 and 12, powering up to two different $3C$-elements, and then in the group generated by these two 3-elements, found an element conjugating one to the other. (This could be done either by a random search, or by pure thought, since the group generated by these elements turned out to be $2^3{:}7{:}3$.) We thus obtain elements of orders 21 and 12 powering up to the same $3C$-element, and generating the whole of its centralizer, $3 \times U_3(3){:}2$. This $3C$-element may be taken as $\boldsymbol{t}^{\pm 1}$.

We then look for $\boldsymbol{a}$, $\boldsymbol{b}$, $\boldsymbol{c}$, $\boldsymbol{d}$, $\boldsymbol{e}$ satisfying the given presentation for $U_3(3){:}2$. We first found an outer involution, which we can take to be $\boldsymbol{e}$, and used standard methods (see **[Lin95, Bra]**) to find its centralizer $2 \times S_4$. There are two subgroups isomorphic to S_4 in here, one of which is inside $U_3(3)$. We take $\boldsymbol{a}$, $\boldsymbol{b}$, $\boldsymbol{c}$ to be the transpositions (12), (23), (34) of the latter. It remains to find $\boldsymbol{d}$, which is unique up to conjugation by $\boldsymbol{e}$. We do this by searching in the centralizer of another involution, such as $\boldsymbol{a}$ (or $\boldsymbol{b}$). Since this centralizer is small, this does not take long.

The next step is to find $\boldsymbol{s}$ inside ${}^3D_4(2){:}3$. Now $\boldsymbol{s}$ is an element of order 7 commuting with $\langle \boldsymbol{a}, \boldsymbol{b}, \boldsymbol{c}\rangle \cong S_4$, so we work inside the normalizer of the fours-group $\langle \boldsymbol{ac}, (\boldsymbol{ac})^{\boldsymbol{b}}\rangle$, which is a group of shape $2^2.[2^9].(7 \times S_3)$. We take any element $\boldsymbol{s}_0$ of order 7 in this group, and then $\boldsymbol{s}_1 = \boldsymbol{s}_0[\boldsymbol{bc}, \boldsymbol{s}_0]$ commutes with $\boldsymbol{bc}$, so $\boldsymbol{s}$ is a suitable power of $\boldsymbol{s}_1$. Indeed, we may take $\boldsymbol{s} = \boldsymbol{s}_1$ without loss of generality. At this point we can use the relation $\boldsymbol{s}^{\boldsymbol{t}} = \boldsymbol{s}^2$ to determine which central element of $3 \times U_3(3){:}2$ is $\boldsymbol{t}$, and which is $\boldsymbol{t}^{-1}$.

Finally we must find $\boldsymbol{u}$, which we do inside the full centralizer $2_+^{1+8}{\cdot}A_9$ of $\boldsymbol{a}$. This centralizer is generated by $\boldsymbol{c}$, $\boldsymbol{d}$, $\boldsymbol{e}$, $\boldsymbol{s}$, $\boldsymbol{t}$ and one other element, found by standard methods. We chop the representation to find a constituent which represents A_9, and convert everything to the natural permutation representation where we can calculate easily. We are looking for an element which commutes with $\boldsymbol{e}$, and inverts $\boldsymbol{t}$. In the quotient A_9, there are just three involutions with these properties, but only one satisfies the relation $(\boldsymbol{us})^3 = 1$. If $\boldsymbol{u}_0$ is any lift of this element in $2_+^{1+8}{\cdot}A_9$, then $\boldsymbol{u}_1 = \boldsymbol{u}_0[\boldsymbol{t}, \boldsymbol{u}_0]$ inverts $\boldsymbol{t}$.

We now have $N(\boldsymbol{t}) \cong (3 \times G_2(3)){:}2$ generated by $\boldsymbol{t}, \boldsymbol{a}, \boldsymbol{b}, \boldsymbol{c}, \boldsymbol{d}, \boldsymbol{e}$ and $\boldsymbol{u}_1$, but $\boldsymbol{u}_1$ still does not satisfy all the required relations, because we have not yet specified precisely which preimage of the element of A_9 to take. Inside $G_2(3){:}2$ generated by $\boldsymbol{a}$, $\boldsymbol{b}$, $\boldsymbol{c}$, $\boldsymbol{d}$, $\boldsymbol{e}$ and $\boldsymbol{u}_1$ we find that the normalizer of the fours-group $\langle \boldsymbol{a}, \boldsymbol{c}\rangle$ is generated by $\boldsymbol{bacb}$, $\boldsymbol{dcd}$, $\boldsymbol{e}$ and $\boldsymbol{u}_1$, and we search in this small group for the required element $\boldsymbol{u}$.

5. The large coset enumeration

The enumeration to determine $|T : H|$ requires the definition of more than 140 million cosets. Thus this step requires substantial resources. Recent descriptions of relevant aspects of coset enumeration are given in **[Sim94, HR]**. Of particular importance here is that the space required (memory locations) for a standard coset enumeration to complete is at least the maximum number of cosets defined times the number of columns in the coset table. The number of columns in the coset table is the number of involutory group generators plus twice the number of noninvolutory generators.

In some circumstances (eg, see **[HS99]**) this means we need to consider finding presentations with small generating sets. In our case here the given presentation for T leads to eleven columns, which is not too many in the context of memory resources available on some supercomputers. Our enumerations were performed on an SGI Origin 2000 computer which has 16 gigabytes of memory, and we were able to obtain access to some 8 gigabytes. This is enough to store over 180 million cosets with the given presentation.

For difficult coset enumeration problems there is sometimes the need for substantial experimentation to determine how to complete an enumeration with given resources (see **[HR]**). However our experience with the various subgroups of T discussed in this paper led us to believe that the enumeration in this case would be relatively easy, and so it turned out.

Using the ACE3 coset enumerator **[ACE]**, with group presentation and subgroup generators as in this paper, we can readily compute the 143,127,000 cosets of H in T. It is worth noting that the final coset table (in standard internal representation) is too big to fit in a computer with 32 bit addressing architecture.

The ACE3 enumerator has a large number of parameters, with a wide choice of settings and enumeration styles (see **[ACE]**). We have completed the enumeration using various strategies without undue difficulty. Felsch-based enumerations seem best. They complete in as little as about 62 cpu hours. For example, with the styles `Hard`, `Felsch:0`, `Felsch:1` and `Sims:3` in each case the maximum number of cosets defined is the index. The total numbers are:

`Hard`	`Felsch:1`	`Felsch:0`	`Sims:3`
145,251,396	151,273,730	155,354,391	240,046,124

The `HLT` style (also available in **[ACE]**) seems less attractive, using a maximum of 179,631,904 cosets and a total of 473,745,756 in an enumeration given 8 gigabytes of memory.

Acknowledgements. The work of the first author was supported by the Australian Research Council. We are grateful to The University of Queensland's High Performance Computing Unit for access to the Origin 2000 supercomputer.

References

[Bra] J. N. Bray, An improved method for generating the centralizer of an involution, Arch. Math. 74 (2000), 241–245.

[BN95] T. Breuer and S. Norton, Improvements to the ATLAS, Appendix 2 of: An ATLAS of Brauer Characters (by C. Jansen, K. Lux, R. Parker and R. Wilson), London Math. Soc. Monogr. (N.S.) 11, Clarendon Press, Oxford, 1995, 297–327.

[ATLAS] J. H. Conway, R. T. Curtis, S. P. Norton, R. A. Parker and R. A. Wilson, An ATLAS of Finite Groups, Clarendon Press, Oxford, 1985.

[GAP4] The GAP Group, GAP — Groups, Algorithms, and Programming, Version 4.2; Aachen, St Andrews, 2000 (`http://www-gap.dcs.st-and.ac.uk/~gap`).

[ACE] G. Havas and C. Ramsay, Coset enumeration: ACE version 3, 1999; available as `http://www.csee.uq.edu.au/~havas/ace3.tar.gz`.

[HR] G. Havas and C. Ramsay, Experiments in coset enumeration, in: Groups and Computation III (W. M. Kantor and Á. Seress, eds.), Ohio State Univ. Math. Res. Inst. Publ. 8, Walter de Gruyter, Berlin–New York, 2001, 183–192.

[HS99] G. Havas and C. C. Sims, A presentation for the Lyons simple group, in: Computational Methods for Representations of Groups and Algebras, Progr. Math. 173, Birkhäuser, Basel, 1999, 241–249.

[HM69] G. Higman and J. McKay, On Janko's simple group of order 50,232,960, Bull. London Math. Soc. 1 (1969), 89–94.

[Iva92] A. A. Ivanov, A presentation for J_4, Proc. London Math. Soc. 64 (1992), 369–396.

[Iva99] A. A. Ivanov, Geometry of Sporadic Groups I: Petersen and Tilde Geometries, Cambridge University Press, 1999.

[JW96] C. Jansen and R. A. Wilson, The minimal faithful 3-modular representation for the Lyons group, Comm. Algebra 24 (1996), 873–879.

[Lin89] S. A. Linton, The maximal subgroups of the Thompson group, J. London Math. Soc. (2) 39 (1989), 79–88.

[Lin91] S. A. Linton, Corrections to "The maximal subgroups of the Thompson group", J. London Math. Soc. (2) 43 (1991), 253–254.

[Lin95] S. A. Linton, The art and science of computing in large groups, in: Computational Algebra and Number Theory (W. Bosma et al., eds), Kluwer, Dordrecht, 1995, 91–109.

[PS97] C. E. Praeger and L. H. Soicher, Low Rank Representations and Graphs for Sporadic Groups, Cambridge University Press, 1997.

[Sim94] C. C. Sims, Computation with Finitely Presented Groups, Cambridge University Press, 1994.

[Soi92] L. H. Soicher, On simplicial complexes related to the Suzuki sequence graphs, in: Groups, Combinatorics and Geometry (M.W. Liebeck and J. Saxl, eds), London Math. Soc. Lecture Note Ser. 165, Cambridge University Press, 1992, 240–248.

[SW88] G. Stroth and R. Weiss, Modified Steinberg relations for the group J_4, Geom. Dedicata 25 (1988), 513–525.

[Wei91] R. Weiss, A geometric characterization of the groups M_{12}, *He* and *Ru*, J. Math. Soc. Japan 43 (1991), 795–814.

[Wil88] R. A. Wilson, Some subgroups of the Thompson group, J. Austral. Math. Soc. (A) 44 (1988), 17–32.

[Wil98] R. A. Wilson, An ATLAS of sporadic group representations, in: The ATLAS 10 Years On (R. T. Curtis and R. A. Wilson, eds), London Math. Soc. Lecture Note Ser. 249, Cambridge University Press, 1998, 261–273.

[W+99] R. A. Wilson et al., An ATLAS of Finite Group Representations, 1999, `http://www.mat.bham.ac.uk/atlas/`.

Centre for Discrete Mathematics and Computing
Department of Computer Science and Electrical Engineering
The University of Queensland
Queensland 4072, Australia
havas@csee.uq.edu.au

School of Mathematical Sciences
Queen Mary and Westfield College
Mile End Road, London E1 4NS, U.K.
L.H.Soicher@qmw.ac.uk

School of Mathematics and Statistics
The University of Birmingham
Edgbaston, Birmingham B15 2TT, U.K.
R.A.Wilson@bham.ac.uk

Computing automorphism groups of finite groups

Derek F. Holt

Abstract. A new method for computing the automorphism group $\mathrm{Aut}(G)$ of a reasonably small finite group G will be described in outline. The method is compared with earlier ones, and with current more specialized algorithms that handle the cases when G is a finite p-group, or a finite soluble group.

1991 Mathematics Subject Classification: primary 20D45; secondary 20B25, 20-04, 20B40.

1. Introduction

In this article, we describe in outline a new method for efficiently computing the automorphism group of a finite group G. Full technical details of the algorithm will be described elsewhere **[2]**. It has been implemented for finite permutation groups within the language of the MAGMA computational algebra system **[1]**, and its performance is encouraging. The MAGMA code is available from the author. It is currently practical for most groups of order up to 1000000, and for many much larger groups. Similar methods for testing pairs of finite groups for isomorphism have also been implemented.

The method used is a generalization of that described for soluble groups by Michael Smith in **[19]**. The idea is to find a series of characteristic subgroups

$$1 = N_r \leq N_{r-1} \leq \cdots \leq N_1 = L \leq G$$

of our given group G, such that each N_i/N_{i+1} is elementary abelian and G/L has no non-trivial soluble normal subgroup. Of course, L is uniquely defined as the largest normal soluble subgroup of G, and $G/L = 1$ if G itself is soluble. We compute $\mathrm{Aut}(G/N_1)$, $\mathrm{Aut}(G/N_2)$, $\ldots$, $\mathrm{Aut}(G/N_r) = \mathrm{Aut}(G)$ successively, where $\mathrm{Aut}(G/N_{i+1})$ is computed by testing which elements of $\mathrm{Aut}(G/N_i)$ can be lifted to automorphisms of G/N_{i+1}. If G/N_1 is non-trivial, then its automorphism group is calculated from information stored in a database of automorphism groups of almost simple groups. The basic strategy of this approach is very similar to that described in **[3]** for computing the subgroup lattice of a finite group.

In Section 2 we discuss some earlier general methods that have been used to compute automorphism groups, together with the more specialized methods of O'Brien and Smith that are applicable to finite p-groups and to finite soluble groups, respectively. In Section 3, we give a brief outline of our algorithm, identify bottlenecks, and suggest possible directions for future development. In Section 4, we provide some performance statistics of the current MAGMA implementation.

2. Various methods for computing automorphisms groups

The earliest implementation of a general purpose automorphism group algorithm was probably that described by Felsch and Neubüser in **[10]**. They started by computing the full subgroup lattice of the group G.

A more refined approach that required less preprocessing was developed by Neubüser in the early 1970's; it required the initial computation of the conjugacy classes of G but not the full subgroup lattice. An automorphism must map an element to one of the same order and in a conjugacy class of the same size. By finding a suitable generating set for G, and using a variety of other tricks, it is usually possible in groups of moderate order to restrict the number of possible mappings of the generators in potential automorphisms to a reasonably small number. To check whether such a map really is an automorphism of G it is necessary to verify that it preserves a system of defining relators for G on the generators.

This method is described in detail in **[16]**. It has proved remarkably enduring and is still used in the GAP system **[7]**. It can be applied equally easily to power-conjugate presentations of soluble groups or to permutation representations of general groups, and is very reliable for most groups of order up to about 10000 and indeed for many larger groups. However there are some quite small examples that it is unable to handle easily, and its effectiveness diminishes rapidly with increasing order. For example, using Version 3.4.4 of GAP on a Sun Ultra 5 WorkStation, the automorphism group of an extraspecial group of order 32 was computed in about 214 seconds, but the calculation for an extraspecial group of order 128 did not complete in 50 hours of cpu-time.

The special case of finite p-groups, which includes the most difficult examples in which the group G has small order, was handled separately in an algorithm due to O'Brien that is described in **[13]**. As mentioned earlier, a method for finite soluble groups given by a power-conjugate presentation is described by Smith in **[19]**. O'Brien's implementation is available from both MAGMA and GAP, and Smith's is available as the GAP share package `autag`.

In common with the author's method for general groups, these both work by lifting automorphism groups through the quotients of a series of characteristic subgroups. In all three cases, the most time-consuming part of the algorithm is to solve the following lifting problem. Suppose that M and N are characteristic subgroups of G, with $N < M$ and M/N elementary abelian. If $A_M = \mathrm{Aut}(G/M)$ is already known, then find the subgroup of A_M containing those elements that lift to automorphisms of G/N. The problem arises when A_M is large; for example, if G/M is elementary abelian, then $A_M = GL(r, p)$ for some r and p; but the subgroup that lifts might be relatively small, and so testing all elements of A_M for lifting in a naïve fashion is not practical.

This lifting problem appears to be considerably more tractable in the p-group case than in the soluble or general cases. For a p-group G, the most suitable series is the lower p-central series. There is a group $\widehat{G}$, known as the *p-covering group* of G/M, which has central elementary abelian subgroups $\widehat{M}$ and $\widehat{N}$ such that $\widehat{G}/\widehat{M} \cong G/M$ and $\widehat{G}/\widehat{N} \cong G/N$, and has the useful property that all elements of A_M lift to automorphisms of $\widehat{G}$. We can regard $\widehat{M}$ as a vector space over the field of order p, and then A_M induces a group T of linear transformations of this space. The subgroup of A_M that lifts to automorphisms of G/N can then be computed indirectly as the subgroup of T that fixes the subspace $\widehat{N}$. This is still a difficult calculation, but it has been the subject of some recent research **[17]**, and it can now be solved considerably more efficiently than the automorphism lifting problem. There is a specific power-commutator presentation of a p-group G known as the *standard presentation* **[14]** with the property that isomorphic groups have the same standard presentation. This is computed simultaneously with $\mathrm{Aut}(G)$, and it can be used to solve the isomorphism problem between G and other groups of the same order.

An issue involved in calculation of $\mathrm{Aut}(G)$ for finite groups G that has (in the author's opinion) not always been adequately addressed is how the resulting automorphism groups should best be represented. This obviously depends on what use the user would like to make of the information once it has been computed. It would be helpful to have more feedback from users on this question; one such instance arose from the investigation of automorphisms of Burnside groups by Sims (see below).

If the group G is quite small, then representing $\mathrm{Aut}(G)$ as a permutation group on the elements of G is the obvious solution, and is likely to be satisfactory for all applications. Cannon and Neubüser tried to find a smaller degree faithful permutation representation on the union of some of the conjugacy classes of G. For larger groups G this can become impractical, because such representations have too large a degree. It is particularly problematic in the p-group and soluble group cases, where one would hope to be able to handle much larger groups. In the published descriptions by O'Brien and Smith, only generators of $\mathrm{Aut}(G)$ are returned, and no attempt is made to compute the group structure, although certain sections of $\mathrm{Aut}(G)$ are explicitly defined as elementary abelian groups or as matrix groups. However, the most recent implementations of O'Brien's algorithm do include some facilities for computing with automorphism groups as groups, such as finding orders of elements of the automorphism group, and these are expected to be developed further in the near future.

As far as the basic lifting problem is concerned, the automorphism group representation problem does not play a rôle in the p-group case, because it reduces to a subspace stabilizer problem. However, in the soluble and general group cases there is no such reduction, and it is often essential to have an efficient concrete representation of $A_M = \mathrm{Aut}(G/M)$.

We cannot offer a completely satisfactory solution to these problems. The author has chosen, in his implementation, to return $\text{Aut}(G)$ as a finitely presented group in which the subgroup $\text{Inn}(G)$ of inner automorphisms is generated by an initial subset of the generators. It is sometimes possible to use this to compute a nice permutation representation by trial and error, perhaps starting from such a representation of G. Another possibility is to apply a soluble quotient algorithm (see **[12]** or **[15]**) if $\text{Aut}(G)$ is soluble or has a soluble subgroup of small index. Of course, such possibilities rely on the knowledge and experience of the user. The permutation action on the elements of G, or the holomorph of G, can also be computed if required.

In **[18]**, Sims considered the problem of computing the order of subgroups of $\text{Aut}(G)$ defined by generating automorphisms supplied by the user, and of membership testing in such subgroups, and he described a GAP implementation of a proposed solution to these problems. He was particularly interested in the case when G is a large finite p-group (for example, quotients of the Burnside group $B(2,5)$), where the automorphisms groups are large. Sims' method also uses a series of characteristic subgroups of G, and the automorphism groups of the associated factor groups. Although we have not done so yet, it would be straightforward to incorporate this idea into the author's implementation.

3. A brief description of the method and its implementation

Since we wish to deal with general finite groups, and the implementations of O'Brien and Smith are available for p-groups and soluble groups defined by power-conjugate presentations, we decided to implement our algorithm for finite permutation groups. Finding an appropriate series

$$1 = N_r \leq N_{r-1} \leq \cdots \leq N_1 = L \leq G$$

of characteristic subgroups can be done by using standard functions available in both GAP and MAGMA. We first find L, for which an algorithm is described in **[11]**, and then find the N_i by repeatedly taking commutator subgroups and refining to enforce prime exponent. It usually proves best to refine this series as much as possible so as to make the factors as small as possible. Computing in quotients of permutation groups is not straightforward in general. However, it is proved in **[5]**, and again in algorithmic terms in **[11]** and by the author in **[8]**, that the soluble radical quotient G/L has a faithful permutation representation of degree at most that of G, and the layer quotients N_i/N_{i+1} can be explicitly computed as G/N_i-modules. All other computations in the quotients G/N_i are actually done by working in their inverse images in G, and so there are no computational bottlenecks involving quotients of permutation groups.

When the soluble radical quotient G/L is non-trivial, its socle is a direct product of nonabelian simple groups S_i, and G/L itself can be embedded in a natural way into a direct product of wreath products of type $\text{Aut}(S_i) \wr \text{Sym}(d_i)$. From this embedding, the computation of $\mathrm{Aut}(G/L)$ can be reduced to that of the $\mathrm{Aut}(S_i)$. The automorphism groups of the simple groups of order up to one million (together with some other larger examples, such as alternating groups) are stored in a database. In order to extract this information, the isomorphism types of the S_i must first be recognized (which is usually possible just from their orders), and then an explicit isomorphism has to be constructed from S_i to the corresponding group in the database, which is stored as a finite presentation. This isomorphism is found by a search for elements in S_i that satisfy the defining relators of this presentation. The stored data on $\mathrm{Aut}(S_i)$ includes its order, and a presentation of the required form having the inner automorphism group as the subgroup generated by the first two generators. There appear to be no particular computational bottlenecks in this part of the computation.

The remainder of the algorithm involves computing $A_N = \mathrm{Aut}(G/N)$ from $A_M = \mathrm{Aut}(G/M)$, where $M = G_i$ and $N = G_{i+1}$ for some i. Now A_N has subgroups C and B with $C \leq B$, where B consists of those automorphisms that induce the identity map on G/M, and C consists of those inducing the identity on both G/M and M/N. Let K be the field of order p, where p is the prime dividing $|M/N|$.

The subgroup C is an elementary abelian p-group isomorphic to the group of derivations from G/M to M/N, and its computation reduces to solving a reasonably small system of linear equations over K. The quotient B/C is isomorphic to a subgroup of the group of module automorphisms of the vector space M/N regarded as a KG/M-module. Finally G/B is isomorphic to the subgroup of A_M of those elements that lift to automorphisms of G/N.

The computations are relatively straightforward when G/N splits as an extension of M/N by G/M. In that case B/C is the full group of module automorphisms. A method of computing this was described by Smith in **[19]**. The MAGMA function for this together with other necessary functions for KG-modules, such as the Jacobson Radical, was implemented by Alan Steel with advice from the author and C. R. Leedham-Green; details are not yet published. Furthermore, the lifting subgroup of A_M consists of those automorphisms which fix M/N as a KG/M-module, and again this can be computed quickly.

The non-split case is more difficult, but we can begin by making an extra reduction. We can refine the series of characteristic subgroups of G so as to make M/N a semisimple KG/M-module contained in the Frattini subgroup of G/N. This is done by computing the Jacobson Radical of M/N and the Frattini subgroup of G/N, and introducing their inverse images into the series. It has been proved by L. G. Kovács under these assumptions that, for an automorphism $\alpha \in G/N$, the restriction $\alpha_{M/N}$ of α to M/N is uniquely determined by the induced action $\alpha_{G/M}$ of α on G/M. (This was a private communication, and the proof is included in **[2]**.) In particular, we get $B = C$ in this case. Unfortunately Kovács' proof is non-constructive and we do not currently know of any general algorithmic method of computing $\alpha_{M/N}$ from $\alpha_{G/M}$, and so for each $\alpha_{G/M} \in A_M$ which we are trying to lift to A_N, we are reduced to trying all possible module automorphisms as candidates for $\alpha_{M/N}$. For a given pair $(\alpha_{G/M}, \alpha_{M/N})$, testing whether there is a corresponding lifting reduces to straightforward linear algebra. Of course, if we do find a pair that lifts, then we need try no further $\alpha_{M/N}$ for that particular $\alpha_{G/M}$, but the vast bulk of the computation time for difficult examples consists in running through large numbers of $\alpha_{G/M}$ for which there is no lifting.

Judging from our current implementation, it seems that all components of this algorithm can be carried out quickly and efficiently apart from the search through the potentially large group A_M for automorphisms that lift. This could be speeded up considerably in some examples if we knew how to compute $\alpha_{M/N}$ from $\alpha_{G/M}$, and devising a practical method to do this must be an immediate priority. However, the search can still be a large one, even for quite small groups. For example, if G is extraspecial of order 128 defined as a central product of three quaternion groups of order 8, $N = 1$ and M is the central subgroup of order 2, then A_M is the group $GL(6, 2)$ of order 20158709760, whereas the lifting subgroup of A_M has order only 51840. This computation took nearly two hours of cpu-time with the author's implementation; to carry out this search, A_M was represented as a permutation group of degree 63 on the non-identity elements of G/M, and standard backtrack searching methods were used for permutation groups.

O'Brien's p-group automorphism took only about 2 seconds to compute the automorphism group of this extraspecial group also in MAGMA on the same machine, which indicates emphatically just how much improvement can be gained by using special properties of p-groups and treating it as a subspace stabilizer problem.

This raises the question whether O'Brien's methods, or even his code, could profitably be inserted somewhere in a general group automorphism program to handle sections of the group that happen to be p-groups. For example, it should be relatively straightforward to deal with p-quotients of the group for all primes p in this manner. There are however still some types of examples that are slower than they ought to be, such as D_6^n, of which the automorphism group is actually just $D_6 \wr \mathrm{Sym}(n)$, but which, with the author's program, is slow for $n = 5$ and intractable for $n \geq 6$.

4. Some current performance figures

We conclude with a few timings of the current MAGMA implementation.

The table contains some sample performance figures on a Sun Ultra 5 WorkStation. The columns headed $|A|$ and $|O|$ list the orders of the automorphism group and outer automorphism group of the group G, respectively. The layer sizes are from the top downwards. All times are in seconds.

In general, the notation for groups follows that in **[4]**. Thus 11_a^{1+2} and 11_b^{1+2} denote the two types of extraspecial groups of order 11^3, where the former is the one with exponent 11. The group $SL_2(13) \vee Q_8$ is a central product of the two factors. The group $T(m, n)$ is the n-th transitive group of degree m, and is returned by the MAGMA function `TransitiveGroup(m,n)`. A full catalogue of all transitive

G	$\|G\|$	$\|A\|$	$\|O\|$	Deg.	TF-Gp.	Layer Sizes	Time
11_a^{1+2}	1331	1597200	13200	121	1	121 11	2.5
11_b^{1+2}	1331	13310	110	121	1	121 11	6.9
$D_8 \times D_8$	64	2048	128	8	1	16 4	9.4
2^{1+6}	128	3317760	51840	128	1	64 2	5886
D_6^5	7776	933120	120	15	1	32 243	718
$2^4 SL_2(5)$	1920	23040	24	40	A_5	60 2 16	0.8
$2^8 A_5$	15360	88473600	5760	32	A_5	60 256	0.6
$C_3 \wr S_5$	29160	38880	4	15	S_5	120 81 3	0.3
$AGL_2(7)$	98784	98784	1	49	$PGL_2(7)$	336 3 2 49	0.9
$SL_2(13) \times Q_8$	17472	209664	48	64	$L_2(13)$	1092 4 2 2	1.7
$SL_2(13) \vee Q_8$	8736	52416	12	224	$L_2(13)$	1092 4 2	1.6
$3L_3(4)$	60480	241920	12	63	$L_3(4)$	20160 3	1.7
$3A_6 \times SL_2(5)$	129600	172800	8	42	$A_6 \times L_2(5)$	2160 3 2	1.5
$SL_2(5)^6$	120^6	720.120^6	46080	30	A_5^6	60^6 64	222
$T(16, 1767)$	12288	98304	16	16	1	3 64 8 4 2	17
$T(16, 1840)$	40320	40320	1	16	A_7	2520 16	0.7
$T(20, 893)$	320000	3840000	12	20	1	32 8 2 625	3727

permutation groups of degree up to 30 has been computed by Hulpke **[9]**, and these provide convenient test examples. In fact, we have successfully run the automorphism group program on all of these groups, and it failed to complete within 24 hours cpu-time only on a handful of 2-groups of degree 16, and on one example of degree 30. The median time for these runs was less than two seconds in all degrees; the example $T(20, 893)$ given in the table was the slowest example of degree 20.

The special purpose method for soluble groups described in **[19]** is faster on some of the soluble examples, but not hugely so, whereas on others, such as $D_8 \times D_8$ it seems to be significantly slower for some reason.

References

[1] W. Bosma and J. J. Cannon, Magma Handbook, Sydney, 1993.

[2] J. J. Cannon and D. F. Holt, Automorphism group computation and isomorphism testing in finite groups, preprint, available from the author.

[3] J. J. Cannon, B. Cox and D. F. Holt, Computing the subgroup lattice of a permutation group, to appear in J. Symbolic Comput.

[4] J. H. Conway, R. T. Curtis, S. P. Norton, R. A. Parker and R. A. Wilson, ATLAS of Finite Groups, Clarendon Press, Oxford, 1985.

[5] D. Easdown and C. E. Praeger, On minimal faithful permutation representations of finite groups, Bull. Australian Math. Soc. 38 (1988), 207–220.

[6] B. Eick, C. R. Leedham-Green & E. A. O'Brien, Constructing automorphism groups of p-groups, to appear.

[7] The GAP Group, GAP — Groups, Algorithms, and Programming, Version 4.2; Aachen, St Andrews, 2000 (`http://www-gap.dcs.st-and.ac.uk/~gap`).

[8] D. F. Holt, Representing quotients of permutation groups, Quart. J. Math. (Oxford) 48 (1997), 347–350.

[9] A. Hulpke, Konstruktion transitiver Permutationsgruppen, Aachener Beiträge zur Mathematik, Band 18, Verlag der Augustinus Buchhandlung, Aachen, 1996.

[10] V. Felsch & J. Neubüser, Über ein Programm zur Berechnung der Automorphismengruppe einer endlichen Gruppe, Numer. Math. 11 (1968), 277–292.

[11] E. Luks & Á. Seress, Computing the Fitting subgroup and solvable radical of small-base permutation groups in nearly linear time, in: Groups and Computation II (L. Finkelstein and W. M. Kantor, eds.), DIMACS Ser. Discrete Math. Theoret. Comput. Sci. 28, Amer. Math. Soc., Providence, RI, 1997, 169–181.

[12] A. C. Niemeyer, A soluble quotient algorithm, J. Symbolic Comput. 18 (1994), 541–561.

[13] E. A. O'Brien, Computing automorphisms of p-groups, in: Computational Algebra and Number Theory (Wieb Bosma and A. J van Poorten, eds.), Proceedings of CANT '92, Kluwer, Dordrecht, 1995, 83–90.

[14] E. A. O'Brien, Isomorphism testing for p-groups, J. Symbolic Comput. 17 (1994), 133–147.

[15] W. Plesken, Towards a soluble quotient algorithm, J. Symbolic Comput. 4 (1987), 111–122.

[16] H. Robertz, Eine Methode zur Berechnung der Automorphismengruppe einer endlichen Gruppe, Diplomarbeit, RWTH Aachen, 1976.

[17] R. Schwingel, Two Matrix Group Algorithms with Applications to Computing the Automorphism Group of a Finite p-Group, PhD Thesis, Queen Mary and Westfield College, University of London, 2000.

[18] C. C. Sims, Computing with subgroups of automorphism groups of finite groups, ISSAC 97, Proceedings of the 1997 Internat. Conference on Symbolic and Algebraic Computation, ACM Press, New York 1997.

[19] Michael J. Smith, Computing automorphisms of finite soluble groups, Australian National University Doctoral Thesis, 1995.

Mathematics Institute
University of Warwick
Coventry, CV47AL, U.K.
`dfh@maths.warwick.ac.uk`

Regular orbits and the $k(GV)$-problem

Christoph Köhler and Herbert Pahlings

Abstract. In this paper all finite quasisimple groups having no regular orbits on finite modules are determined and it is shown that the $k(GV)$-problem has a positive answer for primes different from 3,5,7,11,13,19,31.

1991 Mathematics Subject Classification: primary 20C20; secondary 20D30.

1. Introduction

The $k(GV)$-problem is to show that the number $k(GV)$ of conjugacy classes of the semidirect product GV of a finite p'-group G and a faithful finite $\mathbb{F}_pG$-module V is less than or equal to $|V|$ for any prime p. The interest in this problem comes mainly from the fact, that for p-solvable groups it is known to be equivalent to Brauer's $k(B)$-problem which is to prove that the number $k(B)$ of ordinary irreducible characters of a block B of a finite group G is less than or equal to the order of the defect group of B.

Robinson and Thompson **[RT 96]** have proved that the $k(GV)$-problem has a positive answer for all but finitely many primes p. It follows from the results of this paper, that the asserted inequality is true if $p \notin \{3, 5, 7, 11, 13, 19, 31\}$. The second author learned at the conference in Columbus that the same result has been obtained using different methods by Gluck and Magaard (for the symplectic case) and Riese (for the quasisimple case). These cases occur in the Reduction Theorem of Robinson and Thompson. In fact, it was shown in **[RT 96]** and **[Ro 97]** that if G is a finite p'- group and V a faithful $\mathbb{F}_pG$-module such that the restriction $\mathrm{Res}^G_{C_G(v)} V$ of V to the stabilizer $C_G(v)$ of some vector $v \in V$ has a faithful selfdual summand (such a vector will be called "real" for short) then $k(GV) \leq |V|$. Furthermore in a minimal example of a faithful $\mathbb{F}_pG$-module without a real vector the group G would have to be of symplectic type, i.e. $F^\star(G) = F(G)$ is a central product of the center $Z(G)$ and an extraspecial q-group for some prime $q \neq p$, or of quasisimple type with $F^\star(G)$ being a central product of $Z(G)$ and a quasisimple group $E(G)$.

It is clear that any vector in a regular orbit is real in the above sense, since $C_G(v) = \{1\}$. In **[HLS 92]** it was shown, that with finitely many exceptions any finite quasisimple group E operating faithfully on an $\mathbb{F}_pE$-module V for p not dividing $|E|$ has regular orbits on the vectors of V. In this paper we will explicitly find all these exceptions, there are very few, see Theorem 2.3. Liebeck (see **[Li 96]**) has extended the above result by showing that if G is a p'-group with a faithful $\mathbb{F}_pG$-module V such that G has a quasisimple normal subgroup E operating irreducibly on V then $E \cong A_n$, an alternating group of degree n and V_E is the natural irreducible module of dimension $n-1$ or $p < 5^{30}$. The first case does not present an obstacle to the $k(GV)$-problem, since in this case there is always a real vector, as Robinson and Thompson had shown in **[RT 96]**. The above bound (of Liebeck) has been lowered to 271 by Goodwin (**[Go 00]**). Using this Robinson **[Ro 97]** showed that the $k(GV)$-problem has a positive answer for all primes $p > 211$.

In this paper we will deal mainly with the quasisimple case and the existence of regular orbits. The last section gives a summary of the work of the first author on the symplectic case, for more details concerning the programs and the computations we refer to his diploma thesis **[Ko 99]**. Almost all of the computations have been carried out using GAP (**[GAP 00]**). The data-libraries provided by this system, in particular the library of tables of marks proved to be extremely useful.

Acknowledgement. Thanks are due to D. Gluck, K. Magaard and U. Riese for sending us preprints of their papers, also to M. Liebeck for providing us with a copy of **[Go 00]**. We would also like to thank G. Robinson for valuable suggestions and F. Lübeck, J. Müller and M. Neunhöffer for computing a regular orbit for 6.Suz using their new orbit algorithm.

2. The quasisimple case

Goodwin's main result **[Go 00]** is the following

Theorem 2.1. *Suppose p is a prime, G is a p'-group and V is a faithful n-dimensional $\mathbb{F}_pG$-module such that*

(1) *G has a quasisimple normal subgroup E which is irreducible on V, and*

(2) *G has no regular orbit on the vectors of V.*

Then one of the following holds:

(a) *$E = A_c$, the alternating group of degree c and $c < p$ and V is the natural irreducible module for E of dimension $c-1$.*

(b) *E, n and p are as in the following table*

E	n	p	E	n	p
A_5	3	11	$U_4(2)$	5	7, 13, 19, 31, 37, 43
	4	7, 11, 13		6	$p \leq 43$
$2.A_5$	2	11, 19, 29, 31, 41, 61		10	7
	4	7, 11	$2.U_4(2)$	4	7, 13, 19, 31, 37, 43
A_6	5	7, 11, 13, 17, 19		8	7
$2.A_6$	4	7, 11, 13	$6_1.U_4(3)$	6	$p \leq 151$
$3.A_6$	3	19	$U_5(2)$	10	7, 13
	6	7		11	7
$2.A_7$	4	11	$\mathrm{Sp}_6(2)$	7	$p \leq 79$
$L_2(7)$	3	11, 23	$2.\mathrm{Sp}_6(2)$	8	11
	6	5	$2.\Omega_8^+(2)$	8	$p \leq 271$
$6.L_3(4)$	6	11, 13, 17	$2.J_2$	6	11, 19
$U_3(3)$	6	5, 11	6.Suz	12	19
	7	5			

Theorem 2.2. *Under the assumptions of Theorem* 2.1 *the only exceptions in Case* (b) *are the following:*

G	n	p	minimal stabilizer	real
$A_5 \times Z$	3	11	C_2	
$A_5.2 \times Z$	4	7	C_2	
$2.A_5 \star Z$	2	29, 41, 61	C_2, C_2, C_2	
$2.A_5 \star Z$	2	11, 19, 31	C_5, C_3, C_3	no
$Z.(8 \star 2.A_5).2$	4	7	V_4	
$A_6.2 \times Z$	5	7	C_2	
$2.A_6.2 \star Z$	4	7	C_3	no
$3.A_6 \star Z$	3	19, 31	C_2, C_2	
$2.A_7 \star Z$	4	11	C_3	
$L_2(7) \times Z$	3	11	C_2	
$Z.(6 \times L_2(7)).2$	6	5	C_2	
$U_3(3) \times Z$	7	5	C_2	
$U_3(3).2 \times Z$	7	5	C_2	
$(U_3(3) \times Z).2$	6	5	S_3	
$U_4(2) \times Z$	5	7, 13, 19	S_4, V_4, C_2	
$U_4(2).2 \times Z$	6	7, 11, 13	D_{12}, V_4, C_2	
$2.U_4(2) \star Z$	4	7, 13, 19	$U_{72}, U_{18}, (C_3^2, C_9)$	no
$2.U_4(2) \star Z$	4	31, 37	C_3, C_2	
$6_1.U_4(3).2_2 \star Z$	6	13, 19, 31, 37	$W(B_3), S_3 \times C_2, V_4, C_2$	
$U_5(2) \times Z$	10	7	V_4	
$\mathrm{Sp}_6(2) \times Z$	7	11, 13, 17, 19	C_2^3, V_4, C_2, C_2	
$2.\Omega_8^+(2) \star Z$	8	11, 13, 17, 19, 23	$W(B_3), S_4, S_3, V_4, C_2$	
$2.J_2 \star Z$	6	11	S_3	

In this table in the column headed by G always the largest possible group is listed and $Z \cong \mathbb{F}_p^\times$. In the column headed by "minimal stabilizers" the isomorphism type of a stabilizer of a vector having smallest order is displayed. It is always unique except for $2.U_4(2) \star Z$ in dimension 4 for $p = 19$, where there are two such types. U_{72}, U_{18} are certain subgroups of $U_4(2)$ of order 72 and 18 respectively. "no" in the column headed by "real" means that in the cases of that particular line there are no real vectors. In all other cases real vectors exist in the modules listed.

Theorem 2.3. *If G is a finite quasisimple group, V a faithful $\mathbb{F}_pG$-module for some prime p not dividing $|G|$. Then G has a regular orbit on V or G, p and $n = \dim_{\mathbb{F}_p} V$ are as in the following list:*

G	n	p	G	n	p
$3.A_6$	3	19	$6.U_4(3)$	6	13, 19
$U_3(3)$	7	5	$Sp_6(2)$	7	11, 13
$U_4(2)$	5	7	$2.O_8^+(2)$	8	11, 13, 17, 19, 23
$U_4(2)$	6	7	$2.J_2$	6	11
$2.U_4(2)$	4	7, 13			

3. Regular orbits on vector spaces and projective spaces

If G is a finite group and V an FG-module for some finite field F then V is also a G-set and one can study how this decomposes into orbits and, if F is finite, one can count the multiplicities with which a certain transitive G-set shows up as an orbit (up to G-isomorphism). In **[PP 87]** polynomials were defined which count these multiplicities by specializing the indeterminate to the cardinality of the field F, if the characteristic of F does not divide $|G|$. They depend just on the Brauer character of V. In a special case these polynomials have also been considered by A. Turull **[Tu 88]** and similar polynomials show up in the work of Orlik and Solomon (**[OS 81]**, **[OS 82]**, **[OS 83]**) which give the required polynomials in the case of a reflection group with the natural representation. For example, let V_{n-1} be the natural irreducible $\mathbb{F}_q S_n$-module of dimension $n-1$ as above for the symmetric group S_n over some finite field $\mathbb{F}_q$ =GF(q) of characteristic $> n$. Then the number of regular orbits of S_n on V_{n-1} is just

$$\frac{1}{n!}\prod_{i=1}^{n-1}(q-i).$$

This can easily be verified by observing that a vector in the natural module has a trivial stabilizer if and only if all its components with respect to the natural basis are different. Observe that the roots are just the so-called exponents of the Weyl group $W(A_{n-1}) \cong S_n$, which are one less than the degrees of the basic invariants, a phenomenon common to the finite reflection groups and their natural operation (see **[PP 87]**, **[OS 82]**). In general the number of regular orbits of a finite group G on an $\mathbb{F}_q G$ module V as above with Brauer character χ will always be a polynomial f_χ of degree $n = \dim_{\mathbb{F}_q} V$ in q with a factor $(q-1)$. In particular there will always be regular orbits with just finitely many (at most $n-1$) possible exceptional fields $\mathbb{F}_q$. Observe that there are of course infinitely many modules V over various finite fields yielding the same Brauer character χ.

The above mentioned polynomials can be explicitly computed from the Brauer character corresponding to the $\mathbb{F}_q G$- module V if the table of marks of G is known (see **[PP 87]**), in fact sometimes one can get away with less information. We recall that the table of marks of a finite group G is a square matrix $[m_{ij}]$ indexed by the set of conjugacy classes of subgroups of G with

$$m_{ij} = \mid \mathrm{Fix}_{G/H_i} H_j \mid,$$

where the H_i are representatives of the conjugacy classes of subgroups of G and $\mathrm{Fix}_{G/H_i} H_j = \{gH_i \mid hgH_i = gH_i \text{ for all } h \in H_j\}$ is the set of fixed points of H_j on the transitive G-set G/H_i. If one orders the conjugacy classes of subgroups of G in such a way that $|H_i| \leq |H_j|$ for $i \leq j$ then the table of marks m_{ij} is a lower triangular matrix (with $[N_G(H_i) : H_i]$ on the diagonal). If M is any finite G-set then

$$\mid \mathrm{Fix}_M(H_j) \mid = \sum_{i=1}^{r} n_i m_{ij} \quad \text{for} \quad 1 \leq \mathrm{j} \leq \mathrm{r}$$

provided that

$$M \cong \bigcup n_i \cdot G/H_i$$

as a G-set, i.e. that G has n_i orbits on M with stabilizer conjugate to H_i.

The following lemma shows that the computation of stabilizers of vectors of an absolutely simple $\mathbb{F}_q G$-module can be reduced to the computation of certain stabilizers of $G/Z(G)$. This is essential since the computation of the table of marks of some of the covering groups appearing in Theorem 2.1 would be rather hopeless.

Lemma 3.1. *Let G be a finite group and V be a faithful absolutely irreducible $\mathbb{F}_q G$-module and let $Z = Z(G)$ have order r. Then r divides $q-1$ and the elements of Z operate as scalars ($z \cdot v = \lambda_z v$ with $z \mapsto \lambda_z$ defining an injection $Z \to \mathbb{F}_q^\times$) . Let $\bar{V}_Z = \{Z \cdot v \mid 0 \neq v \in V\}$. (If $r = q-1$ then $\bar{V}_Z$ is just the projective space $P(V)$ of V.) Then G and $\bar{G} = G/Z$ operate on $\bar{V}_Z$ in a natural way. For $v \in V$ and $g \in C_G(Z \cdot v)$ we have $g \cdot v = z_v(g)v$ for some uniquely determined $z_v(g) \in Z$. Then the map*

$$\psi : C_G(Z \cdot v) \to C_G(v), \qquad g \mapsto z_v(g)^{-1} g$$

is a homomorphism with kernel Z. Thus

$$C_G(v) \cong C_{\bar{G}}(Z \cdot v).$$

The proof is straight forward.

Lemma 3.2. *Keeping the notation and assumptions of Lemma* 3.1 *assume in addition that q is coprime to $|G|$. Let χ be the Brauer character of V and $U \leq G$. Then the number of fixed points of U and also of $\bar{U} = Z(G)U/U \leq \bar{G}$ on $\bar{V}_Z$ is*

$$|\operatorname{Fix}_{\bar{V}_Z}(U)| = \sum_{\lambda \in \operatorname{Lin}_r(U)} \frac{q^{(\chi_U, \lambda)_U} - 1}{r},$$

where $\operatorname{Lin}_r(U)$ *is the set of linear characters of U of order dividing r. Also*

$$|\operatorname{Fix}_{P(V)}(U)| = \sum_{\lambda \in \operatorname{Lin}_s(U)} \frac{q^{(\chi_U, \lambda)_U} - 1}{q - 1}$$

with $s = \text{g.c.d}(q - 1, \exp(G))$, *where* $\exp(G)$ *denotes the exponent of G.*

Proof. If $\lambda \in \operatorname{Lin}_r(U)$ and $V_\lambda = \{v \in V \mid u \cdot v = \lambda(u)v \text{ for all } u \in U\}$ is the λ-homogeneous component of V_U then for any $0 \neq v \in V_\lambda$ and any $u \in U$ one has $u \cdot v = \lambda(u)v \in Z \cdot v$. Thus

$$\operatorname{Fix}_{\bar{V}_Z}(U) \supseteq \dot{\bigcup}_{\lambda \in \operatorname{Lin}_r(U)} \{Z \cdot v \mid v \in V_\lambda \setminus \{0\}\}.$$

The converse inclusion is also easily seen. Since $\dim_{\mathbb{F}_q} V_\lambda = (\chi_U, \lambda)_U$ and since all Z-orbits have length r the result follows. The last assertion follows in exactly the same way. □

We will apply the above lemmas in the following situation: Let E be a quasi-simple group, V a faithful absolutely irreducible $\mathbb{F}_q E$-module (of dimension n) for some q coprime to $|E|$. Then embedding E into $\operatorname{GL}_n(\mathbb{F}_q)$ we put $N_{\operatorname{GL}_n(\mathbb{F}_q)}(E) = G$; this is the largest group we have to consider and $Z(G) = C_G(E) \cong \mathbb{F}_q^\times$ because V is an absolutely irreducible $\mathbb{F}_q E$-module. Thus $G/Z(G)$ is isomorphic to a subgroup of $\operatorname{Aut}(E)$ acting trivially on $Z(E)$. We can consider V as an $\mathbb{F}_q G$-module. It is clear that if G has a regular orbit on V then also every subgroup of G. By Lemma 3.1 the stabilizers of vectors $v \in V \setminus \{0\}$ in G are isomorphic to the stabilizers in $G/Z(G)$ of the points $\langle v \rangle$ of the projective space $P(V)$. Observe that $G/Z(G)$ is isomorphic to a subgroup of $\operatorname{Aut}(E)$ — in fact, in all cases we have to consider $G/Z(G) \cong \bar{E} = E/Z(E)$ or $\bar{E}.2$ — and that V defines a projective representation for this subgroup, which can be linearized using a representation group H of $\bar{E}$ or $\bar{E}.2$, i.e. we can work independently of q. We have $\bar{H} \cong G/Z(G)$ and Lemma 3.2 gives the mark of $P(V)$ as an $\bar{H}$-set. Writing this mark as a linear combination of the marks of the transitive $\bar{H}$-sets it follows, that the number of orbits of $\bar{H}$ on $P(V)$ with stabilizer conjugate to a specific subgroup $U \leq \bar{H}$ is $f^{(s)}_{\chi,U}(q)$ for a rational polynomial $f^{(s)}_{\chi,U}$, which depends on $s = \text{g.c.d}(q-1, \exp(H))$.

We will use the following

Notation: If χ is a character of H, $U \leq \bar{H}$ and $s = \text{g.c.d}(q-1, \exp(H)) = q_1 \cdot \dots \cdot q_m$ with prime powers q_i being pairwise relatively prime, we shall write $f^{q_1,\dots,q_m}_{\chi,U}(q) = f^{(s)}_{\chi,U}$. Since s will always be even we will omit q_i if $q_i = 2$ and write $f^{\emptyset}_{\chi,U}(q) = f^{(2)}_{\chi,U}$. Sometimes the polynomials depend only on $s = \text{g.c.d}(q-1, d)$ for a divisor d of the exponent of H, the relevant $q_1, \dots, q_r$ will be clear from the context.

4. Proof of Theorem 2.2

We have to go through the list of groups in Goodwin's Theorem 2.1. We will always assume that $\mathbb{F}_q$ is a field of characteristic p not dividing $|G|$ and that $V = V_\chi$ is a faithful $\mathbb{F}_q G$-module with Brauer character χ of degree n such that $\text{Res}^G_E V$ is irreducible. In addition, we assume (as we may) that $G = N_{\text{GL}(V)}(E)$ embedding $E = E(G)$ (which is supposed to be quasisimple) into $\text{GL}(V)$ using the given representation. In particular G contains the center of $\text{GL}(V)$ isomorphic to $\mathbb{F}_q^\times$.

4.1. A_5. We will treat the case A_5 in detail as an example. The table of marks of A_5 is

	C_1	C_2	C_3	V_4	C_5	S_3	D_{10}	A_4	A_5
A_5/C_1	60								
A_5/C_2	30	2							
A_5/C_3	20	.	2						
A_5/V_4	15	3	.	3					
A_5/C_5	12	.	.	.	2				
A_5/S_3	10	2	1	.	.	1			
A_5/D_{10}	6	2	.	.	1	.	1		
A_5/A_4	5	1	2	1	.	.	.	1	
A_5/A_5	1	1	1	1	1	1	1	1	1

Here C_n (and D_n) denotes a cyclic (dihedral) group of order n and V_4 the Klein four group.

We first consider one of the algebraically conjugate 3-dimensional characters χ of $E = A_5$. By restricting χ to the various subgroups and computing the scalar products with the linear characters of these subgroups which are realized over the field $\mathbb{F}_q$ one obtains according to Lemma 3.2 the following marks of V_χ (giving the numbers of fixed points of the subgroups of E on the projective space $P(V_\chi)$ — provided that χ can be realized over $\mathbb{F}_q$), depending on whether or not $\mathbb{F}_q$ contains the 3rd or 5th roots of unity, i.e. depending on $s = \text{g.c.d}(q-1, 15)$:

s		s	
1	$(1+q+q^2, 2+q, 1, 3, 1, 1, 1, 0, 0)$	3	$(1+q+q^2, 2+q, 3, 3, 1, 1, 1, 0, 0)$
5	$(1+q+q^2, 2+q, 1, 3, 3, 1, 1, 0, 0)$	15	$(1+q+q^2, 2+q, 3, 3, 3, 1, 1, 0, 0)$

Writing these vectors as linear combinations of the rows of the table of marks we get the number of orbits with stabilizer conjugate to the various subgroups:

U :	C_1	C_2	C_3	V_4	C_5	S_3	D_{10}	A_4	A_5
$f^{\emptyset}_{\chi,U}$:	$\frac{1}{60}(q-9)(q-5)$	$\frac{1}{2}(q-5)$	.	1	.	1	1	.	.
$f^{3}_{\chi,U}$:	$\frac{1}{60}(q^2-14q+25)$	$\frac{1}{2}(q-5)$	1	1	.	1	1	.	.
$f^{5}_{\chi,U}$:	$\frac{1}{60}(q-11)(q-3)$	$\frac{1}{2}(q-5)$	.	1	1	1	1	.	.
$f^{3,5}_{\chi,U}$:	$\frac{1}{60}(q-13)(q-1)$	$\frac{1}{2}(q-5)$	1	1	1	1	1	.	.

Here the relevant prime powers are 3 and 5; recall our convention about the polynomials $f^{q_1,\dots,q_m}_{\chi,U}$. So the third line gives for example the number of orbits with various stabilizers for the case $q \equiv 1 \bmod 5$ and $q \not\equiv 1 \bmod 3$. So one concludes that there is always a regular orbit if $q \neq 11$ and for $q = 11$ there are 3 orbits with stabilizers of order 2. Observe that $\mathbb{F}_{13}$ does not have a 5-th root of unity and that q must be coprime to $|E|$.

$E = A_5$ has an irreducible character χ_4 of degree $n = 4$ which can be extended to $H = S_5$ and we are dealing with the natural irreducible module of S_5. We list only the polynomials $f_{\chi_4,\{1\}}$ and $f_{\chi_4,\langle(1,2)\rangle}$ counting regular orbits and orbits with stabilizer conjugate to $\langle(1,2)\rangle$ (of H on $P(V_{\chi_4})$ and thus by Lemma 3.1 of G on V_{χ_4}, if we replace $\langle(1,2)\rangle$ by the corresponding subgroup of G). We get

U :	$\{1\}$	$\langle(1,2)\rangle$
$f^{\emptyset}_{\chi_4,U}$:	$\frac{1}{120}(q-7)(q-3)(q+1)$	$\frac{1}{6}(q-4)(q-1)$
$f^{5}_{\chi_4,U}$:	$\frac{1}{120}(q-1)(3-8q+q^2)$	$\frac{1}{6}(q-3)(q-2)$

Here the polynomials depend only on $s = \text{g.c.d}(q-1, 5)$. So we conclude that there is always a regular orbit unless $q = 7$, in which case there are orbits with stabilizer conjugate to a subgroup of order 2 of G.

We next consider $E = \mathrm{SL}_2(5)$ with one of its characters χ of degree 2. Here also $\bar{E} = E/Z(E) \cong A_5$ is operating on the projective space, so that we can use the table of marks copied above. In order to find the mark of the action we use Lemma 3.2, the preimages of the subgroups of $\bar{E}$ in E and the character table of $\mathrm{SL}_2(5)$. The result is the following:

$U:$	C_1	C_2	C_3	V_4	C_5	S_3	D_{10}	A_4	A_5
$f^{\emptyset}_{\chi,U}:$	$\frac{1}{60}(q+1)$	.	.	.	.	.	.	.	.
$f^{3}_{\chi,U}:$	$\frac{1}{60}(q-19)$	.	1	.	.	.	.	.	.
$f^{4}_{\chi,U}:$	$\frac{1}{60}(q-29)$	1	.	.	.	.	.	.	.
$f^{5}_{\chi,U}:$	$\frac{1}{60}(q-11)$	.	.	.	1	.	.	.	.
$f^{3,4}_{\chi,U}:$	$\frac{1}{60}(q-49)$	1	1	.	.	.	.	.	.
$f^{3,5}_{\chi,U}:$	$\frac{1}{60}(q-31)$	.	1	.	1	.	.	.	.
$f^{4,5}_{\chi,U}:$	$\frac{1}{60}(q-41)$	1	.	.	1	.	.	.	.
$f^{3,4,5}_{\chi,U}:$	$\frac{1}{60}(q-61)$	1	1	.	1	.	.	.	.

Hence we conclude that there is always a regular orbit except for $q = 11, 19, 29, 31, 41, 49, 61$. We also see that there is no vector $v \in V$ with $V_{C_G(v)}$ self-dual if and only if $q = 11, 19, 31$. Recall that $V_{C_G(v)}$ contains the trivial module and that $\dim V = 2$.

For $E = \mathrm{SL}_2(5)$ and $n = 4$ we have two possibilities: We have an absolutely irreducible character χ_4 and over some fields an irreducible but not absolutely irreducible character χ_4' which split into the two 2-dimensional characters over an extension field. χ_4 and χ_4' extend to $H = \mathrm{SL}_2(5).2$ and of course also to the other group isoclinic to H. For χ_4 we find that $f_{\chi_4,\{1\}}(q)$ is always irreducible in $\mathbb{Q}[q]$, hence there are always regular orbits. The other case is different. By Theorem 2.1 we just have to consider the case $q = 7$, since χ_4' is not irreducible over $\mathbb{F}_{11}$. Here $G/Z(G) \not\cong H/Z(H)$, in fact, $G/Z(G) \cong (C_8 \star 2.A_5).C_2$. But $V_{\chi_4'}$ is absolutely irreducible as $\mathbb{F}_qG$-module, so we can again use Lemma 3.2. We find

$U:$	$\{1\}$	C_2	C_2	C_3	V_4	C_6
$f^{3}_{\chi_2',U}$	$(q-7)(-73+8q+q^2)$	$\frac{1}{16}(q-7)$	$\frac{1}{6}(q-7)$	$\frac{1}{16}(q-7)$	1	1

Thus there is no regular orbit for $q = 7$ but stabilizers isomorphic to V_4 and C_6.

4.2. A_6. For $E(G) \cong A_6$ and $n = 5$ we are again dealing with the natural irreducible character χ, which can be extended to $H = S_6$. The only root > 5 of the polynomials $f^{q_1,\ldots,q_m}_{\chi,\{i\}}$ is 7 and $f^{3}_{\chi,\langle(1,2)(3,4)(5,6)\rangle} = \frac{1}{24}(q-5)(q-3)(q-1)$. So we have regular orbits except for $q = 7$ in which case we have orbits with stabilizer of order 2.

For $E = 2.A_6$ we have a pair of characters χ, χ' of degree $n = 4$ which can be extended to $H = 2.A_6.2_1$ (with the notation of the Atlas **[CCN+85]**) or the group isoclinic to it. In all cases we get the same polynomials with 7 being the only positive root coprime to $|E|$. If we choose H as above and χ to be a non-real extension to H

which is realized over $\mathbb{F}_7$ we get as the only values of $f^3_{\chi,U}$ different from 0:

U	$\{1\}$	C_2	C_3	C_3	C_6	$3\times S_3$
$f^3_{\chi,U}$	$\frac{1}{720}(q-7)(q-3)(q+11)$	$\frac{1}{12}(q-7)$	$\frac{1}{12}(q-7)$	$\frac{1}{6}(q-1)$	1	1

Restricting χ to the last three subgroups U_4, U_5, U_6 we find $\chi_{U_4} = 2\cdot 1 + 2\cdot 1a_f$, $\chi_{U_5} = 1 + 1a^s + 1b + 1c_f$, $\chi_{U_6} = 1 + 1a + 2a_f$. Here and in the following we use the following *convention* for characters of subgroups: characters are denoted by their degrees followed by a letter $a, b, \ldots$ to distinguish between different characters of the same degree. Faithful characters carry a subscript f, real characters a superscript s. The trivial character is denoted just by 1. So we see that there are no real vectors for $p = 7$.

For $E(G) \cong 3.A_6$ and $n = 3$ we have four algebraically conjugate characters χ requiring 3rd roots of unity and a root of $X^2 + X - 1$. We get

$U:$	$\{1\}$	$\langle(3,4)(5,6)\rangle$
$f^3_{\chi,U}(q):$	$\frac{1}{360}(q-25)(q-19)$	$\frac{1}{4}(q-15)$
$f^{3,5}_{\chi,U}(q):$	$\frac{1}{360}(q-31)(q-13)$	$\frac{1}{4}(q-17)$

Thus there are regular orbits if and only if $q \neq 19, 31$, and in these cases there are vectors with stabilizer of order 2. ($\mathbb{F}_{13}$ contains no 5th roots of unity and χ is not realized over $\mathbb{F}_{13}$.)

For $E(G) \cong 3.A_6$ and $n = 6$ there is a pair of algebraic conjugate characters of degree 6 which extend to $H \cong 3.A_6.2_3$ using the notation of the Atlas (**[CCN+85]**) and in addition for some q a pair (V, V') of irreducible but not absolutely irreducible $\mathbb{F}_q E(G)$-modules with Brauer character χ_{3a+3b} which splits into two characters of degree 3, but also extending to H. For an extension χ_6 of one of the absolutely irreducible characters of E the only root of $f^{q_1,\ldots,q_m}_{\chi_6,\{1\}}$ is 1, so there are always regular orbits in this case. We have to consider the other case only for $p = 7$ by Theorem 2.1. We embed $E(G)$ into $\mathrm{GL}_6(7)$ using one of the faithful representations and let $G = N_{\mathrm{GL}_6(7)}(E)$. For this group we find $f^3_{\chi_1,\{1\}}(q) = \frac{1}{[G:Z(G)]}(q-5)(q+1)(q+5)(q^2-19)$, so there are always regular orbits.

4.3. A_7. For $E(G) \cong 2.A_7$ and $\chi(1) = n = 4$ we find $1, 7, 9, 11$ as the only positive roots of the polynomials $f^{q_1,\ldots,q_m}_{\chi,\{1\}}$. So there exist regular orbits if and only if $p \neq 11$. Furthermore we list f^5_χ on the subgroups where it does not vanish

U	$\{1\}$	C_3	C_5	$C_7.C_3$
$f^5_{\chi,U}(q)$	$\frac{1}{2520}(q-11)(-7+12q+q^2)$	$\frac{1}{6}(q-5)$	1	1

We compute $\chi_{C_3} = 2\cdot 1 + 1a_f + \overline{1a_f}$, hence we have real vectors.

4.4. $L_2(7)$. For $E = L_2(7)$ and $\chi(1) = n = 3$ we find

U:	$\{1\}$	C_2
$f^{\emptyset}_{\chi,U}(q)$:	$\frac{1}{168}(q-11)(q-9)$	$\frac{1}{4}(q-7)$
$f^{3}_{\chi,U}(q)$:	$\frac{1}{168}(43-20q+q^2)$	$\frac{1}{4}(q-7)$
$f^{7}_{\chi,U}(q)$:	$\frac{1}{168}(q-15)(q-5)$	$\frac{1}{4}(q-9)$
$f^{3,7}_{\chi,U}(q)$:	$\frac{1}{168}(43-20q+q^2)$	$\frac{1}{4}(q-9)$

So there are regular orbits if and only if $p \neq 11$, in which case there is an orbit with stabilizer of order 2. Observe that χ is not realized over $\mathbb{F}_5$.

For $n = 6$ we have an absolutely irreducible character χ_6 and over some fields an irreducible but not absolutely irreducible character χ_{3a+3b}, both are extendible to $H = L_2(7).2$. The only roots of the polynomials $f^{q_1,\dots,q_m}_{\chi_6,\{1\}}$ are divisors of $|H|$, so there are always regular orbits. By Theorem 2.1 we have to consider the other case only for $p = 5$. For $G = N_{\mathrm{GL}_6(5)}(E)$ we have $G/Z(G) \cong (C_6 \times L_2(7)).C_2$ and we get denoting the extension of χ_{3a+3b} to G by χ : $f^{5}_{\chi,\{1\}}(q) = \frac{1}{2016}(q-5)(-311 - 138q + 10q^2 + 6q^3 + q^4)$, $f^{5}_{\chi,C_2}(q) = \frac{1}{48}(3 - 21q + q^2 + q^3)$. We conclude that there is no regular orbit for G and that the smallest stabilizer of a vector has order 2.

4.5. $L_3(4)$. For $E = 6.L_3(4)$ we have a pair of characters χ with $n = \chi(1) = 6$, which are extendible to $H = 6.L_3(4).2_1$ for some q. The tables of marks of $L_3(4)$ and $L_3(4).2_1$ are contained in the GAP-library. The computation shows that the polynomials for E (counting regular orbits of $E \star Z$) are all irreducible, so there are always regular orbits. For H one finds a faithful permutation representation of degree 1512. Using standard generators of $L_3(4).2_1$ (see **[Wi 96]**, **[WWT$^+$]**) one obtains an epimorphism from H onto $L_3(4).2_1$ and preimages of the representatives U_i of the conjugacy classes of subgroups of the latter group in order to compute the fixed point numbers of U_i on $P(V)$ using Lemma 3.2, where V denotes the $\mathbb{F}_q H$- module affording the Brauer character χ. The character χ requires 3rd roots of unity. One finds that the only root of any of the polynomials $f^{3,q_2,\dots,q_m}_{\chi,\{1\}}$ is 13. But over $\mathbb{F}_{13}$ the character χ is not extendible to H and since $G_1 = \mathbb{F}_{13}^{\times} \star E$ has regular orbits, so does the normalizer $G = N_{\mathrm{GL}_6(13)}(E)$ since G cannot be a nontrivial split extension of G_1.

4.6. $U_3(3)$. For $E = U_3(3)$ and $n = 7$ there is up to algebraic conjugacy one irreducible character χ_{7b} which is not extendible to $E.2$. It requires 4th roots of unity. We find $f^{4}_{\chi_{7b},\{1\}}(q) = \frac{1}{|H|}(q-5)(-63 + 256q - 96q^2 - 32q^3 + 6q^4 + q^5)$ and $f^{4}_{\chi_{7b},C_2}(q) = \frac{1}{48}(-1 + 34q - 16q^2 - 2q^3 + q^4)$. The other polynomials $f^{q_1,\dots,q_m}_{\chi_{7b},\{1\}}$ have no roots $\neq 1$. So there are regular orbits if $q \neq 5$ and in this case a minimal stabilizer has order 2. We have also a rational character χ_{7a} of degree 7 extendible to H with $f^{4}_{\chi_{7a},\{1\}}(q) = \frac{1}{|H|}(q-5)(2457 - 1760q - 159q^2 + 31q^3 + 6q^4 + q^5)$ and

$f^4_{\chi_{7a},C_2}(q) = \frac{1}{24}(q-1)(-38+3q+q^2)$ for a cyclic subgroup $C_2 \leq U_3(3)$. So there are regular orbits if $q \neq 5$ and in this case a minimal stabilizer has order 2.

E has also a character χ_6 of degree 6 extendible to H over suitable fields but not over $\mathbb{F}_5$, and χ_6 is not realized over $\mathbb{F}_{11}$. For $\chi_6 \in \mathrm{Irr}(E)$ we find that 5 is the only positive root of the polynomials counting regular orbits of $Z \times E$ so there are no regular orbits for $q = 5$. It turns out that the smallest stabilizer of a vector in $Z \times E$ is a C_4 and in $G = N_{\mathrm{GL}_6(5)}(E) \cong (Z \times E).C_2$ the smallest one is isomorphic to S_3.

4.7. $U_4(2)$. For $E = U_4(2)$ and $n = 5$ we get

$$\begin{aligned}
f^{\emptyset}_{\chi,\{1\}}(q) &= \frac{1}{[H:Z(H)]}(q-15)(q-13)(q-9)(q-7)\\
f^{3}_{\chi,\{1\}}(q) &= \frac{1}{[H:Z(H)]}(q-19)(q-13)(q-7)(q-5)\\
f^{5}_{\chi,\{1\}}(q) &= f^{\emptyset}_{\chi,\{1\}}(q) - \tfrac{1}{5}\\
f^{3,5}_{\chi,\{1\}}(q) &= f^{3}_{\chi,\{1\}}(q) - \tfrac{1}{5}
\end{aligned}$$

So there are regular orbits if and only if $q \notin \{7, 13, 19\}$. Looking at the values of the polynomials for larger groups (not reproduced here) we find that the minimal stabilizers in these exceptional cases are: S_4 , V_4 and C_2 , respectively.

For $E = U_4(2)$ and the absolutely irreducible character χ of degree $n = 10$ all polynomials $f^{q_1,\ldots,q_m}_{\chi,\{1\}}$ are irreducible, so there are always regular orbits. The character of degree $n = 6$ is extendible to $H = U_4(2).2$ and we get

$$\begin{aligned}
f^{\emptyset} &= \frac{1}{[H:Z(H)]}(q-13)(q-11)(q-7)(q-5)(q+1)\\
f^{3} &= \frac{1}{[H:Z(H)]}(q-13)(q-7)(q-1)(25-14q+q^2)
\end{aligned}$$

So there are regular orbits if and only if $q \notin \{7, 11, 13\}$. We get as minimal stabilizers in these cases D_{12}, $C_2 \times C_2$, C_2.

For $E = 2.U_4(2)$ and $n = 4$ the module requires 3rd roots of unity and we get

$$\begin{aligned}
f^{3} &= \frac{1}{[H:Z(H)]}(q-19)(q-13)(q-7)\\
f^{3,4} &= \frac{1}{[H:Z(H)]}(q-37)(q-13)(q+11)\\
f^{3,5} &= \frac{1}{[H:Z(H)]}(q-31)(223-8q+q^2)\\
f^{3,4,5} &= \frac{1}{[H:Z(H)]}(q-1)(-107-38q+q^2)
\end{aligned}$$

So there are regular orbits if and only if $q \notin \{7, 13, 19, 31, 37\}$. For $q = 7$ the group $H/Z(H)$ has 2 orbits on $P(V)$ with stabilizers U_{72}, U_{648} of order 72 and 648, respectively. Both of these groups have no faithful real representation of degree 3. Observe that in the notation of Lemma 3.1 $\mathrm{Res}^G_{\psi(U_i)}(V)$ contains the trivial module, if V is an $\mathbb{F}_qG$-module with Brauer character χ. For $q = 13$ we find stabilizers $U_1, \ldots, U_4$ of orders 18, 48, 72, 648 (U_1, U_2 with multiplicity 2). U_1, U_3, U_4 have no faithful real representation of degree 3. For U_2 we find that $\chi_{\psi(U_2)} = 1 + 1a^r + 2a_f$.

So there is no real vector also in this case. For $q = 19$ we get stabilizers $U_1, \ldots, U_5$ of orders 9, 9, 24, 72, 648. We compute $\chi_{\psi(U_i)}$ and find $1 + 1 + 1a + 1b$, $1 + 1a + 1b + 1c$, $1 + 1 + 2a$, $1 + 1a + 2a$. In none of these cases we obtain a faithful self-dual summand. For $q = 31$ we have stabilizers $U_1, \ldots, U_6$ of order 3, 5, 9, 24, 72, 648 respectively. We find $\chi_{\psi(U_2)} = 1 + 1a_f + \overline{1a_f} + 1b_f$ and we have a faithful self-dual summand. Here the smallest stabilizer has order 3.

For $H = 2.U_4(2).2$ and $n = 8$ all polynomials $f_{\chi,\{1\}}^{q_1,\ldots,q_m}$ are irreducible, so there are always regular orbits.

4.8. $U_4(3)$. Let $E = 6_1.U_4(3)$ the characters χ with $\chi(1) = n = 6$ which are realized over $\mathbb{F}_q$ can be extended to $H = 6_1.U_4(3).2_2$. The table of marks of $U_4(3)$ is available in GAP and using this it is not difficult to compute the table of marks of $U_4(3).2_2$. This group has 972 conjugacy classes of subgroups. To speed up the computation it is only necessary to compute the partial table of marks corresponding to the (531) conjugacy classes of those subgroups $\bar{U} \leq U_4(3).2_2$ which have a fixed points on $P(V_\chi)$ The module V_χ requires the third roots of unity and we find that the number of regular orbits over any such field is

$$f_{\chi,\{1\}}^{3} = \frac{1}{[H : Z(H)]}(q-37)(q-31)(q-25)(q-19)(q-13).$$

So there are regular orbits if and only if $p \notin \{13, 19, 31, 37\}$. In these cases we find as minimal stabilizers $W(B_3)$, $S_3 \times C_2$, V_4, C_2. All these are groups with only rational characters.

4.9. $U_5(2)$. If $E = U_5(2)$ and $n = 11$ all polynomials $f_{\chi,\{1\}}^{q_1,\ldots,q_m}$ are irreducible, so there are always regular orbits. For $n = 10$ there is a rational character χ_{10} which can be extended to an irrational character of $U_5(2).2$, but this extension is not a Brauer character of a module over $\mathbb{F}_7$. The only positive roots of the polynomials $f_{\chi_{10},\{1\}}^{q_1,\ldots,q_m}$ are 1, 5, 7 and we get

$$\begin{aligned}
f_{\chi_{10},\{1\}}^{3} &= \tfrac{1}{|H|}(q-7)(q+5)g_1(q),\\
f_{\chi_{10},C_2}^{3} &= \tfrac{1}{41472}(q-7)g_2(g)\\
f_{\chi_{10},C_3}^{3} &= \tfrac{1}{51840}(q-19)(q-13)(q-7)(q-5),\\
f_{\chi_{10},C_3'}^{3} &= \tfrac{1}{432}(q-7)(q-1)^2\\
f_{\chi_{10},V_4}^{3} &= \tfrac{1}{1152}(q-1)g_3(q)
\end{aligned}$$

with irreducible polynomials $g_1, g_2, g_3 \in \mathbb{Z}[X]$. So there are regular orbits except for $p = 7$ where the smallest stabilizer of a vector is a Klein 4-group.

For $H = U_5(2).2$ the table of marks is not available in GAP but can be constructed using the table of $U_5(2)$ and those of the other maximal subgroups of H for which the computation is not excessive. We have to consider one of the irrational characters χ

with $\chi(1) = n = 10$. The only roots of the polynomials $f_{\chi,\{1\}}^{q_1,\dots,q_m}$ are 7, 5 and 1. Since we can dismiss $p = 7$ as explained above, we have regular orbits in all cases to be dealt with here.

4.10. Sp$_6$(2). For E =Sp$_6$(2) and $n = 7$ we get

$$\begin{aligned} f^{\emptyset} &= \tfrac{1}{[H:Z(H)]}(q-17)(q-13)(q-11)(q-9)(q-7)(q-5) \\ f^{3} &= \tfrac{1}{[H:Z(H)]}(q-19)(q-13)(q-7)(-325+187q-23q^2+q^3) \\ f^{7} &= f^{\emptyset} - \tfrac{1}{7} \\ f^{3,7} &= f^{3} - \tfrac{1}{7} \end{aligned}$$

We conclude that there are regular orbits if and only if $q \notin \{11, 13, 17, 19\}$. Looking at the values of the polynomials for larger groups (not reproduced here) we find that the minimal stabilizers in these four exceptional cases are: $C_2 \times C_2 \times C_2$, $C_2 \times C_2$, C_2 and C_2, respectively.

For $H = 2.\mathrm{Sp}_6(2).2$ and $n = 8$ we find that the only roots of the polynomials $f_{\chi,\{1\}}^{q_1,\dots,q_m}$ are 5 and 7, which are divisors of the group order. So there are always regular orbits.

4.11. $\Omega_8^+(2)$. For $E = 2.\Omega_8^+(2) \cong W(E_8)'$ we use the table of marks of $\bar{E} = E/Z(E)$ which is available in GAP4 and which has been computed by T. Merkwitz ([**Me 98**]). This group has 11171 conjugacy classes of subgroups. We get

$$f^{\emptyset} = \frac{1}{[E:Z(E)]}(q-23)(q-19)(q-17)(q-13)(q-11)(q-7)(q+91)$$

and

	roots of f
$f^3 = f^{\emptyset} - \frac{1}{38880}(q-7)(q-13)(q-19)$	7, 13, 19
$f^4 = f^{\emptyset} - \frac{1}{11520}(q-13)(q-17)(q-29)$	13, 17
$f^5 = f^{\emptyset} - \frac{1}{150}(q-11)$	11
$f^{3,4} = f^3 - \frac{1}{11520}(q-9)(q-13)(q-37)$	13
$f^{3,5} = f^3 - \frac{1}{150}(q-31)$	
$f^{4,5} = f^4 - \frac{1}{150}(q-41)$	
$f^{q_1,\dots,q_m,7} = f^{q_1,\dots,q_m} - \frac{1}{7}$	

We conclude that G has a regular orbit on V if and only if $q \notin \{11, 13, 17, 19, 23\}$. Further inspection shows that in these cases the smallest stabilizer of a vector is a Weyl-group $W(B_3)$ of type B_3 (a rational group of order 48), a S_4, a S_3, a Klein

4-group V_4 and a C_2, respectively.

In all cases there are vectors $v \in V$ with $V_{C_G(v)}$ self-dual.

4.12. J_2. For $E = 2.J_2$ and $n = 6$ we get

	roots of f
$f^{\emptyset} = \frac{1}{[H:Z(H)]}(q-11)(q+1)(q+11)(q^2-193)$	11
$f^{3} = f^{\emptyset} - \frac{1}{1080}(265 - 14q + q^2)$	
$f^{4} = f^{\emptyset} - \frac{1}{120}(45 - 14q + q^2)$	
$f^{5} = f^{\emptyset} - \frac{7}{300}(q - 11)$	11
$f^{3,4} = f^{3} - \frac{1}{120}(25 - 14q + q^2)$	19
$f^{3,5} = f^{3} - \frac{1}{300}(7q - 97)$	−5
$f^{4,5} = f^{4} - \frac{1}{300}(7q - 107)$	11
$f^{3,4,5} = f^{3,4} - \frac{1}{300}(7q - 127)$	
$f^{q_1,\ldots,q_m,7} = f^{q_1,\ldots,q_m} - \frac{1}{7}$	

We see that there are always regular orbits except for $q = 11$. In this case the smallest stabilizer is isomorphic to S_3.

4.13. Suz. In this case $E = 6.\mathrm{Suz}$ and $n = 12$. The table of marks of Suz was not available. F. Lübeck, J. Müller and M. Neunhöffer were able to find a regular orbit for $p = 19$ (the only case left open by Goodwin) using a new orbit algorithm.

5. Proof of Theorem 2.3

Let $V = V_{n-1}$ be the natural irreducible $\mathbb{F}_q S_n$-module of dimension $n-1$. Since S_n has regular orbits on V (see Section 3) if the characteristic of $\mathbb{F}_q$ is greater than n, so does A_n. But it is also easy (see **[PP 87]**) to calculate in this case the number of regular A_n-orbits in V: it is

$$\frac{2}{n!}\prod_{i=1}^{n-2}(q-i)\cdot\left(q+\frac{(n-1)(n-2)}{2}\right).$$

We have now to consider the exceptional cases in Theorem 2.2 where there had been no regular orbits for the maximal group operating on the corresponding module. For any of these cases we list the polynomial counting the number of regular orbits. Here we used the linear operation in case that $Z(G) = \{1\}$ and the operation of the simple group $G/Z(G)$ on the orbits of $Z(G)$ according to Lemma 3.2 in the other case.

G	χ	$\lvert G\rvert \cdot f_\chi(q)$
A_5	χ_3	$(q-1)(q-5)(q+6)$
$2.A_5$	χ_2	$(q-1)(q+1)$
	χ_{2a+2b}	$(q-1)(q+1)(q^2+1)$
$2.A_6$	χ_4	$(q-1)(q+1)(q^2-19)$
$3.A_6$	χ_3	$(q-1)(q-19)(q-25)$
$2.A_7$	χ_4	$(q-1)(q-7)(q^2+8q-83)$
$L_2(7)$	χ_3	$(q-1)(q^2+q-48)$
$U_3(3)$	χ_6	$(q-1)(q+1)(q^4+q^2-398)$
$U_3(3)$	χ_{7a}	$(q-1)(q^6+q^5+q^4+q^3-398q^2-398q+3024)$
$U_3(3)$	χ_{7b}	$(q-1)(q-2)(q-4)(q-5)q(q+4)(q+8)$
$U_4(2)$	χ_5	$(q-1)(q-7)(q-9)(q^2+17q-300)$
	χ_6	$(q-1)(q-4)(q-5)(q-7)(q-8)(q+25)$
$2.U_4(2)$	χ_4	$(q-1)(q-7)(q-13)(q+21)$
$6.U_4(3)$	χ_6	$(q-1)(q-13)(q-19)(q-25)(q^2+58q-2255)$
$U_5(2)$	χ_{10}	$(q-1)(q-5)(q+1)(q+5)(q^2-37)(q^4+63q^2-3508)$
$Sp_6(2)$	χ_7	$(q-1)(q-5)(q-7)(q-9)(q-11)(q-13)(q+46)$
$2.\Omega_8^+(2)$	χ_8	$(q-1)(q-7)(q-11)(q-13)(q-17)(q-19)(q-23)(q+91)$
$2.J_2$	χ_6	$(q-1)(q-11)(q+1)(q+11)(q^2-193)$

6. The symplectic case

We close by reporting briefly about the results of the first author on the symplectic case referring for the details to his Diploma thesis **[Ko 99]**. To simplify the notation we use the following

Definition 6.1. Assume that k is a finite field of characteristic p, G a finite p'-group and V a faithful kG-module. If for some prime $q \neq p$ all abelian characteristic subgroups of $Q = O_q(G)$ are central in Q and $\mathrm{Res}_Q^G(V)$ is absolutely irreducible then (k, G, V) will be called a *symplectic p-triple.* Since V is a faithful kG-module we may embed G into $\mathrm{GL}(V)$ and we call (k, G, V) a *maximal symplectic p-triple* if G is a maximal p'-subgroup of $N_{\mathrm{GL(V)}}(Q)$.

The aim is to investigate, for which primes p it is true that for any symplectic p-triple (k, G, V) the module V contains real vectors (i.e. vectors v such that $\mathrm{Res}_{C_G(V)}^G V$ contains a faithful selfdual summand). Since the largest possible groups have to be considered we may assume that (k, G, V) is a maximal symplectic p-triple. Then G contains all scalar matrices, so that $Z(G) \cong k^\times \cong C_G(Q)$, because Q is supposed to act absolutely irreducibly. Hence $F(G) = Z(G)Q$ and in fact Q is a central product of $Z(Q)$ with a group Q_1 of extraspecial type (see **[As 86]**). If $q \neq 2$ then Q_1 is extraspecial of exponent q and order q^{1+2n}. If $q = 2$ and $|k| \equiv 1 \bmod 4$ then $Q_1 \cong 4 \star 2^{1+2n}$, whereas if $|k| \equiv 3 \bmod 4$ one has $Q_1 = Q$ is extraspecial. Furthermore $G/Z(G)$ is isomorphic to a subgroup of $\mathrm{Aut}(Q_1)$ leaving all elements of $Z(Q_1)$ fixed, in particular to a subgroup of $\mathrm{Sp}_{2n}(q)$. The representations of Q_1 are well known, the non-linear ones have degree q^n, and its extensions to G (often called the Weil representation) have been investigated by many authors. The description given in **[GH 92]** and **[Gl 95]** is particularly suitable for concrete computations. We have to consider the following cases:

type	q	$G/Z(G) \leq$
	odd	$\mathrm{Sp}_{2n}(q)$
$+$	even	$O^+_{2n}(q)$
$-$	even	$O^-_{2n}(q)$
S	even, $\equiv 1 \bmod 4$	$\mathrm{Sp}_{2n}(q)$

Observe that $\dim_k(V) = q^n$ and that $|k| = r$ must be coprime to $|G|$. Since the $k(GV)$-problem has been solved for groups of odd order (c.f. **[Gl 84]**, **[Kn 84]**) we may assume that $r = p^m$ is odd.

Theorem 6.2. *Let (k, G, V) be a maximal symplectic p-triple. Then V contains real vectors with the following possible exceptions:*

q	n	type	r		q	n	type	r
2	1	$-$	7		2	4	$+$	3, 5, 7
	1	S	5, 13			4	$-$	3, 5
	2	$-$	3, 7			4	S	5
3	1		7, 13		2	5	$+, -$	3
						7	$+, -$	3

In the cases listed for $n = 1, 2$ there are no real vectors.

The proof of Theorem 6.2 proceeds in three steps.

1) If V contains no real vectors, then for any non-zero vector $v \in V$ the stabilizer $C_G(v)$ cannot be an elementary abelian 2-group, so that $C_G(v)$ must contain an element of $\delta = \delta(G)$, the set of elements of G of order 4 or of odd prime order. A simple counting argument shows that in this case

$$2 \mid V \mid \leq \sum_{g \in \delta \setminus Z(G), C_V(g) > 1} \mid C_V(g) \mid - 2.$$

Using estimates for $|C_V(g)| = r^{(\chi(g),1)}$ if χ is the Brauer character of V similarly as in **[Ro 97]** and bounds for the number of elements of G of prime power order obtained from the corresponding numbers of $\mathrm{Sp}_{2n}(q)$ one obtains a contradiction, except for the cases in the following list:

q	n	r	q	n	r	q	n	r
2	1	≤ 23	2	5	≤ 25	3	3	≤ 157
2	2	≤ 83	2	6	$3, 5, 7, 9, 11$	3	3	≤ 25
2	3	≤ 139	2	7	3	5	1	11
2	4	≤ 71	3	1	≤ 211			

2) Next the description of the elements of the automorphism groups of the extraspecial groups and the Weil representation and thus the representation of G afforded by V described in principle in **[GH 92]** and **[Gl 95]** and implemented by the first author is used to find explicitly the set $\delta(G) \setminus Z(G)$ together with the dimensions of the fixed spaces of its elements. This is used in the inequality above and yields far better bounds for r. The cases left over at this stage are listed in the following table

q	n	type	r	q	n	type	r	q	n	type	r
2	1	+	3, 7	2	3	+	3, 7, 11	2	5	+	3
	1	−	3, 7		3	−	3, 7		5	−	3
	1	S	5, 9, 13		3	S	5, 9, 13, 17		7	+	3
	2	+	2, 7		4	+	3, 5, 7		7	−	3
	2	−	2, 7		4	−	3, 5	3	1		7, 13
	2	S	5, 9, 13, 17, 25		4	S	5, 9				

3) For the cases $n \leq 3$ left it is feasible to compute the action of G on V or $P(V)$, which is obviously sufficient, using GAP and the Meat-Axe (**[Ri 98]**). The following table displays the cases in which using a random search a vector $v \in V$ was found having a stabilizer $C_G(v)$ such that $\chi_{C_G(v)}$ is real.

q	n	type	r	stabilizer
2	1	+	3, 7	C_2, C_2
2	1	−	3	S_3
2	1	S	9	U_1
2	2	+	3, 7	U_1, D_{12}
2	2	S	5, 9, 13, 17, 25	$S_5, U_1, D_{12}, S_3, D_8$
2	3	+	3, 7, 11	$U_2, S_3, \leq C_2$
2	3	−	3, 7	U_3, S_4
2	3	S	5, 9, 13, 17	$U_4, C_2 \times S_4, S_4, U_5$

Here $U_1, \ldots, U_5$ are certain groups of order 72, 1072, 1296, 1344, 96 respectively. In the remaining cases it was possible for $n \leq 3$ to determine the complete orbit structure of G on V.

From Theorem 2.2 and Theorem 6.2 and the reduction theorems of Robinson and Thompson (**[RT 96]**, **[Ro 97]**) one obtains immediately

Corollary 6.3. *For all primes $p \notin \{3, 5, 7, 11, 13, 19, 31\}$ the $k(GV)$-problem has a positive answer.*

References

[As 86] M. Aschbacher, Finite Group Theory, Cambridge University Press, 1986.

[CCN+85] J. H. Conway, R. T. Curtis, S. P. Norton, R. A. Parker, R. A. Wilson, Atlas of finite groups, Clarendon Press, Oxford, 1985.

[GAP 00] The GAP Group, GAP — Groups, Algorithms, and Programming, Version 4.2; Aachen, St Andrews, 2000 (`http://www-gap.dcs.st-and.ac.uk/~gap`).

[Gl 95] S. P. Glasby, On the faithful representations, of degree 2^n, of certain extensions of 2-groups by orthogonal and symplectic groups, J. Austral. Math. Soc. Ser. A 58 (1995), 232-247.

[GH 92] S. P. Glasby, R. B. Howlett, Extraspecial towers and Weil representations, J. Algebra 151 (1992), 236–260.

[Gl 84] D. Gluck, On the $k(GV)$-problem, J. Algebra 89 (1984), 46–55.

[GM 99] D. Gluck, K. Magaard, The extraspecial case of the $k(GV)$-problem, preprint.

[Go 00] D. P. M. Goodwin, Regular orbits of linear groups with an application to the $k(GV)$-problem, I, II, J. Algebra 227 (2000), 395–432, 433–473.

[HLS 92] J. I. Hall, M. W. Liebeck, G. M. Seitz, Generators for finite simple groups, with applications to linear groups, Quart. J. Math. 43 (1992), 441–458.

[Kn 84] R. Knörr, On the number of characters in a block of a p-solvable group, Illinois J. Math. 28 (1984), 181–210.

[Ko 99] C. D. Koehler, Über das $k(GV)$-Problem, Diplomarbeit 1999, Lehrstuhl D für Mathematik, RWTH Aachen.

[Li 96] M. W. Liebeck, Regular orbits of linear groups, J. Algebra 184 (1996), 1136–1142.

[Me 98] T. Merkwitz, Markentafeln endlicher Gruppen, Diplomarbeit 1998, Lehrstuhl D für Mathematik, RWTH Aachen.

[OS 81] P. Orlik, L. Solomon, Complexes for reflection groups, in: Algebraic Geometry, Lecture Notes in Math. 862, Springer-Verlag, 1981, 193–207.

[OS 82] P. Orlik, L. Solomon, Arrangements defined by unitary reflection groups, Math. Ann. 261 (1982), 339–357.

[OS 83] P. Orlik, L. Solomon, Coxeter arrangements, Proc. of Symposia in Pure Math. 40 (1983), 269–291.

[PP 87] H. Pahlings, W. Plesken, Group actions on Cartesian powers with applications to representation theory, J. Reine Angew. Math. 380 (1987), 178–195.

[Ri 99] U. Riese, Further reductions on the quasisimple case of the $k(GV)$-conjecture, preprint.

[Ri 98] M. Ringe, The C Meataxe, User's guide and technical reference, Lehrstuhl D für Mathematik, RWTH Aachen.

[Ro 97] G. R. Robinson, Further reductions for the $k(GV)$-problem, J. Algebra 195 (1997), 141–150.

[RT 96] G. R. Robinson, J. G. Thompson, On Brauer's $k(B)$-problem, J. Algebra 184 (1996), 1143-1160.

[Tu 88] A. Turull, Polynomials associated to characters, Proc. Amer. Math. Soc. 103 (1988), 463–467.

[Wi 96] R. A. Wilson, Standard generators for sporadic simple groups, J. Algebra 184 (1996), 505–515.

[WWT$^+$] R. A. Wilson, P. Walsh, J. Tripp, I. Suleiman, S. Rogers, R. A. Parker, S. Norton, S. Linton, J. Bray, An Atlas of group representations, `http://for.mat.bham.ac.uk./atlas/`, 1996.

Lehrstuhl D für Mathematik
RWTH-Aachen
Templergraben 64, 52062 Aachen, Germany
c.koehler@math.rwth-aachen.de
h.pahlings@math.rwth-aachen.de

The computational matrix group project

Charles R. Leedham-Green

Abstract. The matrix recognition project seeks to produce computational tools that will deal effectively with matrix groups defined over finite fields. This paper describes a recursive procedure for constructing a composition series and a presentation for a matrix group defined in terms of a generating set. The procedure assumes solutions to many hard problems, such as the constructive recognition of simple groups, finding short presentations for simple groups, producing a constructive version of Aschbacher's theorem, and producing random elements of groups. Sufficient progress has been made on these problems for the implementation of the procedure described here to be practical. The problem of computing the normaliser of a subspace of the natural module and similar problems are also considered.

1991 Mathematics Subject Classification: primary 20H30, secondary 20P05, 20C40

1. Introduction

I should like in this note to describe from my point of view the current state of the matrix group project, and to outline where I think we should be heading.

By 'the computational matrix group project' I mean the attempt to use a computer to answer questions about the group G generated by a given subset X of $\mathrm{GL}(d, q)$ for some $d > 0$ and prime power q.

This is an ambitious project, calling heavily on the work of many mathematicians, some of whom have probably never heard of the project, so it would be foolish to attempt to acknowledge all their contributions. However, I must first and foremost acknowledge the importance of the group theory packages GAP **[GAP]** and MAGMA **[BCP]** (alphabetical order) without which we could not move.

My most long standing collaboration has been with Eamonn O'Brien. We have collaborated with Derek Holt and Sarah Rees, and I have collaborated with Frank Celler. More recently my student Ruth Schwingel has contributed important algorithms and code. Even more recently I have persuaded Marston Conder to join the band. The project was started by Peter Neumann and Cheryl Praeger; they first showed in **[NP]** that the dream of using Aschbacher's classification of matrix groups in **[As]** and **[KL]** to design algorithms could be achieved. To all these and many others my warmest thanks.

The fact that the primary aim of being able to compute a composition series for G is approaching fruition is largely due to the recent pioneering work that has been done in black box recognition of simple and almost groups; in this context one should particularly note the work of Sergei Bratus, Gene Cooperman, Bill Kantor, Steve Linton, Kay Magaard, Larry Finkelstein and Ákos Seress who have produced black box algorithms for the finite groups of Lie type; see **[BKPS]**, **[CFL]**, **[KM]**. Sergei Bratus and Igor Pak have produced a black box algorithm for the Symmetric and Alternating groups; see also **[BLNPS]**.

A *black box group* is a group whose elements are the classes under some equivalence relation on a set of bit-strings of fixed length. Four oracles are provided. One provides a set of bit-strings representing the elements of a generating set for the group. One takes as input an ordered pair of representative bit-strings and returns a representative for their product. One takes as input a single representative bit-string and returns a representative for its inverse. One takes a single representative bit-string and returns 'true' or 'false' depending on whether or not the bit-string represents the identity.

The rationale behind the use of black box algorithms is their great generality. The reason for using equivalence classes of bit-strings to represent group elements rather than bit-strings is that this makes the class of black box groups closed under quotients by those normal subgroups for which one has a membership test. There is a cost to be paid for this freedom; and as we shall not need this extra freedom we will be able to make easy adjustments to most black box algorithms, gaining in efficiency by having each group element represented by a unique bit-string.

There is of course no need to insist that all bit-strings have the same length; a known bound to the length will suffice.

A black box group has no error checking to ensure that the input does consist of valid bit-strings; as input to an oracle will always be taken as the output from an oracle, if the oracles are reliable this is not needed.

To those insufficiently versed in these matters it should be pointed out that the input consists of bit-strings because everything consists of bit-strings.

The questions about the matrix group $G = \langle X \rangle$ to be answered, where X is a subset of $\mathrm{GL}(d, q)$, might be placed in four categories.

(i) Find a composition series for G.

(ii) Answer questions about G as an abstract group; for example find Sylow subgroups, conjugacy classes *etc.*

(iii) Answer questions about G as a matrix group; for example, find the normaliser in G of a subspace of the natural module.

(iv) Find the intersection of G with another subgroup of $\mathrm{GL}(d, q)$.

Note that finding the normaliser in G of a subspace of the natural module is a special case of (iv). Many natural problems in category (iii) above, such as determining blocks of imprimitivity, can be answered using algorithms developed to determine constructively which Aschbacher categories G lies in, as discussed below.

We shall concentrate on (i) and (iii). This will involve various other special cases of (iv).

This report being written after the conference, we cannot ignore all the exciting ideas that were presented there; however, these new ideas will take time to digest.

It seems inevitable that any algorithm to compute the order of the group G generated by an arbitrary subset of $\mathrm{GL}(d, q)$ will need an algorithm to recognise simple matrix groups explicitly. Although special techniques will be needed (and have to some extent been devised) for dealing with the computationally hardest cases, the default algorithms for recognising simple groups will be black box recognition algorithms.

A crucial concept in the theory of black box groups is that of a straight line program defined as follows:

A *straight line program* on a set X consists of a finite linearly ordered set S, each element of which consists either of a pointer or an ordered pair of pointers. Each pointer in an element of S points either to an element of X or to an earlier element of S.

Given a straight line program S on a subset X of a group G, each element of S corresponds to a uniquely defined element of G as follows. An element of S pointing to an element of X corresponds to that element. An element pointing to an earlier element of S corresponds to the inverse of the element of G corresponding to that earlier element of S. An element consisting of an ordered pair of pointers corresponds to the product in that order of the elements of G corresponding to the images of the two pointers, be they elements of X or earlier elements of S.

The reason for using a straight line program is that we may need to record how an element of $\langle X \rangle$ may be obtained as a word in $X \cup X^{-1}$. Clearly a straight line program gives a method for writing the element corresponding to any element of the straight line program as a word in $X \cup X^{-1}$. Since the length of this word may grow exponentially in terms of the size of S, and for other reasons, it is generally more efficient to use straight line programs rather than words.

As the term 'straight line program' seems to us rather distracting, for the purposes of this essay we will use the expression 'structured word' as a synonym.

Roughly speaking, a black box algorithm to recognise some 'standard' group H (normally H will be an almost simple group in some standard representation) will take as input a generating set X for some group G, and either construct computable inverse isomorphisms between G and H, or assert that no such isomorphisms exist. In practice the conclusion that no such isomorphisms exist is sometimes only made with a certain known degree of confidence. Then further testing will either find that the groups are in fact isomorphic, or prove that they are not, or increase our confidence that they are not. A computable isomorphism from H to G will consist of an algorithm for writing the image of any element of H as a structured word in $X \cup X^{-1}$. A computable isomorphism from G to H will specify the image in H of each element of X. It must also be possible to solve the membership problem. This works in a slightly more general situation, in which the black box group has a given subset Y, and we are asked to solve the recognition problem for the group G_0 generated by Y. This is almost the same problem as before, but another complication arises. Given an element g of G we may be asked to determine whether or not g lies in G_0; if so find this element as a structured word on Y, and hence its image in H, and if not to return 'false'.

For a trivial example, to recognise that $G = \langle X \rangle$ is isomorphic to H where H is the symmetric group on three symbols, we might proceed as follows:

1. Find by random search elements a and b of G where a has order 2 and b has order 3.
2. Check that $b^a = b^2$.
3. Compute the elements $1, b, b^2, a, ab, ab^2$ of G.
4. Check that each element of X is equal to an element on this list (using the appropriate oracle).

It is clear that this solves the problem.

If G is indeed isomorphic to H then this algorithm should rapidly lead to success.

Another possibility is that G does not contain elements of order 2 and 3, or at least contains a sufficiently small proportion of such elements that we give up after trying step 1 for a sufficient length of time, and report that G is probably not isomorphic to H, where we can reach any degree of confidence, provided that we trust our random element generator.

Another possibility is that we find a pair of elements a and b but they fail to satisfy the given relations, so we return 'false' with certainty.

A final possibility is that step 4 fails, so we have a proper overgroup of S_3.

Of course this is a ridiculous example, but it illustrates the point that a black box recognition algorithm, based on random sampling of G, may prove that the groups G and H are isomorphic, and construct the required isomorphisms, or it may prove that no such isomorphism exists, or it may simply give evidence that G is not isomorphic to H. The last case will arise if G is isomorphic to a proper subgroup of H, but the probability that the conclusion that G is not isomorphic to H is wrong can be made to tend to 0 exponentially fast by repeated sampling.

Thus explicit recognition algorithms will in general be one-sided Monte Carlo algorithms. They may return 'true' or 'false' with a proof. They may also return 'false' without proof, but with a known degree of confidence, which can be made as near to 1 as one wishes.

It seems highly likely, after listening to the talks at the conference, particularly Babai's description of an algorithm to name the non-abelian composition factors of a matrix group in polynomial time, that one could devise a reasonable matrix group algorithm avoiding Aschbacher's theorem, but using the full power of the black box method; but I believe that an Aschbacher based algorithm will perform better, partly because we will not have to use black boxes with ambiguously encoded group elements. Also, determining whether or not G lies in a particular Aschbacher category may be a question of independent interest.

2. Constructing a composition series

2.1. The algorithm in outline. In fact we do rather more than construct a composition series, so let us start this discussion by a precise statement of what is to be done. Let us first produce a definition of 'composition series' that is perhaps close enough to the original concept to share the same name, though we will sometimes speak of a 'composition tree'.

A *composition series* for a non-trivial finite group G consists of the group G itself if G is simple. If G is not simple, it consists of the ordered pair consisting of a composition series for N and a composition series for G/N for some non-trivial normal subgroup N of G.

Thus a composition series can be thought of as a labelled rooted binary tree. The nodes correspond to certain sections of G, the root corresponding to G itself. A node that corresponds to a section K of G, and is not a leaf, has a left descendant corresponding to a proper normal subgroup N of K and a right descendant corresponding to K/N. The leaves correspond to the composition factors of G.

Now Aschbacher's classification of matrix groups states, roughly speaking, that every matrix group G is either nearly simple, or has a normal subgroup N that has a natural geometric definition. More precisely the image of G in $\mathrm{PGL}(d, q)$ is either an almost simple group, that is to say a group contained between a non-abelian simple group and its automorphism group; or contains a non-trivial normal subgroup that has a natural geometric definition; or is trivial. What is meant by a 'natural geometric definition' will be explained later. In the mean time, the crucial point is that not only will N have a natural definition as the subgroup of G generated by some explicitly computed subset of G, but G/N will also come with a natural faithful representation, generally a matrix or permutation representation. The exceptions suggested by 'generally' arise in some cases when G/N is a small cyclic group. If this representation for G/N is as a permutation group we leave the further analysis of G/N to the permutation group algorithms; the degree of the permutation representation will be small enough for the construction of a composition series in this case to be a triviality. Thus the situation that we will have to consider in general will be as follows. We will have a subset X of $\mathrm{GL}(d, q)$ generating G, and will prove that G lies in some subgroup M, namely the subgroup of $\mathrm{GL}(d, q)$ that preserves an Aschbacher configuration; so M will have a natural action, not in general a faithful action, on some space or set S. If K is the kernel of the action of M on S then N is defined as $G \cap K$, so G/N acts faithfully on S. If we could find a reasonably small generating set for N we could then proceed recursively; the problem with finding such a set makes for complications in the algorithm.

While the composition tree is being constructed, nodes will be formed containing some or all of the information listed immediately below. When the algorithm terminates, all this information will be correctly computed. During the calculation some of this information is likely to be incorrect, in that a subset of a normal subgroup N of a group G that is expected to generate N, as a subgroup, may in fact generate a proper subgroup of N. Such an error will eventually be detected, and the consequences of this error will be corrected. To simplify exposition, we will in general use language that assumes that the conjectural information being used is correct.

0. A binary string b that is the name of the node. If the node is the root of the tree then b is the empty string. If b is not a leaf the left and right immediate descendants of b have names $(b, 0)$ and $(b, 1)$ respectively. In this case G_b will be supplied with a homomorphism onto $G_{(b,1)}$ with kernel $G_{(b,0)}$.

1. A parent group U_b containing the section G_b of G corresponding to the node b. If b is the empty string, so $G_b = G$, then U_b is either $\mathrm{GL}(d, q)$ or $\mathrm{PGL}(d, q)$.

2. A subset (strictly a sequence) X_b of elements of G_b.

3. If $b = (c, 0)$ or $b = (c, 1)$ for some bit-string c, one of the following:

(a). If $b = (c, 1)$ a bijection between X_b and X_c.

(b). If $b = (c, 0)$ an expression for each element of X_b as a structured word in X_c.

4. One of the following:

(a). If b is a leaf, explicit inverse isomorphisms between G_b and a standard copy of some simple group H. The isomorphism from G_b to H may be given by the image in H of every element of X_b; the H-recognition code will include an algorithm that takes any element $g \in U_b$ and computes its image in H if $g \in G_b$, or asserts that $g \notin G_b$. The inverse isomorphism will be given by an algorithm to write the image of any given element h of H as a structured word in X_b. In practice this will be given as an algorithm to write h as a structured word in some canonical generating set for H whose image in G_b is known.

(b). If b is not a leaf, a subgroup M_b of U_b containing G_b, a normal subgroup K_b of M_b, a membership test in U_b for M_b and K_b, and a natural isomorphism of M_b/K_b onto, or sometimes into, $U_{(b,1)}$. Thus the node $(b, 1)$ is set up at the same time that M_b is defined.

5. A constructive membership test for $G_b \leq U_b$, returning for any element g of U_b an expression for g as structured word on X_b if $g \in G_b$, and 'false' if not.

6. A presentation for G_b on the generating set X_b, the relations being given as structured words.

In outline the algorithm will proceed as follows.

Stage A consists of constructing the composition series with the possibility of error, inserting items 1 to 5 in the tree. This stage is one-sided Monte-Carlo in that it may underestimate the order of G. The need for the Monte Carlo approach arises from the difficulty of deterministically finding reasonably small generating sets for the groups $G_{(b,0)}$.

Stage B inserts step 6 for all nodes, and verifies that the final answer is correct, or finds an error.

Stage A constructs nodes in the 'right depth-first' order. Define the subtree rooted at a vertex b to be the tree consisting of b and all (iterated) descendants of b. The nodes of the subtree rooted at b are constructed in the following order. First construct node b. If this not a leaf, construct recursively the subtree rooted at $(b, 1)$, and then the subtree rooted at $(b, 0)$. In Stage A nodes are visited more than once. When a node is created, items 1 to 4 are put in place.

A node b having been created needs to be completed (if this is not done on creation) by upgrading X_b to a probable generating set, and establishing item 5. The node b is updated whenever a leaf of the subtree rooted at b is created, and is completed when all these leaves have been created. When the subtree rooted at b is completed, item 5 for node b is automatically created as described below.

If b and c are leaves, b is created before c if and only if $c < b$ in the lexicographic order.

When the node b is created we can construct random elements of G_b as follows. If $b = (c, 0)$ then the subtree rooted at $(c, 1)$ has been completed. It follows that we can recursively write any element g of $G_{(c,1)}$ as a structured word w_g in $X_{(c,1)}$. Now let h be a random element of G_c constructed recursively. Then writing the natural image g of h in $G_{(c,1)}$ as w_g, and evaluating this structured word by replacing each element of $X_{(c,1)}$ by the corresponding element of X_c gives an element k of G_c, and hk^{-1} is a random element of G_b. If $b = (c, 1)$ we can take a random element of G_c and apply the natural map. Note that each of these random elements can be described as the image of a structured word on $X = X_\emptyset$ under the composite of a sequence of maps, each being the restriction of a natural map. The natural maps are those that arise in item 4(b), and the restrictions are to normal subgroups arising from zeros in the name of the node.

Stage A proceeds as follows. Consider first the creation of a node b.

Constructing 1. In the cases of main interest, b will be $(c, 0)$ or $(c, 1)$ where G_c has been found to lie in some Aschbacher category. If $b = (c, 1)$ then U_b was constructed in item 4 for the node c. If $b = (c, 0)$ we will take U_b to be either U_c (losing information) or the subgroup K_b of U_c that centralises the geometric structure.

Constructing 2. Case (i) $b = (c, 1)$. Take X_b to be the natural image of X_c.

Case(ii) $b = (c, 0)$. Take X_b to be empty.

Constructing 3(a). This bijection is defined by the homomorphism defined at node c under item 4, and strictly speaking requires X_b and X_c to be sequences rather than sets.

Constructing 3(b). This does not arise at the creation stage.

Constructing 4(a). We may recognise that G_b is almost simple modulo scalars by the fact that it lies in no other Aschbacher category. Reducing from 'almost simple modulo scalars' to 'simple' is easy. We then have to decide which simple group we have, and use a black box or more specialist algorithm to construct the required isomorphisms. Unless b is the unique leaf whose name contains no 0 we will be able to construct random elements of G_b, but we will not have a generating set. This should suffice for dealing with the Aschbacher categories, and deciding which simple group H we have. It should then be easy to find a random pair X_b of elements that generate G_b, and to construct the inverse isomorphisms between G_b and H. This completes the node b. If b is the leaf whose name contains no 0 we already have a well defined choice for X_b.

Constructing 4(b). Now G_b 'preserves some geometric structure'. Such a structure is discovered as discussed below, M_b is the subgroup of U_b that preserves this structure, and K_b is the subgroup that centralises it. The various membership tests all come with the geometric structure.

Now we consider updating a node. Consider a leaf b whose name contains at least one 0. We have seen how this node is completed, with the construction of a generating set X_b. However, this is not the whole story; constructing X_b requires us to append elements to various ansestors of this node. Suppose that $b = (c, \epsilon_1, \epsilon_2, \ldots, \epsilon_k)$ where c contains no 0 and $\epsilon_1 = 0$. Then each element of X_b was defined as the image of a structured word w on X_c, and for every $g \in X_b$ we append an element g_i to $X_{(c,\epsilon_1,\epsilon_2,\ldots,\epsilon_i)}$ for $i = 1, 2, \ldots, k-1$ as follows. If $\epsilon_i = 0$ then g_{i-1} was defined as a structured word w_{i-1} on $X_{(c,\epsilon_1,\epsilon_2,\ldots,\epsilon_{i-1})}$ ($w_0 = w$) which evaluates to an element of $G_{(c,\epsilon_1,\epsilon_2,\ldots,\epsilon_i)} < G_{(c,\epsilon_1,\epsilon_2,\ldots,\epsilon_{i-1})}$. Adjoin $g_i := g_{i-1}$ to $X_{(c,\epsilon_1,\epsilon_2,\ldots,\epsilon_i)}$. If $\epsilon_i = 1$ then define g_i to be the natural image of g_{i-1} in $G_{(c,\epsilon_1,\epsilon_2,\ldots,\epsilon_i)}$, and adjoin g_i to $X_{(c,\epsilon_1,\epsilon_2,\ldots,\epsilon_i)}$. While making these adjunctions it is easy to maintain the relationships defined in item 3. Though we do need to adjoin a copy of the identity in $X_{(c,\epsilon_1,\epsilon_2,\ldots,\epsilon_{i-1},1)}$ as the image of g_{i-1} whenever $\epsilon_i = 0$, and recursively adjoin another copy of the identity as the image of this element until we hit a leaf.

It is now easy to see that when the subtree rooted at any node b has been constructed X_b will generate G_b, as required.

Constructing 5. If b is a leaf this is dealt with by an explicit recognition algorithm for the simple group. Now suppose that b is not a leaf. Given an element g of U_b first check that g lies in the appropriate subgroup M_b; that is to say, check that g preserves the required structure. If this test is passed the image h of g in $U_{(b,1)}$ can be constructed. Now check recursively that h lies in $G_{(b,1)}$. If it does, a structured word w in X_b can be constructed as above such that the element g' of G_b corresponding to w has the same image in $U_{(b,1)}$ as g has. Now determine in a similar way whether or not $h = g^{-1}g'$ lies in $G_{(b,0)}$. If it does a structured word in $X_{(b,0)}$ or other known elements of $G_{(b,0)}$ is found representing h; and now, having structured words that represent g' and $h = g^{-1}g'$ gives a structured word representing g.

It remains to describe Stage B. The nodes of the tree are visited in a different order. In this case before processing a node b that is not a leaf we process the nodes $(b, 0)$ and $(b, 1)$; so we work up the tree from leaves to the root. We assume that we have a presentation for each leaf group H_b on the standard generating set, and hence a presentation for G_b on X_b. The relators will be expressed as structured words.

Constructing 6. Having recursively constructed presentations for $G_{(b,1)}$ on $X_{(b,1)}$ and for $G_{(b,0)}$ on $X_{(b,0)}$ we construct a presentation for G_b on X_b. In the first instance, a presentation is constructed on $X_b \cup X_{(b,0)}$. The relators will include the structured words that express the elements of $X_{(b,0)}$ in terms of X_b. Other relators will express elements of the form $y^x : y \in X_{(b,0)}, x \in X_b$ as structured words in $X_{(b,0)}$, making use of item 5 for the node $(b, 0)$. Other relators will simply be the relators from the presentation of $G_{(b,0)}$ on $X_{(b,0)}$. Finally, for every relator for $G_{(b,1)}$ we will have a relator for G_b by lifting the given relator from $X_{(b,1)}$ to X_b, evaluating this relator in G_b, and expressing the resulting element as a structured word in $X_{(b,0)}$, again using item 5 for the node $(b, 0)$.

In constructing item 6, the applications of item 5 for the node $(b, 0)$ will all return a structured word, rather than 'false', if and only if $X_{(b,0)}$ does generate the whole of $G_{(b,0)}$, so it is at this late stage that our expectation that we have made a valid choice for X_b is confirmed or denied. If it is denied, another choice of $X_{(b,0)}$ must be made, for example by incrementing the old choice by the element that failed the membership test. If this happens the processing of the node at $X_{(b,0)}$ and all its descendants must be started again.

We now discuss how step 4(b) is carried out for the various Aschbacher categories.

2.2. Reducible groups. One of the nine Aschbacher categories is that of reducible groups. Specifically, if U is a proper subspace of V then the normaliser M_U of U in $\mathrm{GL}(V)$ is one of Aschbacher's maximal subgroups of $\mathrm{GL}(V)$.

We should, for reasons that will become apparent, consider the situation in which U_b is a central quotient of $\mathrm{GL}(V)$. To simplify the exposition we shall assume here and in various other cases that U_b is in fact $\mathrm{GL}(V)$.

The extreme situation is when G_b is a unipotent group. In this case it is a triviality to diagonalise G_b and then find a composition series for G_b by elementary linear algebra.

If G_b acts reducibly but is not unipotent there is a proper section W of V on which G_b acts non-trivially. Take M_b to be the subgroup of $\mathrm{GL}(V)$ that normalises this section and K_b to be the subgroup that centralises it. Then $U_{(b,1)}$ will be $\mathrm{GL}(W)$. The simplest choice for $U_{(b,0)}$ will be U_b, the recursion proceeding as $G_{(b,0)}$ will be a proper subgroup of G_b.

The question of deciding constructively whether or not $G = \langle X \rangle$ acts irreducibly on V is one of the oldest techniques in computational matrix group theory. The original algorithm, due to Simon Norton and implemented and used to great effect by Richard Parker, is known as 'the meat axe'. The original version was designed to work very efficiently when the composition factors of V are absolutely irreducible G-modules defined over a small field. Holt and Rees **[HR2]** realised that the basic ideas of the meat axe can be generalised to give an efficient algorithm free from these constraints, with the exception of one rare configuration in which V is an irreducible uniserial module with all composition factors isomorphic. This case has been dealt by John Cannon and the author, and is implemented in MAGMA. A more careful analysis of this exceptional case is due to Ivanyos and Lux **[IK]**. Whether any user has ever needed this extension I do not know.

2.3. Semi-linear and non-absolutely irreducible groups. Another of the Aschbacher categories that can be dealt with efficiently consists approximately of the irreducible but not absolutely irreducible groups. More precisely, let V be a k-dimensional vector space over $GF(q^f)$ where $kf = d$. Define $\Gamma L(k, q^f)$ to be the group of $GF(q)$-linear automorphisms of V that are semi-linear over $GF(q^f)$. Note that this notation requires q^f to be written in this form so that both q and f are parameters. A subgroup of $\mathrm{GL}(d, q)$ that is conjugate to a subgroup of $\Gamma L(k, q^f)$ will be called an f-semi-linear group.

Using meat-axe techniques one can decide as in **[HLOR2]** whether or not a subgroup of $\mathrm{GL}(d, q)$ is f-semi-linear for some f, and in the affirmative case construct the corresponding action of $GF(q^f)$ on V. In this case the geometric structure preserved is this action of $GF(q^f)$; so M_b will be the image of $\Gamma L(q^f, d/f)$ in $\mathrm{GL}(d, q)$ determined by this action, and K_b will be $\mathrm{GL}(d/f, q^f)$. Then $U_{(b,1)}$ will be the Galois group of $GF(q^f)$ over $GF(q)$, a cyclic group of order f, and $U_{(b,0)}$ will be $\mathrm{GL}(q^f, d/f)$.

It should be observed that in going from $\mathrm{GL}(d, q)$ to $\mathrm{GL}(d/f, q^f)$ the dimension of the underlying space is reduced by a linear factor, but the field size is increased exponentially. As the space required to store a field element is logarithmic in the size of the field, a reduction in space by a factor of f is achieved when the $f \times f$ blocks are written as elements of $GF(q^f)$. Difficulties arise at this point if algorithms that are polynomial in q rather than in $\log q$ are used.

2.4. Groups that can be written over smaller fields. Another Aschbacher category that can be dealt with efficiently consists of the groups that can be written, modulo scalars, over a smaller field. That is to say, $q = \ell^f$ for some $f > 1$, and G is conjugate to a subgroup of the product M of $\mathrm{GL}(d, \ell)$ with the group S of scalars over $GF(q)$. Note that even if G is conjugate to a subgroup of $\mathrm{GL}(d, \ell)$ the coefficients of the elements of G will not in general lie in $GF(\ell)$. None the less, an efficient algorithm for dealing with this case is due to S. P. Glasby and R. B. Howlett **[GlHo]**. This algorithm constructs a matrix that conjugates G into M. By abuse of notation assume that G lies in M. There is a natural surjection of M onto $\mathrm{PGL}(d, \ell)$ with kernel S. We take M_b to be M and K_b to be S. Then $U_{(b,1)}$ will be $\mathrm{PGL}(d, \ell)$ and $U_{(b,0)}$ will be S. This is one of the places where we must allow a universal group to be a central quotient of a general linear group rather than a general linear group.

2.5. Imprimitive groups. This is the first of the 'hard' Aschbacher categories we have considered. That is to say, the algorithm described in **[HLOR1]** that either constructs a system of imprimitivity for a matrix group G or proves that no such system exists is in some cases slow. So let us examine the algorithm. We assume of course that G is acting irreducibly; in fact we may assume that G does not lie in any of the above 'easy' categories. Now there is only a small number of possibilities for the dimension of a block, and these are considered in turn. There are four possible cases that can arise (for a given block dimension). It may be that G contains an element (or cyclic subgroup) that does not preserve a block decomposition with blocks of this dimension. It may be that G is imprimitive, with blocks of the given dimension, and the normal subgroup N that normalises each of the blocks contains a non-scalar element. It may be that G is imprimitive, with blocks of this dimension, but acts faithfully, modulo scalars, on the set of blocks. Finally, every element of G may preserve some set of blocks of the given dimension, but no set of blocks of this dimension is preserved by G. The first two cases should be easy in that we expect to find and recognise an element of the given type, and then the algorithm will complete quickly. The other cases are worse, in that a back-track search, that should not be too deep, may be needed. Shorn of various tricks, ultimately the algorithm considers the possible action of various elements of G on the set of putative blocks. If the chosen elements are to act in some assumed way this will require some specific subgroup H of G to fix a block. If H fixes few enough subspaces of the natural module of the dimension of the supposed blocks one can check for each such subspace to see whether or not it is in fact a block. The problem with the algorithm from a theoretical point of view is that we have no bound on the number of elements fixing a block that we need to find before H fixes few enough subspaces of the required dimension.

If U_b is found to be irreducible and imprimitive then define M_b to be the subgroup of $\mathrm{GL}(V)$ that preserves the set of blocks, K_b to be the subgroup of M_b that fixes each block, $U_{(b,1)}$ to be the symmetric group on the set of blocks, and $U_{(b,0)}$ to be U_b. This choice of $U_{(b,0)}$ throws away useful information for the sake of simplicity, but as $G_{(b,0)}$ is a proper subgroup of G_b the recursion will operate.

2.6. Tensor factorisable groups. This is the second difficult Aschbacher category; an algorithm to resolve this case is described in **[LGO]** and **[LGO2]**. Suppose that $U_b = \mathrm{GL}(d, q)$ where $d = uw$ with $u > 1$ and $w > 1$; the question arises of whether or not G preserves a tensor decomposition of V as $U \otimes W$, where $\dim(U) = u$ and $\dim(W) = w$. Again there are four cases that arise; either some element (or cyclic subgroup) of G does not preserve such a tensor decomposition, or such a tensor decomposition exists and some element g of G acts as a scalar on one of the tensor factors and not on the other, or such a tensor decomposition does exist, and G acts projectively faithfully on both factors, or every element of G preserves such a tensor decomposition but G itself does not.

The first case is not always easy to deal with, in that, given a matrix A it is not always easy to determine whether or not A preserves a tensor decomposition with factors of given dimensions. If f is the characteristic polynomial of A, the question arises of whether or not f can be written as a tensor product of two polynomials of given degrees. This is not always an easy question to answer. The second case is easier, in that given such an element g the tensor decomposition can be found using SMASH; see **[HLOR2]**. The last two cases may be hard. For the third case, we need to find a subspace S of V of the form $X \otimes W$ or $U \otimes Y$ for a proper subspace X of U (or Y of W). This is done by constructing a suitable subgroup H of G, and finding S as an H-invariant subspace of V. This requires H to act reducibly on U (or W) and for the lattice of H-subspaces of V to be sufficiently small for it be possible to find a suitable S from among them. To deal with case 4, the same technique is used, but now it is essential to be able to prove that S would act reducibly on at least one of the tensor factors if a tensor factorisation of the required type existed. We construct H as a p-local subgroup of G for some prime p. If p is the characteristic of the representation the reducibility of the action is guaranteed. Of course the characteristic may be prime to the order of G, but in this case the order of G is likely to be not too big, and other simple methods may suffice to prove that H would act reducibly.

If a tensor factorisation $V = U \otimes W$ of the required form is found then G lies in the central product M of $\mathrm{GL}(U)$ with $\mathrm{GL}(W)$. Take M_b to be M and K_b to be the subgroup of M that acts as scalars on U. Then $U_{(b,1)} = M_b/K_b = \mathrm{PGL}(U)$, and $U_{(b,0)} = \mathrm{GL}(W)$. Here again we have a projective linear group appearing as a universal group.

2.7. Tensor induced groups. This is the third difficult Aschbacher category. A theoretical description of the various problems and an implementation of an algorithm to deal with this case by the author and O'Brien is nearing completion. In this case V has a decomposition as a tensor product of k subspaces of equal dimension, and G permutes these k factors transitively. So we take M_b to be the subgroup of $\mathrm{GL}(V)$ that permutes the tensor factors, and K_b to be the subgroup of M_b that fixes each subspace. Then $U_{(b,1)}$ is the symmetric group of degree k and $U_{(b,0)}$ is $\mathrm{GL}(V)$, again throwing away information for the sake of simplicity.

2.8. Symplectic p-groups. This is the fourth difficult Aschbacher category. It may arise when the dimension d of V is a power of some prime p dividing $q - 1$. Then G is a central extension of a p-group P of symplectic type by a subgroup of a symplectic group. No reasonable algorithm has been devised to determine whether or not a subgroup G of $\mathrm{GL}(V)$ is of this type. Alice Niemeyer **[ACN]** has solved the problem in some cases, and SMASH solves the problem if a non-central element of P can be found. It is then easy to construct the subgroup M_b of $\mathrm{GL}(V)$ that normalises the P, and the normal subgroup $K_b = P(Z(\mathrm{GL}(V)) \cap G)$ of M_b. Now M_b/K_b is a symplectic group $\mathrm{Sp}(W)$ or (in the case $p = 2$) $O(W)$ where $W \cong P/Z(P)$ is a $GF(p)$-space. We take $U_{(b,1)}$ to be $\mathrm{GL}(W)$, so in this case $U_{(b,1)}$ will properly contain M_b/K_b. Again we may take $U_{(b,0)}$ to be U_b, the recursion progressing since the image of G_b in $U_{(b,1)}$ is non-trivial.

2.9. Almost simple groups. This is the last Aschbacher category, known as C9, and is certainly difficult. More precisely, G_b modulo scalars will be an almost simple group, that is to say a group containing a simple group S and contained in the automorphism group of S, where S is equated with its group of inner automorphisms. Reduction to the case when G is simple modulo scalars is easy, though not always helpful.

From a theoretical point of view the various techniques developed for the black box recognition of simple groups have been remarkably successful. Some cases have been programmed, but not all; for example, not the algorithms for recognising the exceptional groups of Lie type. Putting together all the known techniques will be a major programming effort, and there are still finitely many simple groups that remain outside the net. It seems unlikely that dealing with these will be harder than the cases already dealt with, but that does not imply that the task will be easy.

2.10. Random group elements. Generating random elements of a group is something of a hot potato, as often happens with statistical techniques.

The basic question is: given a set X that generates a finite group G, find a process that will generate independent uniformly distributed elements of G.

A variant is: given X as above, and another subset Y of G, find a process that will generate independent uniformly distributed elements of the normal closure in G of the subgroup generated by Y.

If in fact G has a permutation representation, and a base and strong generating set for G is constructed, there is a simple 'perfect' solution to this problem. That is to say, given a perfect random number generator it is easy to find a perfect solution to the first problem, and to reduce the second to the first.

In the matrix group context in which we wish to use the algorithm there seems to be no useful way in which the matrix group is any better than a black box group, so this is the context in which we operate.

As a result of a conversation between Leonard Soicher and the author, the product replacement algorithm was born: see **[CLMNO]**. This describes a Markov process for constructing random elements of G. The process requires just one group multiplication. There is then a theoretical observation, namely that if our random process consists of running the Markov process k times then for some large k (depending on presumably unknown features of G) the output will consist of almost independent elements of G that will, under certain hypotheses, be close to the uniform distribution. The practical observation was that, in a wide variety of cases, taking the random process to consist of running the Markov process for a modest number of times, in many cases 100 or so times, and then simply taking successive outputs as our random sequence, although the output was clearly not independent, none the less, gave results that passed a variety of χ^2 tests with flying colours.

This paper received more attention than this author had expected. Diaconis and Saloff-Coste carried out their own tests, agreed that the method was effective, and supplied some extremely sophisticated analysis. At a study group in Perth, a simple variant came to life, in which the Markov process is replaced by one requiring two multiplications rather than one, and in which the limit distribution is precisely the even distribution; so repeating the Markov process k times to construct a random element comes as near as one wishes to having an independent uniformly distributed sequence of elements. Unfortunately, there seems no prospect of finding a theoretical justification for taking k to be some small number such as 1; and the fact that a significant amount of pre-processing is needed to get away from a 'bad' generating system is symptomatic of the difficulties involved.

Weaknesses in the original product replacement algorithm are shown to exist in **[BaPa]**. For a general discussion of the product replacement algorithm see **[Pak]**.

I am grateful to Igor Pak for much helpful advice he has given me on the limitations of product replacement.

A considerable amount of work is being done to obtain a clearer picture. It would be premature to speculate too much on the outcome. But it does seem from **[BLGNF]** that in at least some cases where the product replacement algorithm converges to a seriously non-uniform distribution, the new method converges not only to the uniform distribution, but also converges much faster.

The effect of poor random element selection should not be too serious. As the above algorithm constructs a presentation of the group G that is proved to be correct, poor element selection will only slow down the algorithm, and not cause a wrong answer to be returned. This does assume that the algorithm will complete, and hence that the random element generator will eventually find an element in any non-empty subset of G, and this the new improved product replacement algorithm will surely do.

2.11. Further details. Many details have been suppressed, partly for ease of exposition, and partly because the algorithm has not yet been subjected to the detailed thought and refinement that arises from implementation.

Some of the issues that need to be addressed are as follows.

1. Leaves.

The algorithm has been described as if the leaves are to correspond to simple groups, and this is certainly an option. However Aschbacher's theorem reduces the problem to the following classes.

(a) Groups that are almost simple modulo scalars.

(b) Unipotent groups.

(c) Permutation groups.

(d) Cyclic groups.

It may be more sensible to allow a leaf to be a group of any one of these types. The data structures and algorithms contained in item 4(a) of the definition of a node in the composition tree are easily extended to this case.

2. Direct products.

If the node b corresponds to a group that is found to be an imprimitive subgroup of $\mathrm{GL}(V)$ preserving a direct sum decomposition of V as $\sum_1^k W_i$ then $U_{(b,0)}$ is simply defined as $U_b = \mathrm{GL}(V)$. It is probably better to define $U_{(b,0)}$ to be $\prod_1^k \mathrm{GL}(W_i)$, and then take $U_{(b,0,1)}$ to be $\mathrm{GL}(W_1)$, and $U_{(b,0,0)}$ to be $\prod_2^k \mathrm{GL}(W_i)$, with a similar decomposition of the nodes $(b, 0, 0, \ldots, 0)$ with fewer than k zeros following b. This would give more control over the computation.

3. Generating sets.

The algorithm as described has a single generating sequence X_b at each node b, these generating sequences being perhaps inadequate at intermediate stages of the algorithm. It would be better to have two sequences X_b and R_b of elements of G_b for each node b. The sequence X_b is treated as a generating sequence (though it may be inadequate), and the sequence R_b is a random sequence to be used for test purposes. Thus one would use the sequence X_b to recognise that G_b lies in some specific Aschbacher category; for example preserving some explicit geometric structure, or being some explicit nearly simple group. Then one would check that the elements of R_b all lie in the corresponding Aschbacher group. If one of them does not then it is transferred from R_b to X_b. The elements of $R_\emptyset$ would be constructed from using the random element generator described above. If $b = (c, 1)$ the sequences X_b and R_b would be constructed from X_c and R_c by taking natural images. If $b = (c, 0)$ we have seen how to construct a random element of G_b from a random element of G_c. So we can construct random elements of G_b from R_c. A small number of these, say the first six, could be used for the initial definition of X_b, and the rest could be used to construct R_b. If an element of R_b is transferred to X_b, where $b = (c, 1)$, the corresponding element of R_c must be transferred to R_b. This means of constructing random elements and generators will produce a considerable improvement in efficiency. I have no strong feeling for how big R should be. It could always be incremented during the calculation if necessary, with no overhead for having chosen too small a value initially.

2.12. Conclusion. It is hard to assess at this stage what are the real limitations in matrix group recognition. It seems highly unlikely that computing a composition series for an arbitrary matrix group can be achieved in time that is polynomial in the size of the input.

The most difficult Aschbacher cases other than the C9 case tend to arise with a group that is in fact almost simple modulo scalars. The question of whether or not the group then preserves, for example, a system of imprimitivity does not need to be answered. (Of course a positive answer would be a great help). So if we find that we cannot easily decide whether or not G lies in some other Aschbacher category, we can try to recognise it as an almost simple group modulo scalars, and if successful, avoid the problem.

Thus it seems that recognising C9 groups is the real hurdle. As alternating and symmetric groups are no problem this reduces us to groups of Lie type and a finite number of sporadic groups, some finite number of groups of Lie type having it seems to be treated as sporadic. The main problem here is the contribution of the size of the field to the complexity. The original algorithms were all polynomial in the size of the field, rather than in the logarithm of the size of the field. Success in reducing this to a discrete logarithm problem in the case of SL and Sp as in **[CLG]** gives grounds for hope that other groups of Lie type might also have similar reductions in their complexity. In practice it seems that these discrete log methods, where applicable, are sufficiently powerful to remove the field size from being a problem; but this is a practical observation, and not an asymptotic one.

In an ideal world the project should make full use of Aschbacher's theorem, which analyses subgroups of classical groups, not just general linear groups. Thus we should have versions of the first four hard Aschbacher cases that could make use of G being a subgroup of any classical group.

The possibility of G being a classical group in its natural representation is an Aschbacher category that we have not discussed, as this also comes under the final classification. The Niemeyer–Praeger algorithm **[NiP1]** and **[NiP2]**, the successor of the original Neumann–Praeger algorithm, is one of the most efficient algorithms in the business; but unfortunately it does not construct explicit isomorphisms. More precisely, given an arbitrary element g of the classical group $G = \langle X \rangle$ the algorithm does not give a method for writing g as a structured word in X.

3. Other matrix group problems

If one wishes to solve a question concerning the internal structure of a group G, such as finding a Sylow subgroup, one may hope, as in **[Ka]** to be able to solve the problem by using a composition series for G as described above. This makes no reference to how G was defined in the first place. Thus the same algorithm will work if G is a matrix group or a permutation group. In so far as this is possible, having written code to construct a composition series for a matrix group we should not have to write matrix group code to solve the various problems of interest that concern internal structure. But some questions concern the embedding of G in the general linear group. Of these the most obvious are:

(1.) Compute the intersection of two subgroups of $\mathrm{GL}(V)$.

(2.) Compute the normaliser in a subgroup G of $\mathrm{GL}(V)$ of a subspace U of V.

(3.) Compute the stabiliser in a subgroup G of $\mathrm{GL}(V)$ of a vector in V.

The second is a special case of the first, and the third is almost a special case of the second.

Problem 1 in general seems too hard to contemplate; but there are countless tricks that one can try and special cases where progress can be made. For example, Ruth Schwingel **[RS]** has produced a polynomial time algorithm to compute the intersection in $\mathrm{GL}(V)$ of the normalisers of any set of subspaces of V.

Problem 2 arises for example in computing the automorphism group of a finite p-group. In general the larger the orbit of U the harder the problem. The obvious way to reduce the size of the orbit is to look for some way to reduce G to a smaller group. If there is a subgroup K of modest index in G that contains the required normaliser then computing K may be feasible. The perennial problem then arises that G may have no subgroup of modest index. This is particularly likely to happen if G is a classical group, so we need specialist algorithms to deal with this problem with compact representations of classical groups. For example the central issue in computing the automorphism group of a finite p-group P is the problem of finding the subgroup of $\mathrm{GL}(V)$ that normalises a given subgroup of $\wedge^2 V$ where $V = P/\Phi(P)$. O'Brien and the author have made some progress with this problem without a definitive solution having appeared.

If the subspace U in problem 2 is of small enough dimension, then a solution to problem 3 can be a great help in solving problem 2. For example the stabiliser of a vector in $\wedge^2 V$ under $\mathrm{GL}(V)$ is the subgroup of $\mathrm{GL}(V)$ that preserves a (possibly degenerate) alternating form, so generators for this group can be written down immediately.

Another point to be made is that before attempting to solve problem 2 in any potentially difficult case it seem wise to compute first a number of random images of U under G, hashing these images. When twenty co-incidences have been recorded (that is to say, twenty pairs of random elements of G have mapped U to the same subspace), one can estimate, to a 95% level of confidence, how big the orbit of U under G will be, within a factor of about 2. This enables one to decide whether or not a direct attack on the problem will succeed within the resources available.

The symmetric group S_n has a natural embedding in $\mathrm{GL}(n-1, q)$ for any n, and it is easy enough to find an algorithm to write down generators for the subgroup that fixes a vector. However, to find generators for the subgroup of S_{100} embedded in $\mathrm{GL}(99, 5)$ that fixes an arbitrary given space of dimension 20 is presented as a challenge problem.

References

[As] M. Aschbacher, On the maximal subgroups of the finite classical groups, Invent. Math. 76 (1984), 469–514.

[BaPa] L. Babai and I. Pak, Strong bias of group generators: an obstacle to the "product replacement algorithm", Proc SODA 2000.

[BLGNF] Adrian Baddeley, C. R. Leedham-Green, Alice C. Niemeyer and Martin Firth, Measuring the Performance of Random Element Generators in Large Algebraic Structures, in preparation.

[BCP] W. Bosma, J. Cannon and C. Playoust, The Magma algebra system I: The user language, J. Symbolic Comput. 24 (1997), 235–265.

[BKPS] L. Babai, W. M. Kantor, P. Pálfy and Á. Seress, Black-box recognition of finite simple groups of Lie type by statistics of element orders, in preparation.

[BLNPS] R. Beals, C. R. Leedham-Green, A. C. Niemeyer, C. E. Praeger, and Á. Seress, A mélange of black box algorithms for recognising finite symmetric and alternating groups I, II, in preparation.

[CFL] G. Cooperman, L. Finkelstein and S. Linton, Constructive recognition of a black box group isomorphic to $GL_n(2)$, in: Groups and Computation II (L. Finkelstein and W. Kantor, eds.), DIMACS Ser. Discrete Math. Theoret. Comput. Sci. 28, Amer. Math. Soc., Providence, RI, 1997, 85–100.

[CLMNO] F. Celler, C. R. Leedham-Green, S. H. Murray, A. C. Niemeyer and E. A. O'Brien, Generating random elements of a finite group, Comm. Algebra 23 (1995), 4931–4948.

[CLG] M. Conder and C. R. Leedham-Green, Fast recognition of classical groups over large fields, in: Groups and Computation III (W. M. Kantor and Á. Seress, eds.), Ohio State Univ. Math. Res. Inst. Publ. 8, Walter de Gruyter, Berlin–New York, 2001, 113–121.

[GAP] The GAP Group, GAP — Groups, Algorithms, and Programming, Version 4.2; Aachen, St Andrews, 2000 (`http://www-gap.dcs.st-and.ac.uk/~gap`).

[GlHo] S. P. Glasby and R. B. Howlett, Writing representations over minimal fields, Comm. Algebra 25 (1997), 1703–1711.

[HLOR1] D. F. Holt, C. R. Leedham-Green, E. A. O'Brien and S. Rees, Testing matrix groups for primitivity, J. Algebra 184 (1996), 795–817.

[HLOR2] D. F. Holt, C. R. Leedham-Green, E. A. O'Brien and S. Rees, Computing matrix group decompositions with respect to a normal subgroup, J. Algebra 184 (1996), 818–838.

[HR2] D. F. Holt and S. Rees, Testing modules for irreducibility, J. Austral. Math. Soc. Ser. A 57 (1994), 1–16.

[IK] G. Ivanyos and K. Lux, Treating the exceptional cases of the MeatAxe. Experiment. Math., to appear.

[Ka] W. M. Kantor, Sylow's theorem in polynomial time, J. Comput. System Sci. 30 (1985), 359–394.

[KL] P. B. Kleidman and M. W. Liebeck, The subgroup structure of the finite classical groups, London Math. Soc. Lecture Note Ser. 129, Cambridge Univ. Press, 1990.

[KM] W. M. Kantor and K. Magaard, Black box exceptional groups of Lie type, in preparation.

[KS] W. M. Kantor and Á. Seress, Black box classical groups, Mem. Amer. Math. Soc. 149, Nr. 708, Amer. Math. Soc., Providence, RI, 2001.

[LGO] C. R. Leedham-Green and E. O'Brien, Tensor products are projective geometries, J. Algebra 189 (1997), 514–528, 1997.

[LGO2] C. R. Leedham-Green and E. A. O'Brien, Recognising tensor products of matrix groups, Intern. J. Algebra Comp. 7 (1997), 541–559.

[NP] P. M. Neumann and C. E. Praeger, A recognition algorithm for special linear groups, Proc. London Math. Soc. 65 (1992), 555–603.

[ACN] A. C. Niemeyer, A constructive recognition algorithm for normalisers of small extra-special groups as matrix groups, submitted.

[NiP1] A. C. Niemeyer and C. E. Praeger, Implementing a recognition algorithm for classical groups, in: Groups and Computation II (L. Finkelstein and W. Kantor, eds.), DIMACS Ser. Discrete Math. Theoret. Comput. Sci. 28, Amer. Math. Soc., Providence, RI, 1997, 273–296.

[NiP2] A. C. Niemeyer and C. E. Praeger, A recognition algorithm for classical groups over finite fields, Proc. London Math. Soc. 77 (1998), 117–169.

[Pak] I. Pak, What do we know about the product replacement algorithm?, in: Groups and Computation III (W. M. Kantor and Á. Seress, eds.), Ohio State Univ. Math. Res. Inst. Publ. 8, Walter de Gruyter, Berlin–New York, 2001, 301–347.

[RS] Ruth Schwingel, Ph.D. thesis, Queen Mary and Westfield College, University of London, 2000.

School of Mathematical Sciences
Queen Mary College
University of London
Mile End Road, London E1 4NS, U.K.
crlg@maths.qmw.ac.uk

Finding p'-elements in finite groups of Lie type

Frank Lübeck

Abstract. We give estimates for the proportion of elements of order divisible by a given number m in finite groups of Lie type which are defined over finite fields with characteristic prime to m.

1991 Mathematics Subject Classification: primary 20G40; secondary 20D60.

1. Introduction

Let G be a connected reductive algebraic group over an algebraic closure of a finite prime field $\mathbb{F}_p$ with p elements and let F be a Frobenius endomorphism of G. Then some power of F, say F^a, induces on the character group of an F-stable maximal torus of G the map $k \cdot id$, where k is some power of p. We define $q > 0$ by $q^a = k$ and denote $G(q)$ the group of F-fixed points of G. This is a finite group of Lie type. (This definition includes the Suzuki and Ree groups.) We will write W for the Weyl group of G.

Assume that $G(q)$ contains an element of order m. We want to investigate the proportion $c_{G,m}(q) = |M_{G,m}(q)|/|G(q)|$, with

$$M_{G,m}(q) = \{x \in G(q) \mid m \text{ divides the order } |x| \text{ of } x\},$$

in the case where m is prime to p.

Our main statement is as follows (we denote by Φ the Euler Φ-function): Let $\gcd(m, p) = 1$. For each constant $0 \leq c < \Phi(m)/m$ and each $l \in \mathbb{N}$ there exists $q_0 \in \mathbb{N}$, such that for all $G(q)$ as above with $q > q_0$ and rank at most l we have $c_{G,m}(q) > c/(2^l \cdot |W|)$. We will also give an explicit q_0 (which becomes bigger for smaller differences $(\Phi(m)/m) - c$).

We mention one consequence of this statement. If we fix the type of G (i.e., its root datum) and consider the case of a prime m which is different from p, then there is a constant $\varepsilon > 0$ such that for all prime powers q the value of $c_{G,m}(q)$ is either zero or at least ε (take some $c < 1/2$ above and for ε the minimum of $c/(2^l \cdot |W|)$ and all nonzero $c_{G,m}(q)$ with $q \leq q_0$). On the other hand it is not difficult to see that $c_{G,p}(q)$ tends to zero when q becomes large. See **[GL01]** for a more precise statement.

This result has an interesting interpretation in computational group theory, where one is often looking for certain elements by a random search. For a fixed probability

Groups and Computation III
Ohio State Univ. Math. Res. Inst. Publ. 8

$\alpha < 1$ there is a number $n \in \mathbb{N}$, depending just on α and the root datum of G, such that for any prime divisor $m \neq p$ of $|G(q)|$ any set of n random elements from $G(q)$ contains, with probability at least α, an element whose order is divisible by m. On the other hand, for growing q it becomes more and more difficult to find a p-singular element by a random search.

Acknowledgement. I would like to thank Bill Kantor for asking me to write this note. He originally wanted to know an estimate for $c_{G,m}(q)$ in the case of exceptional groups G and m a product of two prime powers.

2. A lower bound for finding elements of given order

In this section we will keep the notation from the introduction. Let G, p, F, q, W, m with $\gcd(m, p) = 1$ as above. Recall that the rank of G, denoted rank(G), is the dimension of a maximal torus of G.

Theorem 2.1. (a) *Let $l \in \mathbb{N}$ and $c < \Phi(m)/m$. There exists $q_0 \in \mathbb{N}$ with the following property: For all $G(q)$ with rank of G at most l and $q > q_0$ which contain an element of order m, the proportion of regular semisimple elements of order divisible by m is at least $c/(2^l \cdot |W|)$ (and so $c_{G,m}(q) > c/(2^l \cdot |W|)$).*

(b) *In* (a) *we can take q_0 such that for all $q > q_0$ we have*

$$2 \cdot l^2 \cdot 2^{l-1} \cdot ((q+1)/(q-1))^{l-1}/((\Phi(m)/m) - c) + 1 < q.$$

For example one can choose for q_0 any number greater than

$$2l^2 \cdot 6^{l-1}/((\Phi(m)/m) - c) + 1.$$

To illustrate the statement we give an example of an application.

Corollary 2.2. *Assume that $m \in \mathbb{N}$ has prime factorization of form $r_1^{a_1} r_2^{a_2} r_3^{a_3}$, with different primes r_1, r_2, r_3 not equal to p and given $a_i \geq 0$.*

(a) *Let G be of rank l and $q > 2^l \cdot 6^{l-1} \cdot 300/77 + 1$. Further assume that $G(q)$ contains an element of order m. Then the proportion of regular semisimple elements of $G(q)$ which have order divisible by m is at least $1/(100 \cdot 8^l \cdot l!)$.*

(b) *Let G be of rank at most 8 and $q > 63848$. Assume that $G(q)$ contains an element of order m. Then the proportion of regular semisimple elements of $G(q)$ which have order divisible by m is at least $1/(1.8 \cdot 10^{13})$.*

Proof. First we note that $\Phi(m)/m \geq (r_1 - 1)/r_1 \cdot (r_2 - 1)/r_2 \cdot (r_3 - 1)/r_3 \geq 1/2 \cdot 2/3 \cdot 4/5$. For the application of Theorem 2.1 we choose $c = 1/100$ and so we have $(\Phi(m)/m) - c \geq 77/300$.

Putting these numbers into the second formula in 2.1(b) and using the estimate $|W| < 4^l \cdot l!$, we get part (a).

Taking now $l = 8$ a simple calculation shows that all $q > 63848$ fulfill the first inequality in 2.1(b). The largest possible Weyl group of some G with rank at most 8 is the one of type E_8, which has a bit less than $7 \cdot 10^8$ elements. Hence, in this case we see that $c/(2^l \cdot |W|)$ in 2.1(a) is at least $1/(1.8 \cdot 10^{13})$. □

Note that in a statement like 2.2 it is necessary to fix an upper bound for the number of different prime divisors of m since the sequence $a_n = \prod_{i=1}^n (r_i - 1)/r_i$, r_i being the i-th prime, tends to zero with growing n. In 2.3 we show that the proportion of elements with order divisible by m can become arbitrarily small, even for G a torus, when m has many different prime divisors.

Now we collect some propositions needed for the proof of the theorem.

Proposition 2.3. *Let A be a finite Abelian group which contains an element of order m. Then A contains at least $\Phi(m)/m \cdot |A|$ elements whose order is divisible by m.*

Proof. In the case where A is a cyclic group of order r^a, r a prime, it contains $\Phi(r^a) = (r-1)r^{a-1}$ elements of order r^a, and hence of order divisible by r^b for all $b \leq a$.

In the general case let $m = \prod_{i=1}^k r_i^{b_i}$ be the prime decomposition of m. The Abelian group is isomorphic to a direct product of cyclic groups of prime power order. For any $r_i^{b_i}$, $i = 1, \ldots, k$, there must be a direct factor of A which is cyclic of order $r_i^{a_i}$ with $a_i \geq b_i$. The proposition follows from the result for the special case above, applied to these factors, and from $\Phi(m)/m = \prod_{i=1}^k (r_i - 1)/r_i$. □

Proposition 2.4. *Let T be an F-stable torus of G of rank a. Then we can estimate the number of elements of $T(q)$ by*

$$(q-1)^a \leq |T(q)| \leq (q+1)^a.$$

Proof. The order $|T(q)|$ is the specialization at q of the characteristic polynomial of a matrix of finite order (see, e.g., **[Ca85]**, Proposition 3.3.8). Such a polynomial is a product of linear terms $X - \zeta$ with ζ on the unit circle. Since q is real and greater than 1 we have for each such factor $q - 1 \leq |q - \zeta| \leq q + 1$. □

Proposition 2.5. *Let T be a maximal torus of G and $t \in T$.*

(a) *Then the connected component C of the centralizer of t in G is generated by T and the root subgroups U_α with $\alpha \in \Psi(t) = \{\alpha \mid \alpha$ root with respect to T, $\alpha(t) = 1\}$. The subgroup C is again a reductive group and it has root system $\Psi(t)$.*

(b) *Let $Z = Z((G^*)')$ be the center of the commutator subgroup of the dual group of G. If T is F-stable and $t \in T(q)$ then the index of $C(q)$ in the whole centralizer of t in $G(q)$ is at most $|Z|$.*

(c) *Two elements of T which are conjugate in G are conjugate under an element of the Weyl group of G with respect to T.*

Proof. For these results we give references to **[Ca85]**. Part (a) is in Theorem 3.5.3 and 3.5.4. Part (b) follows from **[Ca85]**, Section 4.5, similar to the proof of 4.5.8. And (c) is in 3.7.1. □

Proof (of Theorem 2.1). **(1)** An element of $G(q)$ of order m with $\gcd(m, p) = 1$ is contained in an F-stable maximal torus T of G. From Proposition 2.3 we know that at least $\Phi(m)/m \cdot |T(q)|$ elements of $T(q)$ have order divisible by m.

(2) The semisimple part of the dual group of G has a center containing at most 2^l elements: For this it is enough to find an upper bound for the order of the center for all simply connected groups of rank at most l. These groups are direct products of simple simply connected groups. And a simple group of rank k has at most $k+1$ central elements (in case A_k). So the maximal possible order of such a center is that of a direct product of groups of type A_1 which all have centers of order 2.

(3) Let $t \in T(q)$ be non-regular semisimple, i.e., its connected centralizer is not the maximal torus T. It follows from 2.5(a) that there is a root α with respect to T with $\alpha(t) = 1$. Let Ψ be the smallest F-stable root subsystem containing α and consider the subgroup G_Ψ generated by T and the root subgroups U_β with $\beta \in \Psi$. Then t is an element of the center Z of G_Ψ. The connected component of Z is a torus S of rank smaller than $\operatorname{rank}(G)$. As in (2) we see that the index $(Z(q) : S(q))$ is at most 2^{l-1}.

The number of such subgroups G_Ψ is at most the number of positive roots of G. And using the classification of root systems we can estimate this number in all cases by $2 \cdot \operatorname{rank}(G)^2$.

From the upper bound for torus orders in 2.4, applied to the centers of the $G_\Psi(q)$, we find that $T(q)$ contains at most $2 \cdot l^2 \cdot 2^{l-1} \cdot (q+1)^{\operatorname{rank}(G)-1}$ non-regular elements.

(4) We assume now that q fulfills the first inequality in 2.1(b). We subtract 1 and multiply by $(q-1)^{l-1}$ in that inequality. Using the lower bound for $|T(q)|$ in 2.4 and (3) this shows that for such q the proportion of non-regular elements in $T(q)$ is at most $(\Phi(m)/m) - c$. Together with (1) we see that the proportion of regular elements in $|T(q)|$ with order divisible by m is at least c.

(5) We make the same assumption as in (4). We know from 2.5(b) and (2) that each conjugacy class of a regular semisimple element in $T(q)$ has at least $|G(q)|/(2^l \cdot |T(q)|)$ elements. Furthermore (4) and 2.5(c) say that there are at least $c/|W| \cdot |T(q)|$ such conjugacy classes whose elements have order divisible by m. This finishes the proof of the theorem. □

3. A refinement for classical groups

In our quite simple arguments of the last section all the estimates are very rough. In particular we did not take into account that for a given m there can be several non-conjugate maximal tori containing elements of order divisible by m. In this section we will do this for the cases of simple groups G of classical type.

We will use the same notation as in Section 2. Furthermore from now on we assume that G is a simple group and of classical type. Let k be the number of pairwise different prime divisors of m and define $c_k = \frac{1}{2}\prod_{i=1}^{k}\frac{p_i-1}{p_i}$, where p_i is the i-th prime number.

Theorem 3.1. *Let G, m, k as above, G of rank l. Assume that $G(q)$ contains an element of order m and $q > 2l^2 \cdot 6^{l-1}/c_k+1$. Then the proportion of regular semisimple elements of $G(q)$ of order divisible by m is at least $c(k,l) = c_k/(2(2l)^k(l+1))$. If we assume further that G is simply connected the estimate can be improved to $c(k,l) = c_k/(2(2l)^k)$.*

Proof. **(1)** The smallest possible value of $\Phi(m)/m$ for an m as above is $2c_k$. Taking $c = c_k$ in Theorem 2.1, we get the statement as in the theorem with $c'(k,l) = c_k/(2^l \cdot |W|)$ instead of $c(k,l)$. The restriction on q given in the theorem is taken from the second estimate in 2.1(b).

(2) The term 2^l in $c'(k,l)$ comes from estimating the number of connected components of the center of the dual group of G. Under our current assumption that G is simple, this term can be replaced by $(l+1)$ - the worst case being $G = PGL_{l+1}$. If G is simply connected the dual group has trivial center and hence this term can even be replaced by 1. (See part (2) of the proof of Theorem 2.1.)

(3) The $G(q)$-conjugacy classes of F-stable maximal tori of G are parameterized by the F-conjugacy classes of W. Let $T(q)$ be a maximal torus of $G(q)$ containing an element of order m and let T be parameterized by $w \in W$.

The term $|W|$ in $c'(k,l)$ comes from estimating the number of elements in $T(q)$ which are $G(q)$-conjugate to a fixed regular semisimple element in $T(q)$ (see part (5) of the proof of Theorem 2.1). Using the w parameterizing T we can give the exact number of such elements; it is $|C_{W,F}(w)|$, the order of the F-centralizer of w in W. (See **[Ca85]**, 3.3 and 3.7, for more details.)

(4) Let q be as in the theorem and $w_1, \dots, w_r \in W$ be representatives of the F-conjugacy classes of W parameterizing classes of tori $T(q)$ containing elements of order m. Adding up the contributions from the single classes of tori as given in (3) we get that the proportion of regular semisimple elements in $G(q)$ whose order is divisible by m is at least $c_k/(l+1) \cdot \sum_{i=1}^{r} 1/|C_{W,F}(w_i)|$.

Hence, to prove the theorem, we have to show that the proportion of elements of W parameterizing maximal tori $T(q)$ which contain an element of order m is at least $1/(2 \cdot (2l)^k)$. We will show this in the next step case by case.

(5) (Type A_l). We consider W with respect to a maximally split torus T_0, then F acts trivially on W. Here W is isomorphic to the symmetric group S_n on $n = l+1$ letters. Let $w \in S_n$ be of cycle type $(a_1, \ldots, a_r)$. Then a maximal torus $T(q)$ parameterized by w is isomorphic to a direct product of cyclic groups of order $(q^{a_1}-1)/(q-1), q^{a_2}-1, \ldots, q^{a_r}-1$. When $T(q)$ contains an element of order m then any prime power dividing m is a divisor of one of the orders of the cyclic factors. Since m is a product of at most k prime powers, we need to answer the following question: Given $b_1, \ldots, b_k \in \mathbb{N}$. What is the proportion of elements of S_n with cycles whose lengths contain multiples of $b_1, \ldots, b_k$? (We must be a bit careful if a prime power dividing m divides $q-1$. But then all maximal tori contain elements with order divisible by this prime power, except the case where n is prime and $T(q)$ corresponds to an n-cycle.)

By replacing $b_1, \ldots, b_k$ by multiples and deleting b_j which divide others, we may assume that $l+1-\sum_{i=1}^k b_i < b_j$ for all j. In this case the number of elements in S_n having cycles of length $b_1, \ldots, b_k$ is easily counted by:

$$\frac{n!\,(b_1-1)!}{(n-b_1)!\,b_1!} \cdot \frac{(n-b_1)!\,(b_2-1)!}{(n-b_1-b_2)!\,b_2!} \cdots$$

$$\frac{(n-b_1-\ldots-b_{k-1})!\,(b_k-1)!}{(n-b_1-\ldots-b_k)!\,b_k!} \cdot (n-b_1-\ldots-b_k)! = \frac{n!}{b_1\cdots b_k}$$

Hence the proportion we are looking for can be estimated by $1/(b_1\cdots b_k) > 1/(l+1)^k$.

(Type ^2A_l). Here the argument goes as in type A_l, we only have to replace q by $-q$ and adjust signs.

(Type B_l, C_l). Here we consider W as a wreath product of a cyclic group with 2 elements with a symmetric group on l letters. The maximal torus parameterized by a $w \in W$ is a direct product of cyclic groups of order q^a-1 for a positive cycle of length a, respectively q^a+1 for a negative cycle of length a.

The argument is now similar to the case of type A_{l-1}, we only need to adjust the sign of each cycle correctly, which gives the additional factor $1/2^k$.

(Type D_l). The Weyl group W of this type is a normal subgroup of the Weyl group W' of type B_l of index 2. An element $w \in W'$ is in W, if and only if the number of negative cycles is even. The argument is now similar to case B_l, we only have to assure that we only count elements in the subgroup W of W' in those cases where we count more than one conjugacy class. This gives another factor $1/2$ in the estimate.

(Type ^2D_l). Here the F-conjugacy classes of W correspond to the conjugacy classes of W' outside W. The argument is exactly the same as for type D_l. □

References

[Ca85] R.W. Carter, Finite Groups of Lie Type - Conjugacy Classes and Complex Characters, Wiley-Interscience, Chichester, 1985.

[GL01] R. M. Guralnick and F. Lübeck, On p-singular elements in Chevalley groups in characteristic p, in: Groups and Computation III (W. M. Kantor and Á. Seress, eds.), Ohio State Univ. Math. Res. Inst. Publ. 8, Walter de Gruyter, Berlin–New York, 2001, 169–182.

Lehrstuhl D für Mathematik
RWTH Aachen
Templergraben 64, D-52062 Aachen, Germany
Frank.Luebeck@Math.RWTH-Aachen.de

Balanced presentations of the trivial group on two generators and the Andrews–Curtis conjecture

Alexei D. Miasnikov and Alexei G. Myasnikov

Abstract. The Andrews–Curtis conjecture states that every balanced presentation of the trivial group can be reduced to the standard one by a sequence of the elementary Nielsen transformations and conjugations. In this paper we describe all balanced presentations of the trivial group on two generators and with the total length of relators ≤ 12. We show that all these presentations satisfy the Andrews–Curtis conjecture.

1991 Mathematics Subject Classification: Primary 20E05, 20F05, 68T05; Secondary 57M05, 57M20.
This research was partially supported by the NSF, grant DMS-9973233.

1. Introduction

Let $F = F(X)$ be a free group of rank $n \geq 2$ with a basis $X = \{x_1, \ldots, x_n\}$. Consider the following transformations of an n-tuple $W = (w_1, \ldots, w_n)$ of elements from F:

(AC1) replace w_i by $w_i w_j$, $j \neq i$;

(AC2) replace w_i by w_i^{-1};

(AC3) replace w_i by $f w_i f^{-1}$ for some $f \in F$,

and leave w_k fixed for all $k \neq i$.

The transformations (AC1) and (AC2) are usually called elementary Nielsen transformations. We will refer to the transformations (AC1)–(AC3) as the *elementary AC-transformations.*

Let F^n be the cartesian product of n copies of the group F. Two n-tuples V and W from F^n are called *Andrews–Curtis equivalent* (or *AC-equivalent*) if one of them can be obtained from the other by a finite sequence of elementary AC-transformations. In this event, we write $V \sim W$. The relation $\sim$ is an equivalence relation on the set F^n. The following conjecture, which appears to be of interest in topology as well as group theory, was raised by J. J. Andrews and M. L. Curtis in [AC].

The Andrews–Curtis conjecture. The elements $w_1, \ldots, w_n \in F(X)$ generate $F(X)$ as a normal subgroup if and only if $(w_1, \ldots, w_n) \sim (x_1, \ldots, x_n)$.

One can formulate this conjecture in terms of presentations. A group presentation $\langle x_1, \ldots, x_m \mid r_1, \ldots, r_n \rangle$ is called *balanced* if $m = n$. The balanced presentation $\langle x_1, \ldots, x_n \mid x_1, \ldots, x_n \rangle$ of the trivial group is called *standard.* We shall say that two presentations with the same generators are AC-equivalent if the tuples of relators in these presentations are AC-equivalent. Plainly, the AC-conjecture is equivalent to the following one: *every balanced presentation of the trivial group is AC-equivalent to the standard one with the same set of generators.*

There is a survey **[BM]** by R. G. Burns and O. Macedonska on the AC-conjecture from group theory viewpoint. For relevant topological results we refer to a survey **[HM]** by C. Hog-Angeloni and W. Metzler. The prevailing opinion seems to be that the AC-conjecture is false. Moreover, several potential counterexamples are known. We say that a balanced presentation of the trivial group $\langle x_1, \ldots, x_n \mid w_1, \ldots, w_n \rangle$ is a *potential counterexample* to the AC-conjecture if, firstly, it is not known to be AC-equivalent to the standard one, and, secondly, no one of the elementary AC-transformations decreases the total length $|w_1| + \cdots + |w_n|$ of the relators. Below we list some of the shortest and the most established potential counterexamples:

(1) $\langle x, y \mid x^{-1}y^2x = y^3,\ y^{-1}x^2y = x^3 \rangle$.

(2) $\langle x, y, z \mid y^{-1}xy = x^2,\ z^{-1}yz = y^2,\ x^{-1}zx = z^2 \rangle$.

(3) $\langle x, y \mid x^3 = y^4,\ xyx = yxy \rangle$.

The first two presentations have been known for almost 20 years; for a discussion we refer to the survey **[BM]**. Example (3) is the second presentation in the series

(4) $\langle x, y \mid x^n = y^{n+1},\ xyx = yxy \rangle,\ n \geq 2,$

which is due to S. Akbulut and R. Kirby **[AK]**. This series has been known for 15 years. Notice that for $n = 2$ the series (4) gives the presentation

(5) $\langle x, y \mid x^2 = y^3,\ xyx = yxy \rangle,$

which, until recently, has also been considered as a potential counterexample. In **[MA]** the first author proved that this presentation satisfies the AC-conjecture. For this purpose he designed a genetic algorithm to search for corresponding sequences of the elementary AC-transformations (see **[MA]** for details). This algorithm is also instrumental for the following theorem which is the main result of this paper.

Theorem 1.1. *Every balanced presentation of the trivial group on two generators with the total length of the relators at most* 12 *satisfies the Andrews–Curtis conjecture.*

The idea of the proof of this theorem is very simple: we list all balanced presentations of the trivial group on two generators where the total length of relators is at most 12:

$$\mathcal{P} = \{P_1, P_2, P_3, \ldots\}$$

and then apply the genetic algorithm to each of the listed presentations. However, there are two issues to address here. The first one concerns the listing of the trivial presentations. Indeed, it is known that there is no algorithm to decide whether a group given by a finite presentation is trivial or not **[Adjan]**, **[Rabin]**. Whether such an algorithm exists for balanced presentations is an open and difficult problem **[BMS]**. In the particular case when the total length of relators in the presentations is at most 12, the following result enables one to list the set $\mathcal{P}$.

Theorem 1.2. *Let G be a group defined by a presentation*

(6) $\langle\, x, y \mid r(x, y),\ s(x, y)\,\rangle$, *where* $|r(x, y)| + |s(x, y)| \leq 12$.

If the abelianization of G is trivial then G is either the trivial group or G is isomorphic to the following finite group of order 120

$$\langle x, y \mid yxy = x^2,\ xyx = y^4\,\rangle.$$

The proof of this theorem is based on computer computations with the software package Magnus. We discuss the proof in Section 2.

The second issue is related to the real time required to carry out the computations with the genetic algorithm. It turns out that there are about 10^6 presentations in the set $\mathcal{P}$. To run the genetic algorithm on each of them would take too much time. So one has to exploit tricks and shortcuts to decrease the time. We discuss this in Section 2. However, it is worthwhile to note here that the presentations from $\mathcal{P}$ with total length up to 10 are relatively easy to reduce to the standard one by AC-transformations. So, the minimal total length of relators in non-trivial examples from $\mathcal{P}$ appears to be 11. Surprisingly enough, all the "difficult" presentations from $\mathcal{P}$ with total length 11 (which cannot be easily reduced to the standard one) are readily seen to be AC-equivalent to the presentation **(5)**: $\langle x, y | \, x^2 = y^3,\ xyx = yxy\rangle$. The most difficult examples to "crack" are those with total length 12, which cannot be easily reduced neither to the standard one nor to the presentation **(5)** (of length 11). We list these most interesting presentations in the corollary below.

Corollary 1.3. *The following presentations of the trivial group are AC-equivalent to* $\langle x, y | \, x, y\rangle$*:*

$$\langle x, y \mid x^{-1}y^2x = y^3,\ x^2 = y^{\epsilon}xy^{\delta}\rangle,$$

where $\epsilon, \delta \in \{1, -1\}$.

One can find in **[MA]** the corresponding sequences of AC-transformations that reduce the presentations above to the standard presentation of the trivial group.

It follows now from Theorem 1.1 that the minimal total length of relators in potential counterexamples, that still stand, is 13. This allows us to formulate the following

Corollary 1.4. *The presentation* (3),

$$\langle x, y \mid x^3 = y^4,\ xyx = yxy\rangle,$$

is, at the present time, a minimal potential counterexample to the Andrews–Curtis conjecture.

2. Description of the algorithms

We start with few known algorithms, which play an important part in our proofs.

In 1936 J. H. C. Whitehead **[Whit]** gave an algorithm which for given m-tuples $U = (u_1, \dots, u_m)$ and $V = (v_1, \dots, v_m)$ from F^m decides whether there exists an automorphism $\phi \in \mathrm{Aut}(F)$ such that $\phi(u_i) = v_i$, $i = 1, \dots, m$. Furthermore, if such an automorphism exists, then the algorithm finds one. In general, the time-complexity of the Whitehead algorithm is exponential. However, in the particular case when one needs to check whether a given element $f \in F$ can be mapped by an automorphism of F to the element x_1, the Whitehead method is polynomial in time with respect to the length of f (for a given fixed F). Recall that an element $f \in F$ is called *primitive* if $\phi(f) = x_1$ for some $\phi \in \mathrm{Aut}(F)$. It follows that we can recognize primitive elements in a free group of rank two quite effectively.

The other algorithm that we used in our proofs is the Todd–Coxeter algorithm **[TC]**, **[Jon]**. This is a systematic procedure for enumerating cosets of a given finitely generated subgroup of a given finitely presented group. In particular, if the group given by a finite presentation is finite, then the Todd–Coxeter algorithm will eventually recognize this, and it will give a multiplication table of the group. At the present time, the Todd–Coxeter algorithm (and its variations) are among the most powerful application of computers to group theory (see **[Havas]** for details).

The third algorithm we want to mention here allows one to check whether the abelianization $G/[G,G]$ of the group G given by a finite presentation $\langle x_1, \dots, x_n \mid r_1, \dots, r_m\rangle$ is trivial or not.

This algorithm is relatively fast (at least in our case); it calculates the canonical invariants of the abelian group $G/[G,G]$. One can find a complete description of the algorithm in the book **[Sims]** by C. Sims.

Now we explain how these methods can be used in proving Theorems 1.1 and 1.2. We combine both proofs into a single procedure that was carried out by a computer. This procedure consists of the following steps.

1. Generate a list L_1 of balanced presentations on two generators and with the total length of relators ≤ 12. There are about $6 \cdot 10^6$ of such presentations.

2. For each presentation P in the list L_1, check to see whether the abelianization of the group defined by P is trivial or not. Delete, one by one, all presentations from L_1 with a non-trivial abelianization. Denote by L_2 the resulting list. Plainly,

all the trivial groups from L_1 are also in L_2. There are about 10^6 presentations in L_2.

3. Apply the Whitehead algorithm to each of the relators in every presentation P in L_2 to check whether the relator is a primitive element in F or not. If the relator is primitive then the group defined by P is trivial. Moreover, it is not hard to see that, in this event, the presentation P is AC-equivalent to the standard one **[MA]**. Delete, one by one, all the presentations from L_2 with primitive elements among its relators. All the deleted presentations satisfy the AC-conjecture. Denote by L_3 the resulting list. There are 122240 presentations in L_3.

4. Observe that a cyclic permutation of a relator is a particular case of the transformation $(AC3)$. We can use this to reduce the number of presentations in L_3 which belong to the same AC-equivalence class. Compare presentations in L_3, by cyclically permuting their relations, and leave only one presentation from each equivalence class. Denote the resulting list by L_4. Only 1648 presentations are left in the list L_4.

5. At this point we want to sort out the presentations in L_4 which define the trivial group. We apply the Todd–Coxeter algorithm to each presentation P from L_4 to compute the order of the group defined by P. Notice that the Todd–Coxeter algorithm is time-consuming, so we want to apply it to as few presentations as possible. Luckily, all the groups in the list L_4 happened to be finite, so the Todd–Coxeter algorithm eventually stopped and gave the answer. It turns out that all the groups in L_4 are trivial, except 16 groups of order 120. It follows now that all the groups from L_2 are either trivial or of order 120. To finish the proof of Theorem 1.2 it suffices to notice that all these groups are isomorphic to each other, and hence they are isomorphic to this particular one, given by the presentation

$$\langle x, y \mid yxy = x^2,\ xyx = y^4 \rangle.$$

Indeed, every group of order 120 with trivial abelianization is a central extension of a cyclic group of order 2 by the simple group A_5. All such groups are isomorphic. Denote by L_5 the list of all presentations from L_4 which define the trivial group. There are 1632 presentations in L_5.

6. This is the last and the most time-consuming step. We apply the genetic algorithm from **[MA]** (and some of its variations) to check whether the presentations from L_5 satisfy the AC-conjecture. We show that every presentation from L_5 is AC-equivalent either to the standard presentation of the trivial group, or to the presentation **(5)**, or to one of the presentations in Corollary 1.3, which are already known to satisfy the AC-conjecture (see **[MA]**).

At this step, all the presentations of the trivial group on two generators and with total length at most 12 have been shown to satisfy the AC-conjecture. This proves Theorem 1.1.

All routines and procedures which we used here are available via the Internet at `www.grouptheory.org` as a part of the software package Magnus.

Notice that in order to show that two presentations are AC-equivalent we use a modification of the genetic algorithm from **[MA]** in which the fitness function is replaced by a new one. Namely, in this case we used the sum of the Hamming distances between the relators as cyclic words. In most of the occasions the genetic algorithm with this new fitness function worked very fast. We refer to **[Hol]** and **[Mit]** for a general discussion on genetic algorithms.

References

[Adjan] S. I. Adjan, On algorithmic problems in effectively complete classes of groups, Doklady Akad. Nauk SSSR 123 (1958), 13–16.

[AK] S. Akbulut and R. Kirby, A potential smooth counterexample in dimension 4 to the Poincare conjecture, the Schoenflies conjecture, and the Andrews–Curtis conjecture, Topology 24 (1985), 375–390.

[AC] J. J. Andrews and M. L. Curtis, Free groups and handlebodies, Proc. Amer. Math. Soc. 16 (1965), 192–195.

[BMS] G. Baumslag, A. Myasnikov, and V. Shpilrain, Open problems in combinatorial group theory, in: Groups, Languages and Geometry (R. Gilman, ed.), Contemp. Math. 250, Amer. Math. Soc., Providence, RI, 1999, 1–27.

[BM] R. G. Burns and Olga Macedonska, Balanced presentations of the trivial group, Bull. London Math. Soc. 25 (1993), 513–526.

[Havas] John J. Cannon, Lucien A. Dimino, George Havas and Jane M. Watson, Implementation and analysys of the Todd–Coxeter algorithm, Math. Comput. 27, 463–490.

[HM] C. Hog-Angeloni, W. Metzler, The Andrews–Curtis conjecture and its generalizations, in: Two-dimensional homotopy and combinatorial group theory, London Math. Soc. Lecture Note Ser. 197, Cambridge Univ. Press, Cambridge, 1993, 365–380.

[Hol] J. H. Holland, Adaptation in Natural and Artificial Systems, University of Michigan Press, 1975.

[Jon] D. L. Johnson, Topics in the theory of group presentations, London Math. Soc. Lecture Note Ser. 42, Cambridge Univ. Press, Cambridge, 1980.

[MA] A. D. Miasnikov, Genetic algorithms and the Andrews–Curtis conjecture, Internat. J. Algebra Comput. 9 (6) (1999), 671–686.

[Mit] M. Mitchell, An introduction to genetic algorithms, MIT Press, Cambridge, MA, 1996.

[Ne] B. H. Neumann,On some finite groups with trivial multiplicator, Publ. Math. Debrecen 4 (1956), 190–194.

[Rabin] M. O. Rabin, Recursive unsolvability of group theoretic problems, Ann of Math. 67 (1958), 172–194.

[TC] J. A. Todd, H. S. M. Coxeter, A practical method for enumerating cosets of a finite abstract group, Proc. Edinburg Math. Soc. 5 (1936), 26–34.

[Sims] C. Sims, Computation With Finitely Presented Groups, Cambridge Univ. Press, Cambridge, 1994.

[Whit] J. H. C. Whitehead, On equivalent sets of elements in a free group, Ann. of Math. 37 (1936), 782–800.

Department of Mathematics
The City College of The City University of New York
Convent Avenue at 138 Street, New York, NY, 10031, U.S.A.
alex@rio.sci.ccny.cuny.edu
alexei@rio.sci.ccny.cuny.edu

Deterministic algorithms for management of matrix groups

Takunari Miyazaki *

Abstract. We report some results on deterministic algorithms for efficient management of matrix groups over finite fields. Our results generalize Luks's work on solvable matrix groups to wider classes of matrix groups.

In this article, we present algorithms for membership-testing and its closely related problems. The algorithms run deterministically in time polynomial in these three parameters: the input length of given generators for a matrix group G, the largest prime dividing $|G|$ other than the field characteristic, and the maximum order of a nonabelian composition factor of G.

In a future paper, these algorithms will be key procedures in an efficient divide-and-conquer algorithm that will solve matrix-group problems resembling graph-isomorphism testing in the class Γ_d: finite groups all of whose nonabelian composition factors lie in S_d.

1991 Mathematics Subject Classification: Primary 20C40; Secondary 68Q25, 68Q40.

1. Introduction

In this article, we investigate methods of *deterministic* computation for efficient management of matrix groups over finite fields. Polynomial time is our measure of efficiency.

A critical foundation for group computation is the ability to perform testing membership of an element in a group. In matrix groups over a finite field k, the *membership-test* problem formally means: *Given generators for* $G \leq \mathrm{GL}(n, k)$ *and an element* $x \in \mathrm{GL}(n, k)$, *test whether or not* $x \in G$. In general, a polynomial-time method seems unlikely: even for 1×1 matrices, membership-testing already subsumes a version of the well-known *discrete-log* problem. To separate such an obstacle, Luks introduced an additional timing parameter $\mu(G)$: the largest prime dividing $|G|$ other than $\operatorname{char} k$. For solvable matrix groups $G \leq \mathrm{GL}(n, k)$, Luks then devised deterministic

*This work was done as part of the author's Ph.D. dissertation at the University of Oregon under the direction of Professor Eugene M. Luks. The author was supported in part by National Science Foundation Grant CCR-9820945 as a research assistant.

algorithms to solve membership-testing and many important applications, all running in time polynomial in the input length of given generators for G and $\mu(G)$ **[13]**.

Subsequently, randomization has emerged as an indispensable tool in membership-test algorithms for various classes of matrix groups (cf. **[4]**, **[5]**, and **[6]**). For any $G \leq \mathrm{GL}(n, k)$, the Beals–Babai algorithm of **[6]** performs membership-testing in Las Vegas time that depends polynomially on $\mu(G)$ and another parameter $\nu(G)$: the smallest integer ν such that all the nonabelian composition factors of G have faithful permutation representations of degree at most ν (see also **[4]**). For abelian matrix groups, there are other deterministic methods (cf. **[2]**, **[7]**, and **[10]**).

Of great importance among the applications considered in **[13]** are matrix-group problems that generalize graph-isomorphism testing via the problem of finding set-stabilizers in permutation groups (see also **[14**, §10**]**). These problems include finding the stabilizers of vectors and subspaces and finding centralizers and intersections of subgroups.

For an integer constant $d > 0$, let Γ_d denote the class of finite groups all of whose nonabelian composition factors lie in S_d (evidently, Γ_d includes all solvable groups). An earlier result of Luks **[12]** asserts that, for permutation groups in Γ_d, one can find set-stabilizers in polynomial time (see also the work of Babai, Cameron, and Pálfy **[3]** for important related results). In a future paper **[16]**, we will generalize the methods of **[13]** that solve these graph-isomorphism-inspired matrix-group problems from solvable groups to Γ_d (see also **[17]**). The algorithms in **[16]** run deterministically in time polynomial in these three parameters: the input length of given generators for G, the largest prime dividing $|G|$, and $\mathrm{char}\, k$.

In this article, we focus on the deterministic membership-test machinery we have developed, primarily, to enable basic matrix-group manipulation in the class Γ_d for **[16]**. In fact, with $\mu(G)$ in timing, we are able to achieve polynomial time in a broader class of matrix groups than those in Γ_d. These algorithms are indeed direct generalizations of Luks's membership-test methods for solvable groups. In addition to the basic tools of **[13]**, our methods also incorporate a variant of the *distilling* lemma of Beals and Babai **[6**, Lemma 2.1**]** in our deterministic setting. To handle nonsolvable groups, our approach is guided by the basic structure theory of normal series (cf. **[6]**).

Let $k = \mathrm{GF}(q)$ throughout denote a finite field of characteristic p. For an input/output $G \leq \mathrm{GL}(n, k)$, we assume that G is specified by a generating set of matrices S. We also assume some reasonable encoding of the field k so that *polynomial in the input length* throughout means a polynomial in n, $\log q$, and $|S|$, unless it is specified otherwise (here, log denotes the logarithm to base 2).

The principal result of this article is the following theorem.

Theorem 1.1. *Let k be a finite field. Given a matrix group $G \leq \mathrm{GL}(n, k)$, one can solve the following problems deterministically in time polynomial in these three parameters: the input length, the largest prime dividing $|G|$ other than* $\mathrm{char}\, k$*, and the maximum order of a nonabelian composition factor of G.*

(i) *Find $|G|$.*

(ii) *Given $x \in \mathrm{GL}(n, k)$, test whether or not $x \in G$.*

(iii) *Find a generator-relator presentation of G.*

We list immediate applications of Theorem 1.1 in the following corollary.

Corollary 1.2. *The list of problems in Theorem* 1.1 *continues as follows.*

(iv) *Find the kernel of a given homomorphism $G \to M$, where M is a finite group equipped with a polynomial-time algorithm for testing membership and finding presentations.*

(v) *Given a normal subgroup N of G, find the centralizer $C_G(N)$.*

(vi) *Find a composition series of G.*

(vii) *Find a series of G-normal subgroups*

$$G = G_1 > G_2 > \cdots > G_r = 1$$

such that each G_i/G_{i+1} is a nonabelian chief factor of G or an elementary abelian group.

We establish important notational conventions at the outset. Our general reference is **[22]**.

For a finite group G and a subset $R \subseteq G$, the *normal closure* of R in G, denoted by $\langle R^G \rangle$, is the smallest normal subgroup of G containing R; we call R *normal generators* for $\langle R^G \rangle$ in G. To formalize complexity analysis, we define three parameters on groups as follows.

Definition. For a finite group G, define

(i) $\kappa(G)$ to be the maximum order of a nonabelian composition factor of G,

(ii) $\gamma(G)$ to be the maximum order of a nonabelian chief factor of G, and

(iii) if $G \leq \mathrm{GL}(n, k)$, then $\mu(G)$ to be the largest prime dividing $|G|$ other than char k.

As usual, $M(n, k)$ denotes the algebra of all the $n \times n$ matrices over k. If $X \subseteq M(n, k)$, then $\mathrm{Span}(X)$ denotes the subalgebra of $M(n, k)$ spanned by X. If $G \leq \mathrm{GL}(n, k)$, then $k[G] = \mathrm{Span}(G)$ denotes the *linear span* (or *enveloping algebra*) of G.

Implicit to our complexity analysis is the fact that the length of a strictly increasing series of subgroups in $\mathrm{GL}(n, k)$ is polynomially bounded (evidently, at most $n^2 \log q$ by Lagrange's theorem; for better estimates, see **[19]** and **[21]**).

To establish Theorem 1.1, we claim that it suffices to prove for the case when k is of prime order. Indeed, observe that, with some reasonable encoding of the field k (e.g., an irreducible polynomial f over the prime field GF(p) such that $k \cong \mathrm{GF}(p)[X]/(f)$), it is elementary to convert our setting to matrix groups over GF(p) by blowing up by a factor of deg f. Thus, in the sections to follow, we throughout assume that $k = \mathrm{GF}(p)$.

Our approach to the problem of membership-testing follows a general paradigm of *manageability* introduced in **[13]**. In §2, we formalize such a paradigm. We then summarize basic polynomial-time tools in §3 and important facts on normal structure in §4. In §5, we first establish the manageability of matrix groups $G \leq \mathrm{GL}(n, k)$ such that $\mu(G)$ and $\gamma(G)$ are polynomially bounded. In §6, we prove Theorem 1.1 by generalizing the methods of §5 to matrix groups $G \leq \mathrm{GL}(n, k)$ such that $\mu(G)$ and $\kappa(G)$ are polynomially bounded.

Acknowledgment. The author would like to express his sincere gratitude to his adviser Professor Eugene M. Luks for generous guidance.

2. Membership-testing and manageable groups

In this section, we describe a general paradigm of *manageability* and the overall structure of a generic membership-test procedure we will adapt. This method was originally formalized for solvable matrix groups in **[13**, §§4.1–4.2].

We begin with a brief outline. Given $G \leq \mathrm{GL}(n, k)$, the method constructs, in a top-down fashion, a G-normal series

$$G = N_1 > N_2 > \cdots > N_r = 1$$

specified by generator-relator presentations of G/N_i and G-homomorphisms $\pi_i : N_i \to M_i$, where $N_{i+1} = \mathrm{Ker}\,\pi_i$, and M_i are finite groups equipped with certain basic polynomial-time machinery. With such a series, membership-testing can be performed by a *sifting* process: map a candidate x for membership in N_i by π_i to M_i; find an $a \in N_i$ such that $\pi_i(a) = \pi_i(x)$; then test membership of xa^{-1} in N_{i+1} (cf. **[9]** and **[20]**).

In order to formalize our method, we must first define algorithmic notions of presentations (cf. **[13**, §4.2]).

Let $\mathcal{F}(X)$ denote the free group on a set X. Then, for a group G and a function $\phi : X \to G$, there is a natural extension of ϕ to a homomorphism $\hat{\phi} : \mathcal{F}(X) \to G$.

Let G be a group and N a normal subgroup of G. A *constructive presentation of* G mod N is $\Pi = (X, \phi, \psi, \mathcal{R})$ in which X is a set, $\phi : X \to G$ is a function, $\psi : G \to \mathcal{F}(X)$ is a homomorphism, and $\mathcal{R}$ is a subset of $\mathcal{F}(X)$, satisfying the following properties.

(1) $g\hat{\phi}(\psi(g))^{-1} \in N$ for each $g \in G$.

(2) $\hat{\phi}^{-1}(N) = \langle \mathcal{R}^{\mathcal{F}(X)} \rangle$.

For computational purposes, we assume that $\Pi = (X, \phi, \psi, \mathcal{R})$ is specified by $\phi(X)$, $\mathcal{R}$, and a procedure for determining $\psi(g)$ for any given $g \in G$.

It is immediate from the definition that generators for G and a constructive presentation of G mod N for $N \trianglelefteq G$ offer a subset $R \subseteq N$ such that $\langle R^G \rangle = N$ (that is, R forms normal generators for N) **[13**, Lemma 4.2]. Furthermore, for normal subgroups N and K of G such that $K \leq N$, there is also a simple procedure to glue constructive presentations of G mod N and N mod K together to produce a constructive presentation of G mod K **[13**, Lemma 4.3].

We now define our notion of manageability involving finite groups equipped with certain basic polynomial-time machinery.

A finite group M specified by some input string is *manageable* if there is a polynomial-time algorithm for testing membership of a given element and finding a constructive presentation for all the subgroups of M.

Theorem 2.1. *The following finite groups are manageable.*

(i) *Permutation groups* $G \leq \mathrm{Sym}(\Omega)$ (Sims **[20]**; Furst–Hopcroft–Luks **[9]**).

(ii) *Solvable matrix groups* $G \leq \mathrm{GL}(n, k)$ *such that* $\mu(G)$ *is bounded by a fixed polynomial in the input length of* G (Luks **[13]**). □

Let G be a finite group and N a normal subgroup of G. A *manageable representation* for N is a homomorphism $\pi : N \to M$ for some manageable G-group M, where $\pi(N) \neq 1$, such that π is a G-map satisfying $\pi(x)^g = \pi(x^g)$ for $x \in N$ and $g \in G$. Very often, π is defined as the restriction of a homomorphism $G \to M$ on N. For computational purposes, we assume that π is a procedure for determining $\pi(x)$ for any given $x \in N$.

A *manageable series* of a finite group G is a G-normal series

$$G = N_1 > N_2 > \cdots > N_r = 1$$

specified by constructive presentations of G mod N_i and manageable representations $\pi_i : N_i \to M_i$ for manageable groups M_i such that $N_{i+1} = \mathrm{Ker}\, \pi_i$.

The following problem is of our main interest.

Manageable representation.

Input: $G \leq \mathrm{GL}(n, k)$ *and a constructive presentation of* G mod N *for* $N \trianglelefteq G$.
Find: *a manageable representation* $\pi : N \to M$ *for a manageable group* M.

Suppose, for the moment, that a procedure to solve the manageable representation problem is available. Given $G \leq \mathrm{GL}(n, k)$, we construct a manageable series of G, in a top-down fashion, as follows.

Let $N_1 = G$, and start with the trivial constructive presentation of G mod N_1. As a generic step, suppose that a constructive presentation of G mod N_i for $N_i \trianglelefteq G$ is given. Then we may assume that we have normal generators R_i for N_i. Here, find a manageable representation $\pi_i : N_i \to M_i$ for some manageable group M_i. Then, as π_i is a G-homomorphism, obtain, in M_i, generators for and a constructive presentation of $\pi_i(N_i) = \langle \pi_i(R_i)^G \rangle$. This presentation of $\pi_i(N_i)$ can be pulled back to a constructive presentation of N_i mod $\operatorname{Ker} \pi_i$. Now, define $N_{i+1} = \operatorname{Ker} \pi_i$, and use the constructive presentations of G mod N_i and N_i mod N_{i+1} to form a constructive presentation of G mod N_{i+1}. Repeat the procedure until $N_i = 1$.

When $N_r = 1$ is reached, a manageable series of G is complete. Membership-testing can be then performed by the sifting procedure through this series as described before.

There are also important corollaries. The order of G is $\prod_{i=1}^r |\pi_i(N_i)|$. With the ability to perform membership-testing in G, for a subset $Q \subseteq G$, one can also find the normal closure $\langle Q^G \rangle$ by adding conjugates (cf. [**14**, 3.11]). Further, one can find the kernel of a homomorphism ϕ from G into a manageable group M: If $G = \langle S \rangle$, where ϕ is specified by $\phi(S)$, then obtain a constructive presentation of $\phi(G)$ and thus G mod $\operatorname{Ker} \phi$. Form normal generators for $\operatorname{Ker} \phi$, and complete the normal closure to find generators for $\operatorname{Ker} \phi$.

To prove Theorem 1.1 in the sections to follow, we will thus focus on the manageable representation problem.

3. Basic polynomial-time tools

In this section, we review basic polynomial-time tools developed earlier in [**13**]. We also derive a variant of the *distilling* lemma of Beals and Babai [**6**, Lemma 2.1].

The following result is observed in [**13**, Theorem 4.5].

Theorem 3.1. *Given $G \leq \mathrm{GL}(n,k)$ and $R \subset \mathrm{GL}(n,k)$, where $N := \langle R^G \rangle$, in polynomial time one can find a basis B of the linear span $k[N]$ such that $B \subseteq N$.* □

In general, observe that an element $a \in \mathrm{GL}(n,k)$ centralizes $G \leq \mathrm{GL}(n,k)$ if and only if a commutes with a basis of $k[G]$. In particular, given $G \leq \mathrm{GL}(n,k)$ and $R \subset \mathrm{GL}(n,k)$, where $N := \langle R^G \rangle$, in polynomial time one can prove that N is abelian, or find $a, b \in N$ such that $[a,b] \neq 1$.

Theorem 3.1 can be strengthened further by applying the following observation [**15**].

For an n-dimensional vector space V over k, let X be a subset of V. For $G \leq \mathrm{GL}(V)$, let X^G and $X^{k[G]}$ denote the images of X under G and $k[G]$, respectively. Then we have $\mathrm{Span}(X^G) = \mathrm{Span}(X^{k[G]})$.

Proposition 3.2 (Luks). *Given $G \leq \mathrm{GL}(n,k)$, and $R, Q \subset \mathrm{GL}(n,k)$, where $N := \langle R^G \rangle$ and $L := \langle Q^N \rangle$, in polynomial time one can find a basis B of the linear span $k[L]$ such that $B \subseteq L$.*

Proof. Let $\bar{\ }: G \to \mathrm{GL}(M(n,k))$ be the conjugation action of G on $M(n,k)$. Regarding Q as a set of vectors of the vector space $M(n,k)$, we have $k[L] = \mathrm{Span}(Q^{\overline{N}}) = \mathrm{Span}(Q^{k[\overline{N}]})$. Since $\overline{N} = \langle \overline{R}^{\overline{G}} \rangle$, a basis of $k[\overline{N}]$ as elements of $\overline{N}$ is available by Theorem 3.1.

Note that, with respect to normal generators, the depth of L from G is two, and the algorithm squares an upper bound of a dimension twice: Let $m = n^2 = \dim_k M(n,k)$. Since $\overline{N} \leq \mathrm{GL}(M(n,k)) \cong \mathrm{GL}(m,k)$, the linear span $k[\overline{N}]$ is a k-vector space of dimension at most m^2, that is, at most n^4. □

Next, we state a reduction theorem from **[13]**. The following result, implicit in **[13, §4.7]**, is built on the manageability of nilpotent matrix groups.

Theorem 3.3. *Given $G \leq \mathrm{GL}(n,k)$, a constructive presentation of G* mod *N for $N \trianglelefteq G$, and $B < \mathrm{GL}(n,k)$, where B is abelian or class-2 nilpotent, such that $B^G = B$ and $1 < C_N(B) < N$, one can find one of the following in time polynomial in the input length and $\mu(G)$.*

(i) *A manageable representation $\pi : N \to M$ for a manageable group M.*

(ii) *A homomorphism $\phi : G \to \mathrm{GL}(m,k)$ such that $m < n$ and $\phi(N) \neq 1$.*

(iii) *A homomorphism $\phi : G \to \mathrm{GL}(m,k)$ such that $m \leq n$ and $1 < |\phi(N)| \leq |N|/2$.* □

We now derive an important tool for locating a nonidentity element in a proper normal subgroup when it exists. The following result has appeared in another version involving randomization (cf. **[1]**, **[4]**, and **[6]**). Our version is built on Proposition 3.2 using the argument of **[6**, Lemma 2.1**]**.

Proposition 3.4. *Given the following inputs:*

(i) *$G \leq \mathrm{GL}(n,k)$, $R, Q \subset G$, where $N := \langle R^G \rangle$ and $L := \langle Q^N \rangle$ such that $L \leq N$, and L is nonabelian, and*

(ii) *nonidentity elements $a, b \in L$ such that a or b belongs to a proper N-normal subgroup of L,*

in polynomial time one can find a nonidentity element c in a proper N-normal subgroup of L.

Proof. If $[a,b] \neq 1$, then $[a,b]$ suffices for c as $[a,b] \in \langle a^N \rangle \cap \langle b^N \rangle$.

Suppose that $[a,b] = 1$. By Proposition 3.2, one can prove that a centralizes $\langle b^N \rangle$, or find $d \in \langle b^N \rangle$ such that $[a,d] \neq 1$. If such d is found, then $[a,d]$ suffices for c since $[a,d] \in \langle a^N \rangle \cap \langle b^N \rangle$.

Suppose that a centralizes $\langle b^N \rangle$. Again, by Proposition 3.2, test whether or not a centralizes $L = \langle Q^N \rangle$. If $a \in Z(L)$, then return $c = a$ since $Z(L)$ is a proper N-normal subgroup of L. If $a \notin Z(L)$, then $\langle b^N \rangle < L$; thus, return $c = b$. □

4. Normal structure

In this section, we summarize several basic facts about normal closures and normal series. Let G be a finite group.

Let L be a subnormal subgroup of G. Then every composition factor of $\langle L^G \rangle$ is isomorphic to some composition factor of L (see, e.g., [**8**, Lemma 4.4C]). Evidently, if L is solvable, then $\langle L^G \rangle$ is also solvable.

For nonabelian simple groups, we state the following familiar result (see, e.g., [**18**, 13.3.1]).

Lemma 4.1. *Suppose that a finite group G has a subnormal subgroup T such that T is nonabelian simple. Then $\langle T^G \rangle$ is a direct product of G-conjugates of T, where G acts transitively on the set of these conjugates.* □

Now, let N be a normal subgroup of G. Suppose that T is a nonabelian simple normal subgroup of N. Then, from Lemma 4.1, it follows that N is a nonabelian minimal normal subgroup of G if and only if $N = \langle T^G \rangle$. Furthermore, if L is a nonabelian minimal normal subgroup of N, then $\langle L^G \rangle$ is a nonabelian minimal normal subgroup of G.

The next proposition summarizes elementary facts about the structure of chief factors (see, e.g., [**17**, Propositions 3.4.5, 3.4.6]).

Proposition 4.2. *Suppose that a finite group G has a chief factor N/K.*

(i) *If L_1 is a normal subgroup of N such that $L_1 \not\leq K$, then there is a proper N-normal subgroup L_2 of L_1 such that L_1/L_2 is simple.*

(ii) *Suppose that N/K is nonabelian. If $L_1, \ldots, L_r$ are nonabelian normal subgroups of N, where each $L_i \not\leq K$, such that $[L_i, L_j] = 1$ for all pairs $L_i \neq L_j$, then $r \leq \log_{60} |N/K|$.* □

5. Small chief factors

We first prove that matrix groups $G \leq \mathrm{GL}(n, k)$ such that $\mu(G)$ and $\gamma(G)$ are bounded by a fixed polynomial in the input length are manageable. Our main objective is to solve the manageable representation problem defined in §2 to establish the manageability of G.

We begin with the following result, immediate from Proposition 3.4.

Proposition 5.1. *Given $G \leq \mathrm{GL}(n, k)$ and $R \subset G$, where $N := \langle R^G \rangle$ is nonabelian, one can either*

(i) *prove that N is a minimal normal subgroup of G, or*

(ii) *find a nonidentity element in a proper G-normal subgroup of N,*

in time polynomial in the input length and $\gamma(G)$.

Proof. Write $\gamma = \gamma(G)$. If $|N| \le \gamma$, then it is trivial. Suppose that $|N| > \gamma$. First, form a set X of $\gamma + 1$ distinct elements of N and then the set Y of $\gamma(\gamma+1)$ products of the form $x^{-1}y$, where $x, y \in X$ such that $x \ne y$. Now, find $a, b \in N$ such that $[a, b] \ne 1$, and add $[a, b]$ to Y. Observe that Y contains at least one nonidentity element in a proper G-normal subgroup of N. Then we apply Proposition 3.4 $|Y| - 1$ times. □

We are now ready to prove

Proposition 5.2. *Matrix groups $G \le \mathrm{GL}(n, k)$ such that $\gamma(G)$ and $\mu(G)$ are bounded by a fixed polynomial in the input length are manageable.*

Proof. Our goal is to find a manageable representation for a given normal subgroup $N = \langle R^G \rangle$, specified by normal generators R. Since solvable groups are manageable, if N happens to be indeed solvable, one can complete the normal closure to find generators for N and verify the solvability of N (cf. **[13**, Theorem 3.1**]**). Therefore, we may assume that N is nonsolvable.

The following polynomial-time algorithm finds a manageable representation for $N = \langle R^G \rangle$, given by R, when N is nonsolvable, or reduces the input to an instance for Theorem 3.3.

Let $\bar{\ } : G \to \mathrm{GL}(k[N])$ denote the conjugation action of G on $k[N]$.

procedure SMALL_CHIEF
Input: $G \le \mathrm{GL}(n, k)$ and $R \subset G$, where $N := \langle R^G \rangle$ is nonsolvable.
Output: a manageable representation $\pi : N \to \mathrm{Sym}(\Omega)$, where $|\Omega|$ is
  polynomially bounded, or a nilpotent group of class at most two
  $B < N$ such that $B^G = B$ and $1 < C_N(B) < N$.

begin
  let $Q := R$ and X denote $\langle Q^G \rangle$;
  while $\overline{X}$ is nonabelian containing a proper $\overline{G}$-normal subgroup **do**
    begin
      find $1 \ne \bar{a} \in \overline{X}$ such that $\langle \bar{a}^{\overline{G}} \rangle < \overline{X}$;
      let a be a preimage of $\bar{a}$ in G so that $\langle a^G \rangle < X$;
      let $Q := \{a\}$ and X denote $\langle Q^G \rangle$;
    end;
  if $\overline{X}$ is abelian **then return** $B := X$;
  else return the natural action $G \to \mathrm{Aut}(\overline{X})$ restricted to N;
end.

At any point in the above while loop, $\overline{X} \ne 1$ so that X acts nontrivially on $k[N]$ by conjugation; that is, $C_N(X) < N$.

Suppose that $\overline{X}$ is found to be abelian. Then $[\overline{X}, \overline{X}] = 1$ and thus $[X, X] \leq Z(N)$. Therefore, X is nilpotent of class at most two. Also, since $1 \leq [X, X] \leq Z(X) \leq C_N(X)$, it follows that $C_N(X) \neq 1$.

If $\overline{X}$ is found to be a nonabelian minimal normal subgroup of $\overline{G}$, then return π as defined by the conjugation action of $\overline{G}$ on $\overline{X}$. Evidently, $\pi(N) \neq 1$. □

6. Small composition factors

From now on, we relax the condition on the chief factors. That is, we assume that we are dealing with $G \leq \mathrm{GL}(n, k)$ such that $\mu(G)$ and $\kappa(G)$ are polynomially bounded.

As in §5, our goal is to show the manageability of G, and we will maintain the overall structure of the algorithm roughly the same. In this section, we focus on the main difficulty of finding a nonidentity element of a proper G-normal subgroup of a nonabelian normal subgroup $N = \langle R^G \rangle$, given by normal generators R, in case N is not minimal normal in G.

First, note that, if N is a minimal normal subgroup of G, then $\gamma(N) \leq \kappa(G)$. When $\gamma(N) \leq \kappa(G)$ holds, Proposition 5.2 asserts that N is manageable; in particular, membership-testing allows us to complete the normal closure and find generators for N. The following procedure tests whether or not a nonabelian normal subgroup $N = \langle R^G \rangle$, given by R, is a minimal normal subgroup of G.

procedure TEST_MIN_NORM
Input: $G \leq \mathrm{GL}(n, k)$ and $R \subset G$, where $N := \langle R^G \rangle$ is nonabelian.
Output: a nonabelian simple group $T \trianglelefteq N$ if N is a minimal normal
    subgroup of G; otherwise, *false*.

begin
    assuming $\gamma(N) \leq \kappa(G)$, attempt to find generators Q for N and a
        nonabelian simple group $T \trianglelefteq N$ based on Proposition 5.2;
    if the above attempt fails to find Q or T within the assumed polyno-
        mial-time **then return** *false*;
    if the above attempt succeeds, and $\langle T^G \rangle = N$, **then return** T;
    else return *false*;
end.

The algorithm returns the correct answer since N is a minimal normal subgroup of G if and only if $\langle T^G \rangle = N$ (cf. Lemma 4.1).

Suppose that N is not a minimal normal subgroup of G. Our main objective now is to locate a nonidentity element in a proper G-normal subgroup of N. To find such an element, we will exploit the following important structural properties of chief factors.

Consider a chief factor N/K of G. If N/K happens to be abelian, then a nonidentity commutator of elements of N lies in K.

Suppose that N/K happens to nonabelian. Then $N/K = T_1/K \times \cdots \times T_\ell/K$, where the T_i/K are isomorphic to a nonabelian simple group T/K; furthermore, G acts transitively on the set $\{T_1/K, \ldots, T_\ell/K\}$ (cf. Lemma 4.1). Now, suppose that X/K, where $X \neq K$, is a proper normal subgroup of N/K. Then, without loss of generality, $X/K = T_1/K \times \cdots \times T_j/K$ for some $j < \ell$. By the transitivity of G, there is $g \in G$ such that $X/K \cap X^g/K \neq K/K$ and has a fewer than j copies of T/K, or X/K commutes with X^g/K for all $g \in G$ unless $X/K = X^g/K$.

Recall that, in our setting, normal subgroups are represented by normal generators, and we have relied on their linear spans to test commutativity (cf. Proposition 3.2). In the following subroutine PERM_REP, we therefore consider the above procedure with respect to the linear spans of normal subgroups. The subroutine descends inside a chief factor N/K or constructs the conjugation action of G on a set of the linear spans of normal subgroups of N that commute one another.

procedure PERM_REP
Input: $G \leq \mathrm{GL}(n, k)$, $R \subset G$, and $x \in G$, where $N := \langle R^G \rangle$ is nonabelian,
  and $L := \langle x^N \rangle$ is nonabelian properly contained in N.
Output: one of the following:
  (A) $1 \neq y \in L^g$ for some $g \in G$ such that $\langle y^N \rangle < L^g$;
  (B) $1 \neq c \in N$ such that $\langle c^G \rangle < N$;
  (C) the set $\mathcal{E}$ of G-conjugates of $k[L]$, where each $E \in \mathcal{E}$ centralizes
    all $F \in \mathcal{E}$, $F \neq E$, and $|\mathcal{E}| < n^2 \log_{60} p$.

```
begin
    let 𝓔 := {k[L]};
    for each E ∈ 𝓔 and each s ∈ S do
        begin
            if E^s ∉ 𝓔 and E^s does not commute with some F ∈ 𝓔 then
                begin
                    find [a, b] ≠ 1, where a and b are basis elements of
                       E^s and F, respectively;
                    return y := [a, b] for (A);
                end;
            else if E^s ∉ 𝓔 then add E^s to 𝓔;
            if |𝓔| ≥ n^2 log_60 p then
                begin
                    collect the following in a set X: [a, b] ≠ 1, where
                       a, b ∈ L, and a basis of each E ∈ 𝓔;
                    from X, find c ∈ N such that ⟨c^G⟩ < N;
                    return c for (B);
                end;
        end;
end.
```

By Proposition 4.2 (ii), it is routine to verify the correctness of the procedure PERM_REP.

To descend further in N/K when PERM_REP outputs (C), we will make use of a point-stabilizer of G under the permutation representation found in (C). To find such a point-stabilizer, our algorithm will rely on the following fundamental lemma (see, e.g., **[8**, Theorem 3.6A**]**). Note that our method will only construct one point-stabilizer, whereas the well-known Schreier–Sims method **[20]** constructs a series of point-stabilizers recursively.

Lemma 6.1 (Schreier). *Let G be a group and H a subgroup of G. If S is a generating set for G, and R is a complete set of right coset representatives of H in G, then the set $T = \{r_1 s r_2^{-1} \mid r_1, r_2 \in R, s \in S, \text{ and } r_1 s r_2^{-1} \in H\}$ generates H. The set T is called Schreier generators for H.* □

With the procedure PERM_REP and Lemma 6.1, we are now ready to prove the following.

Proposition 6.2. *Given $G \le \mathrm{GL}(n, k)$ and $R \subset G$, where $N := \langle R^G \rangle$ is nonabelian, one can either*

(i) *prove that N is a minimal normal subgroup of G, or*

(ii) *find a nonidentity element in a proper G-normal subgroup of N,*

in time polynomial in the input length and $\kappa(G)$.

Proof. Let ℓ denote the maximum length of a normal series in $\mathrm{GL}(n, k)$; here, recall that ℓ is polynomially bounded (see §1). As usual, assume that $G = \langle S \rangle$. We first outline our main algorithm and then formalize it in the procedure PROPER_NORM presented below.

Step 1. First, if N is solvable, then it suffices to return a nontrivial commutator of two elements of N (cf. the proof of Proposition 5.2).

Step 2. Suppose that N is nonsolvable, containing a proper G-normal subgroup, and there is a G-chief factor N/K. In Step 2, we construct a sequence of N-normal subgroups

$$N = L_1, L_2, \dots, L_r, \tag{1}$$

where $r \le \ell$, each $L_{i+1} < L_i^{g_i}$ for some $g_i \in G$, and each $L = L_i$ is represented by an element x such that $L = \langle x^N \rangle$. Step 2 is a loop, consisting of four sub-steps (a)–(d), that iterates at most ℓ^2 times and forces L to step down the sequence (1) at most ℓ times. First, to begin the sequence (1), find $1 \ne a \in N$ such that $\langle a^N \rangle < N$ using Proposition 3.4, and let $x := a$ and L denote $\langle x^N \rangle$.

Step 2 (a). If L is found to be abelian or a minimal normal subgroup of N, then the loop halts and returns $\langle L^G \rangle$ (cf. §4).

Step 2 (b). Suppose otherwise. Then Step 2 (b) first calls the procedure PERM_REP. If PERM_REP returns an element y for Output (A), then we let $x := y$ and descend the sequence (1) with a new $L := \langle x^N \rangle$. If PERM_REP returns an element c for Output (B), then we halt the loop and return $c \in N$ such that $\langle c^G \rangle < N$.

Steps 2 (c), (d). Now, we consider the case when the procedure PERM_REP returns Output (C). Let $E = k[L]$. Here, G acts on the set $\mathcal{E}$ of polynomially-bounded size. Thus, we may assume that we have Schreier generators U for the point-stabilizer G_E by the Schreier–Sims method **[20]**. Since N normalizes L, we know N stabilizes $E = k[L]$ so that $N \leq G_E$. Thus, $L = \langle x^N \rangle \leq \langle x^{G_E} \rangle$.

Steps 2 (c) and (d) consider the following two cases, respectively.

(I) There is $1 \neq g \in G_E$ such that $L \neq L^g$.

(II) G_E normalizes L.

When Case (I) holds, Step 2 (c) works as follows. As $G_E = \langle U \rangle$, there is $u \in U$ such that $L \neq L^u$. Recall that $k[L^u] = k[L]$ is noncommutative. Hence, there are basis elements a of $k[L]$ and b of $k[L^u]$, where $a \in L$ and $b \in L^u$, such that $[a, b] \neq 1$. Since $[a, b] \in L \cap L^u < L$, we step down the sequence (1).

When Case (II) holds, Step 2 (d) works as follows. We consider the case when $L \not\leq K$. Here, we have $L^{G_E} = L$ so that $L = \langle x^N \rangle \leq \langle x^{G_E} \rangle \leq \langle L^{G_E} \rangle = L$ and thus $L = \langle x^N \rangle = \langle x^{G_E} \rangle$. Therefore, distinct elements of L can be now listed so that we can find $a \in L$ such that $\langle a^N \rangle < L$ (cf. Propositions 3.4, 4.2).

Every time L steps down the tower, we first assume Case (I) and repeat Step 2 (c) for ℓ times. After these ℓ iterations, if L is not found to be a minimal normal subgroup of N, then Case (II) must hold. If $L \not\leq K$, then Step 2 (d) finds a proper N-normal subgroup of L, and we step down the sequence (1).

Finally, when the loop reaches $L = L_\ell$, then we conclude that either L is a minimal normal subgroup of N, or $L \leq K$.

The following procedure PROPER_NORM formalizes our method.

procedure PROPER_NORM
Input: $G \leq \mathrm{GL}(n, k)$ and $R \subset G$, where $N := \langle R^G \rangle$ is nonabelian.
Output: a nonidentity element in a proper G-normal subgroup of N, or
  "N is a minimal normal subgroup of G".

begin
Step 1. (* Halt if N is solvable or a minimal normal subgroup of G. *)
  if N is solvable **then**
    find $a, b \in N$ such that $[a, b] \neq 1$ and **return** $d := [a, b]$;
  if TEST_MIN_NORM(G, R) **then**
    return "N is a minimal normal subgroup of G";
Step 2. (* Form a sequence $N = L_1, \ldots, L_r$, where $r \leq \ell$. *)
  find $a \in N$ such that $\langle a^N \rangle < N$;
  let $x := a$ and L denote $\langle x^N \rangle$;

```
    let i := 1 and j := 1;
    while i ≤ ℓ do
        begin
            let found := false;
        (a). if L is abelian then return d := x;
            if TEST_MIN_NORM(G, {x}) then return d := x;
        (b). PERM_REP(G, R, x);
            if PERM_REP returns y for (A) then
                let found := true, x := y, and L denote ⟨x^N⟩;
            else if PERM_REP returns c for (B) then return d := c;
            else
                begin
                    let E denote k[L];
                    under the G-action on ℰ by conjugation, obtain
                      Schreier generators U for the point-stabilizer G_E;
        (c). (* When L^{G_E} ≠ L, we will find 1 ≠ c ∈ [L, L^u] < L for
              some u ∈ U. *)
                    if j ≤ ℓ then
                        begin
                            let Y := ∅;
                            for each u ∈ U do
                                add [a, b] ≠ 1, where a ∈ L and
                                  b ∈ L^u, to Y;
                            from Y, find c ∈ L such that ⟨c^N⟩ ≤ L
                              (and ⟨c^N⟩ < L when L^{G_E} ≠ L);
                            let x := c and L denote ⟨x^N⟩;
                            let j := j + 1;
                        end;
        (d). (* When j = ℓ, we have L^{G_E} = L and ⟨x^N⟩ = ⟨x^{G_E}⟩. *)
                    else
                        begin
                            as L^{G_E} = L, find a ∈ L such that
                              ⟨a^N⟩ ≤ L (and ⟨a^N⟩ < L if L ≰ K);
                            let x := a and L denote ⟨x^N⟩;
                            let found := true;
                        end;
                end;
            if found then let i := i + 1 and j := 1;
        end;
    return d := x;
end.
```

□

We are finally ready to establish Theorem 1.1. As we have discussed in §2, in order to prove Theorem 1.1, it suffices to establish manageability.

Proposition 6.3. *Matrix groups $G \leq \mathrm{GL}(n, k)$ such that $\kappa(G)$ and $\mu(G)$ are bounded by a fixed polynomial in the input length are manageable.*

Proof. Using the procedures TEST_MIN_NORM and PROPER_NORM, we apply the argument of Proposition 5.2. Here, we either return the conjugation action of Lemma 4.1, or reduce the input to an instance of Theorem 3.3. □

As a final note, we briefly comment on Corollary 1.2.

The method discussed in §2 evidently finds the kernel of a given homomorphism from $G \leq \mathrm{GL}(n, k)$ into a manageable group in the desired timing; thus, Corollary 1.2 (iv) holds. Now, observe that, for $G \leq \mathrm{GL}(n, k)$ and a normal subgroup N of G, the centralizer $C_G(N)$ is the kernel of the conjugacy action of G on the linear span $k[N]$. Therefore, Corollary 1.2 (v) also holds.

Applying the method of Proposition 5.1 on quotient groups, it is straightforward to devise a polynomial-time algorithm for finding composition series for Corollary 1.2 (vi). For Corollary 1.2 (vii), we apply the method of **[11**, §4].

References

[1] L. Babai and R. Beals, A polynomial-time theory of black box groups I, in: Groups St Andrews 1997 in Bath I (C. M. Campbell, E. F. Robertson, N. Ruskuc, and G. C. Smith, eds.), London Math. Soc. Lecture Note Ser. 260, Cambridge Univ. Press, Cambridge, 1999, 30–64.

[2] L. Babai, R. Beals, J. Cai, G. Ivanyos, and E. M. Luks, Multiplicative equations over commuting matrices, in: Proc. 7th Annu. ACM-SIAM Sympos. on Discrete Algorithms, SIAM, Philadelphia, 1996, 498–507.

[3] L. Babai, P. J. Cameron, and P. P. Pálfy, On the orders of primitive groups with restricted nonabelian composition factors, J. Algebra 79 (1982), 161–168.

[4] R. Beals, Towards polynomial time algorithms for matrix groups, in: Groups and Computation II (L. Finkelstein and W. M. Kantor, eds.), DIMACS Ser. Discrete Math. Theoret. Comput. Sci. 28, Amer. Math. Soc., Providence, RI, 1997, 31–54.

[5] —, Algorithms for matrix groups and the Tits alternative, J. Comput. System Sci. 58 (1999), 260–279.

[6] R. Beals and L. Babai, Las Vegas algorithms for matrix groups, in: Proc. 34th Annu. Sympos. on Foundations of Computer Science, IEEE Comput. Soc. Press, Los Alamitos, CA, 1993, 427–436.

[7] J. Cai, R. J. Lipton, and Y. Zalcstein, The complexity of the membership problem for 2-generated commutative semigroups of rational matrices, in: Proc. 35th Annu. Sympos. on Foundations of Computer Science, IEEE Comput. Soc. Press, Los Alamitos, CA, 1994, 135–142.

[8] J. D. Dixon and B. Mortimer, Permutation groups, Grad. Texts in Math. 163, Springer-Verlag, New York, 1996.

[9] M. Furst, J. Hopcroft, and E. Luks, Polynomial-time algorithms for permutation groups, in: Proc. 21st Annu. Sympos. on Foundations of Computer Science, IEEE Comput. Soc. Press, Washington, DC, 1980, 36–41.

[10] G. Ge, Testing equalities of multiplicative representations in polynomial time, in: Proc. 34th Annu. Sympos. on Foundations of Computer Science, IEEE Comput. Soc. Press, Los Alamitos, CA, 1993, 422–426.

[11] W. M. Kantor and E. M. Luks, Computing in quotient groups, in: Proc. 22nd Annu. ACM Sympos. on Theory of Computing, ACM Press, New York, 1990, 524–534.

[12] E. M. Luks, Isomorphism of graphs of bounded valence can be tested in polynomial time, J. Comput. System Sci. 25 (1982), 42–65.

[13] —, Computing in solvable matrix groups, in: Proc. 33rd Annu. Sympos. on Foundations of Computer Science, IEEE Comput. Soc. Press, Los Alamitos, CA, 1992, 111–120.

[14] —, Permutation groups and polynomial-time computation, in: Groups and Computation (L. Finkelstein and W. M. Kantor, eds.), DIMACS Ser. Discrete Math. Theoret. Comput. Sci. 11, Amer. Math. Soc., Providence, RI, 1993, 139–175.

[15] —, Personal communication.

[16] E. M. Luks and T. Miyazaki, in preparation.

[17] T. Miyazaki, Polynomial-time computation in matrix groups, Ph.D. dissertation, Tech. Rep. CIS-TR-99-11, Department of Computer and Information Science, University of Oregon, Eugene, 1999.

[18] D. J. S. Robinson, A course in the theory of groups, 2nd ed., Grad. Texts in Math. 80, Springer-Verlag, New York, 1996.

[19] G. M. Seitz, R. Solomon, and A. Turull, Chains of subgroups in groups of Lie type II, J. London Math. Soc. (2) 42 (1990), 93–100.

[20] C. C. Sims, Computational methods in the study of permutation groups, in: Computational Problems in Abstract Algebra (J. Leech, ed.), Pergamon Press, Oxford, 1970, 169–183.

[21] R. Solomon and A. Turull, Chains of subgroups in groups of Lie type I, III, J. Algebra 132 (1990), 174–184; J. London Math. Soc. (2) 44 (1991), 437–444.

[22] D. A. Suprunenko, Matrix groups, Nauka, Moscow, 1972 (Russian); English transl., Transl. Math. Monogr. 45, Amer. Math. Soc., Providence, RI, 1976.

Department of Computer and Information Science
University of Oregon
Eugene, OR 97403, U.S.A.

Current Address:

Department of Computer Science
Bucknell University
Lewisburg, PA 17837, U.S.A.
tmiyazak@eg.bucknell.edu

Nearly simple matrix groups: desiderata and suggestions

Peter M. Neumann

Abstract. This note is concerned with two related points about matrix groups and matrix-group computations over finite fields. One concerns the last of the Aschbacher classes of subgroups of finite general linear groups, the class of nearly simple groups. The point is that this can be naturally divided into three subclasses, here called 'connected algebraic', 'discrete algebraic' and 'casual' (or 'properly modular') groups. This subdivision is intended partly to impose some structure on a disparate collection of linear groups, partly to respond to a feeling that different computational techniques may be appropriate in the three different cases. The second point concerns the desirability and feasibility of reducing the dependence of matrix-group algorithms on computing orders of elements in favour of more genuinely linear-algebraic computations.

1991 Mathematics Subject Classification: primary 20C40; secondary 20G40.

1. Introduction

This note is a paraphrase of short lectures presented in Dagstuhl 1997 and Columbus 1999. They were intended to focus two suggestions made in August 1993 at a semi-formal session at Groups-St Andrews in Galway and elaborated in informal discussions at various times and places in recent years. The first point is that the last of the Aschbacher classes of subgroups of finite general linear groups, the class of nearly simple groups, can sensibly be divided into three subclasses. This subdivision is intended partly to impose some structure on a disparate collection of linear groups, partly to respond to a feeling that different computational techniques may be appropriate in the three different cases. The second point concerns the desirability and feasibility of reducing the dependence of matrix-group algorithms on computing orders of elements in favour of more genuinely linear-algebraic computations. Both these points have as much to do with attitudes and with personal taste as anything substantially mathematical. Nevertheless (or therefore) they should be worth recording—under a title that echoes one of Cayley's **[6]** from 1878.

Groups and Computation III
Ohio State Univ. Math. Res. Inst. Publ. 8

2. Subdividing the last of the Aschbacher classes

By now the Aschbacher classification **[2]** of (maximal) subgroups of $\mathrm{GL}(d, q)$, the group of $d \times d$ invertible matrices over the finite field $\mathrm{GF}(q)$, is well enough known in Computational Group Theory that it need not be rehearsed here. The parenthetic word maximal was much of the point of Aschbacher's original work, but has now lost most of its relevance. Here we focus on the last of its categories, that of almost simple groups. These are the groups $G \leqslant \mathrm{GL}(V)$, where V is a d-dimensional vector space over $\mathrm{GF}(q)$, with the following properties:

> there exists $G_0 \trianglelefteq G$ such that $Z \leqslant G_0$ where $Z := G \cap \{\text{scalars}\}$ and if $T := G_0/Z$ then T is non-abelian simple and $T \leqslant G/Z \leqslant \mathrm{Aut}\, T$;
>
> as a linear group G_0 is absolutely irreducible, primitive, not definable over a proper subfield, etc.

In the context of **[20, §3]** Cheryl Praeger and I noted that it could be also assumed that G_0 does not preserve a classical form (a non-degenerate sesquilinear, bilinear or quadratic form). That was good for our purposes then, but is not desirable here because very many of the interesting nearly simple groups arise as subgroups of classical groups.

My first point is that there is a useful subdivision of this category into three subclasses. The names attached to them are intended as suggestive labels, no more.

(I) *Connected algebraic groups*: groups G of Lie type in their defining characteristic.

(II) *Discrete algebraic groups*: groups G for which there exists $G^* \leqslant \mathrm{GL}(d, \mathbb{C})$ such that $G \cong G^*$ and $G \equiv G^* \pmod{p}$.

(III) *Casual groups* or *properly modular groups*: all others.

The descriptions are perhaps best explained by examples.

The most obvious examples of subclass (I) are the classical groups $\mathrm{GL}(d, q)$, $\mathrm{U}(d, q)$ (as a subgroup of $\mathrm{GL}(d, q^2)$), $\mathrm{Sp}(d, q)$ (when d is even), $\mathrm{O}^{\varepsilon}(d, q)$ acting on their natural representation space V. But then also there are functorially defined representations on such spaces as the exterior square $\wedge^2 V$, the exterior cube $\wedge^3 V$, the symmetric square $\Sigma^2 V$, the symmetric cube $\Sigma^3 V$, and many others, including, for example, representations of exceptional groups in their natural representations on their Lie algebras. From a purely theoretical point of view all of these must be of interest. From a practical point of view the dimensions are sufficiently large that representations other than V will arise only rarely. If V is m-dimensional then $\dim(\wedge^k V) = \binom{m}{k}$, and even if k is only 3 this is out of range for practical purposes well before m has reached 20.

In (II) the notation $G \equiv G^* \pmod{p}$ means the following. Since G^* is finite it is conjugate to, and therefore may be taken to be, a subgroup of $\mathrm{GL}(d, R)$, where R is the ring of integers of a number field contained in $\mathbb{C}$. Reduction modulo a prime ideal divisor of p is a homomorphism $\mathrm{GL}(d, R) \to \mathrm{GL}(d, F)$, where F is a finite field of characteristic p, say $F = \mathrm{GF}(q)$ where $q = p^b$. If this homomorphism maps G^* to G then I write $G \equiv G^* \pmod{p}$. The simplest and best known examples in this class are:

$2\,\mathrm{Alt}(5) \leqslant \mathrm{GL}(2, q)$ for $q = p \equiv \pm 1 \pmod{10}$, or $q = p^2$ where $p \equiv \pm 3 \pmod{10}$;

$2\,\mathrm{Alt}(7) \leqslant \mathrm{GL}(4, p)$ for $p \equiv 23, 37, 53, 67 \pmod{70}$;

$\mathrm{PSL}(2, r) \leqslant \mathrm{GL}(\frac{1}{2}(r-1), q)$ for $q = p$ where $(\frac{p}{r}) = 1$ or $q = p^2$ where $(\frac{p}{r}) = -1$ (the Jacobi symbol $(\frac{p}{r})$ expresses the quadratic character of p modulo r);

$\mathrm{Sym}(n) \leqslant \mathrm{GL}(n-1, p)$ when $p \nmid n$.

The condition that the field is $\mathrm{GF}(p)$ or $\mathrm{GF}(p^2)$ in these examples reflects the condition that G is not definable over a proper subfield of $\mathrm{GF}(q)$. It is typical of discrete algebraic groups that they appear as subgroups of $\mathrm{GL}(d, F)$ for finite fields F of all except finitely many characteristics p.

By way of contrast, it is typical of the groups of subclass (III) that they occur for only finitely many values of q. It was tempting to call these groups 'sporadic' or 'exceptional', but, mindful of the confusion that such usage would have caused, I scoured a dictionary for something similar. The first two meanings of 'casual' are given as 'Due to, characterized by, or subject to chance; accidental, fortuitous' and 'Occurring unpredictably; irregular; occasional'.

For a given group G, an absolutely irreducible (primitive, etc.) representation of G is genuinely modular of characteristic p precisely when it does not lift to a representation in characteristic 0 and therefore lies in a block of non-zero defect. In particular, of course $p \mid |G|$. Familiar examples in this class are:

$\mathrm{Alt}(7) \leqslant \mathrm{GL}(4, 2)$;

$M_{11} \leqslant \mathrm{GL}(5, 3)$;

$J_1 \leqslant \mathrm{GL}(7, 11)$;

$M_{23} \leqslant M_{24} \leqslant \mathrm{GL}(11, 2)$;

$\mathrm{Sym}(n) \leqslant \mathrm{GL}(n-2, p)$ when $p \mid n$.

Examples may be manufactured at will by taking an almost simple group G and one of its absolutely irreducible, primitive representations over $\mathrm{GF}(q)$ that is not the reduction modulo p of an ordinary representation and which cannot be written over a proper subfield of $\mathrm{GF}(q)$. Very many of the representations catalogued in the Modular ATLAS **[17]** fall into this subclass (see also **[16]**).

3. Recognition algorithms

There are many computational problems in relation to the last of the Aschbacher classes but I propose to focus just on the question of recognition—to recognise T or G, or, preferably, G and the representation.

Recognising connected algebraic groups. Recognition algorithms for the classical groups in their natural representations are now well-established—see, for example, Celler **[7]**, Celler & Leedham-Green **[9, 10]**, Neumann & Praeger **[20]**, Niemeyer & Praeger **[22]**. Black-box algorithms are available to recognise T and sometimes G_0 or G—see, for example, Cooperman et al. **[12]**, Altseimer & Borovik **[1]**, Kantor & Seress **[18]**, and several algorithms described at this workshop. Of course black-box algorithms are not specially focussed on matrix groups, but the point is that they may be used to recognise G as abstract group and the representation can then be identified using the MEATAXE. For example, if a black box tells us that G is abstractly isomorphic to $\mathrm{GL}(m, q)$ and if the originally given representation space V has dimension $\frac{1}{2}m(m-1)$, then one may use the MEATAXE to test for isomorphism between V and the exterior square $\wedge^2 F^m$ of the natural representation of G. There are, however, still some difficulties associated with the cost of black-box methods in this particular context. A matrix group of degree d over $\mathrm{GF}(q)$ requires approximately $d^2 \log q$ bits to describe its elements. Therefore a black-box method which succeeds at a cost $O(n^3\mu)$, where n is the size of the input and μ is the cost of multiplying two elements, costs $O(d^9 \log^5 q)$ to process if matrix multiplication, done in the obvious way, is taken to cost $O(d^3 \log^2 q)$ units, and not much less even if sophisticated multiplication algorithms are used. Moreover, some black-box methods have components whose cost depends on q rather than $\log q$.

Sometimes costly computations are justifiable when they yield valuable extra information. Some recognition algorithms, for example, are 'constructive' in Leedham-Green's sense (see, for example, **[10]**) that they make provision for an arbitrary element of the group to be expressed explicitly as a word in the originally given generators. Most of the black-box algorithms are of this kind, and from time to time the extra information may be worth the extra cost. Nevertheless, as always, it is worth investing effort to seek cheaper methods if they can be devised.

Recognising discrete algebraic groups. In the case of discrete algebraic groups it seems reasonable to hope for algorithms that depend rather little upon the particular field F over which the representation is defined. For example, one rather stringent test of the quality of an algorithm is whether it might work over fields of characteristic 0 such as $\mathbb{Q}$ or algebraic number fields. The procedure given in **[20]** for recognising $\mathrm{PSL}(2, r)$ as subgroup of $\mathrm{GL}(d, F)$ where $d = \frac{1}{2}(r-1)$ meets this criterion. It assumes that irreducible and nearly irreducible elements x and y have been found. The assumption there that $\frac{1}{2}(r-1)$ is prime can easily be circumvented; the assumption that a nearly irreducible element y is available is acceptable since such elements appear with frequency very close to $\frac{1}{2}$. But that an irreducible element x is available is not so clear if we take PROCEDURE PSL(2, r) out of the context in which it is embedded in **[20]**. These are the elements of order r in PSL(2, r). They appear with frequency $2/r$ and therefore the expected number of random trials to find one is r. The methods suggested by Celler *et al* **[11]** allow random elements to be generated by one or two group operations after a pre-processing calculation. If we may take the cost of generating a single random element of the group as being that of a bounded number of matrix multiplications (as we may provided the pre-processing phase is not too costly) then this leads to an expected cost $O(d^4)$ operations in the field F, where d is the degree of the matrix group (that is $d = \frac{1}{2}(r-1)$), to find the element x. Such a cost is acceptable in the context of **[20]**, where the remainder of the procedure also costs $O(d^4)$. And in the context of matrix group algorithms cost $O(d^4)$ is not unreasonable, especially if the constant hiding in the scholarly vagueness of the 'Oh' notation is not too large.

Another example of a method that works for infinite fields is the following simple recognition algorithm for the groups $\mathrm{Sym}(d+1)$ of degree d in sub-class (II) when $p \nmid d+1$. Use random search to find an element g such that $c_g(t) = t^d - 1$ and $\dim \mathrm{Ker}(1-g) = 1$, where $c_g(t)$ is the characteristic polynomial of g. It is not hard to see that if G really is $\mathrm{Sym}(d+1)$ then such elements are d-cycles because they have order dividing d and they have at most two cycles in the set of size $d+1$. If v is a non-zero vector such that $vg = v$ then v is fixed by the stabiliser of a point in $\mathrm{Sym}(d+1)$, and the natural permutation representation can be reconstructed as the orbit of v under the action of G on V. The d-cycles in $\mathrm{Sym}(d+1)$ occur with frequency $1/d$ and therefore the expected number of random elements to be inspected is $2d$. Moreover, inspection of elements g of G costs at most $O(d^3)$ even by unsophisticated methods, so the expected cost of this method is $O(d^4)$. And again, this method is applicable also when F has characteristic 0.

Recognising casual groups. In a lecture at this meeting Gerhard Hiss described the substantial progress that he and Gunter Malle have made with the compilation **[16]** of a data-base of representations of nearly simple matrix groups of degrees $\leqslant 250$. Their work is relevant to all three of the subclasses, but I believe that it will be particularly important for the casual groups. Black-box methods to recognise symmetric and alternating groups have been devised by Bratus & Pak **[4]** and Beals et al. **[3]**, and again, although these are applicable also to the groups in class (II), I believe that they will be most relevant to properly modular representations. Other methods may be based on the observation that many of the casual groups are relatively small. For example, such groups may often be recognised by permutation-group methods. These are exemplified in the procedures for recognising the Mathieu groups M_{11} in GL(5, 3), M_{23} in GL(11, 2) and M_{24} in GL(11, 2) described in **[20]**.

4. Desiderata

Most of the presently available recognition algorithms depend at some point on the order $\mathrm{ord}(g)$ of elements of the group G. The second of my main points is that it seems desirable to develop algorithms that are less dependent on it.

I am not advocating total avoidance of the use of $\mathrm{ord}(g)$: there are, of course, some situations where it is perfectly natural to compute with $\mathrm{ord}(g)$. For example, the algorithm sketched above to recognise $\mathrm{Sym}(d+1)$ of degree d as a discrete algebraic group when $p \nmid d+1$ uses $\mathrm{ord}(g)$ implicitly and quite naturally. Such examples aside, what I would like to see is the development of algorithms that make better use of linear algebra instead of $\mathrm{ord}(g)$. Why? Well, I feel that $\mathrm{ord}(g)$ is an uncomfortable and slightly unnatural parameter in the context of matrix groups. Such a feeling is bound to be subjective and hard to justify. Nevertheless there are at least some rationalisations.

One is the point made above that, at least in the case of discrete algebraic groups, it seems reasonable to hope for algorithms that depend rather little upon the particular field F over which the representation is defined. An algorithm that depends on computation of the orders of elements is unlikely to be adaptable to fields of characteristic 0 because it is not clear how one can efficiently find the order of a matrix over an infinite field (compare the natural method for finite fields sketched in the next paragraph). An example of an algorithm that seems to be restricted for this reason to algebra over finite fields is the procedure proposed by Leedham-Green and O'Brien **[19]** for recognising when a group G preserves a tensor decomposition. Cheryl Praeger and I are engaged in a project to produce a tensor factorisation algorithm which has some hope of being field-independent.

Another is that the order of a matrix is not easy to calculate. The best known method (see **[8]**) is to calculate the characteristic (or the minimal) polynomial; factorise it as $\prod f_i(t)^{m_i}$, where the polynomials f_i are monic and irreducible; define $N := p^m \operatorname{lcm}(q^{d_i} - 1)$, where p^m is a smallish power of the characteristic ($m = \lceil \log_2 d \rceil$ will always suffice) and $d_i := \deg f_i$; factorise N and seek the order as a divisor of N. The problem here is the factorisation of $q^{d_i} - 1$. For computations that arise in practice this does not appear to be a serious difficulty. The Cunningham Project **[5]** to compute and tabulate factorisations of numbers of the form $a^d - 1$ covers a range that reasonably matches the range of feasibility of matrix-group calculations set by other constraints. Nevertheless, the match is not perfect, and there are many numbers of the form $q^{d_i} - 1$ with d_i around 100 and q between 1024 and 65536 that cannot yet be factorised, whereas other aspects of matrix-group computations seem to be feasible for $q \leqslant 65536$ and $d \leqslant 200$. Besides, it is not clear whether the match is serendipitous or whether there is a good reason for it. If there is not then, as computers grow in power and speed and as ambitions grow in Computational Group Theory, there may come a time when the Cunningham Project falls substantially behind what is needed: factorisation seems unlikely to be achievable in polynomial time, whereas other components of matrix-group algorithms are, and as time progresses should therefore outstrip progress on factorisation of the relevant numbers.

A good response to this difficulty, and one which has been the basis of much successful research, is to work with implicit or partial information about $\mathrm{ord}(g)$, namely that g is what is now called a ppd-element of G (see **[23]**). A primitive prime divisor of $q^{d_i} - 1$ is one which does not divide $q^a - 1$ for $1 \leqslant a < d_i$ and a $\mathrm{ppd}(q, d_i)$-element is one whose order is divisible by such a prime. The question whether g is such an element can be answered satisfactorily (see Procedure $\Psi(d, q)$ in **[20**, §6]), and it is this that permits information about $\mathrm{ord}(g)$ to be used effectively in **[20]** and **[22]**. A related idea is exploited by Celler and Leedham-Green, who point out in **[8]** that we rarely need $\mathrm{ord}(g)$ exactly and therefore we do not need a complete factorisation of the numbers $q^{d_i} - 1$. Nevertheless, there are some algorithms, such as the tensor-factorisation procedure of Leedham-Green and O'Brien **[19]** mentioned above, for which the need to know the orders of group elements imposes severe restrictions both on practical implementations and on theoretical analysability.

A second response is probabilistic. It is along the lines: we are working with a random element g of G, and so the numbers d_i above may perhaps be taken to be random; the probability that a random k-bit integer can be factorised by elementary methods in time $O(k)$ is greater than $1 - 1/\log k$; therefore the procedure for finding $\mathrm{ord}(g)$ will almost always succeed. That this is true in practice is a matter of experience—see, for example, **[8, 9]**. From a theoretical point of view, however, there are some difficulties with this response. We simply do not know whether numbers of the special form $q^{d_i} - 1$ behave like random integers as input to factorisations algorithms. Conditioned as I am by half-knowledge of Lucas's tests for primality of Mersenne and Fermat numbers, and by knowledge of the existence of the Cunningham Project, I find myself inclined to believe that they behave better. But my ignorant beliefs are of no real value and I can find no reason why they may not behave worse. This is a serious impediment to a proper theoretical analysis of the probability of success and of the expected computational cost of some of the algorithms which we know from experience work pretty well in implementations.

In what ways might we introduce more linear algebra? One was suggested in Note 11.5 of **[20]**. There a $d \times d$ matrix g was called 'prima' if its characteristic polynomial $f(t)$ has an irreducible factor of multiplicity 1 and degree d_1, where $2 \leqslant d_1 \leqslant d - 2$ and d_1 is coprime with d and with the degrees of all the other irreducible factors of $f(t)$. The hope was expressed that a theory analogous to the long-established theory of 'prima' elements of permutation groups might be developed. I know of no progress along these lines, however.

Another method is to develop recognition algorithms based on sampling. More precisely, these are non-recognition algorithms. We know, for example, that $\mathrm{Prob}[g \in \mathrm{GL}(d, q)$ is eigenvalue-free$]$ is very close to e^{-1} (see **[21]**). Testing for eigenvalue-freeness can be done very quickly, so it is feasible to become confident that G is *not* the general linear group (in fact, does not contain $\mathrm{SL}(d, q)$) by random sampling. This rather crude idea is presently being tested and refined by Andrew Spencer, a research student at Oxford.

Work has also been initiated on a project to factorise tensor products and to extract exterior-square roots using linear algebra alone. If a fast exterior-square root algorithm could be designed then it would permit the replacement of a linear group G of degree $\frac{1}{2}m(m-1)$ with one of degree m, leading to considerable savings for most computations with G. In particular, one very narrowly focussed hope and expectation, is that $\mathrm{GL}(V)$ acting on $\wedge^2 V$ might be recognisable rather more cheaply than by black-box methods. First steps towards the extraction of exterior-square roots were taken by Catherine Greenhill in her Oxford DPhil thesis (1996: see **[14, 15]**) and further work is being done by Duncan Brydon at Oxford. It remains to be seen, however, whether these ideas can be made to succeed and to generalise naturally to the many other functorial factorisation problems connected with matrix representations.

Acknowledgements. I am very grateful to Gerhard Hiss, to Cheryl Praeger, and to Bill Kantor for their helpful comments on drafts of this note.

References

[1] Christine Altseimer and A. Borovik, Probabilistic recognition of orthogonal and symplectic groups, in: Groups and Computation III (W. M. Kantor and Á. Seress, eds.), Ohio State Univ. Math. Res. Inst. Publ. 8, Walter de Gruyter, Berlin–New York, 2001, 1–20.

[2] M. Aschbacher, On the maximal subgroups of the finite classical groups, Invent. Math. 76 (1984), 469–514.

[3] Robert Beals, C. R. Leedham-Green, Alice C. Niemeyer, Cheryl E. Praeger and Ákos Seress, A mélange of black box algorithms for recognising finite symmetric and alternating groups, in preparation.

[4] Sergey Bratus and Igor Pak, Fast constructive recognition of a black box group isomorphic to S_n or A_n using Goldbach's conjecture, J. Symbolic Comput. 29 (2000), 33–57.

[5] John Brillhart, D. H. Lehmer, J. L. Selfridge, Bryant Tuckerman, and S. S. Wagstaff, Jr., Factorizations of $b^n \pm 1$, $b = 2, 3, 5, 6, 7, 10, 11, 12$ up to high powers, second edition, Amer. Math. Soc., Providence, RI, 1988; updated at `http://www.cs.purdue.edu/homes/ssw/cun/`; see also `ftp://ftp.comlab.ox.ac.uk/pub/Documents/techpapers/Richard.Brent/factors`.

[6] A. Cayley, Desiderata and suggestions: (1) The theory of groups, Amer. J. Math. 1 (1878), 50–52 = Mathematical Papers, Vol. X, Paper 694, 401–403.

[7] Frank Celler, Konstruktive Erkennungsalgorithmen klassischer Gruppen in GAP, PhD Dissertation, RWTH Aachen, 1997.

[8] Frank Celler and C. R. Leedham-Green, Calculating the order of an invertible matrix, in [13], 55–60.

[9] Frank Celler and C. R. Leedham-Green, A non-constructive recognition algorithm for the special linear and other classical groups, in [13], 61–67.

[10] F. Celler and C. R. Leedham-Green, A constructive recognition algorithm for the special linear group, in: The atlas of finite groups: ten years on (Birmingham, 1995), London Math. Soc. Lecture Note Ser. 249, Cambridge Univ. Press, Cambridge, 1998, 11–26.

[11] Frank Celler, Charles R. Leedham-Green, Scott H. Murray, Alice C. Niemeyer and E. A. O'Brien, Generating random elements of a finite group, Comm. Algebra 23 (1995), 4931–4948.

[12] Gene Cooperman, Larry Finkelstein and Steve Linton, Constructive recognition of a black box group isomorphic to $\mathrm{GL}(n, 2)$, in [13], 85–100.

[13] Larry Finkelstein and William M. Kantor, eds., Groups and Computation II, DIMACS Ser. Discrete Math. Theoret. Comput. Sci. 28, Amer. Math. Soc., Providence, RI, 1997.

[14] Catherine Greenhill, An algorithm for recognising the exterior square of a matrix, Linear andMultilinear Algebra 46 (1999), 213–244.

[15] Catherine Greenhill, An algorithm for recognising the exterior square of a multiset, London Math. Soc. J. Comput. Math. 3 (2000), 96–116.

[16] Gerhard Hiss and Gunter Malle, Low degree representations of quasi-simple groups, preprint 1999.

[17] Christoph Jansen, Klaus Lux, Richard Parker and Robert Wilson, An atlas of Brauer characters, London Math. Soc. Monogr. (N.S) 11, Clarendon Press, Oxford 1995.

[18] W. M. Kantor and Á. Seress, Black box classical groups, Mem. Amer. Math. Soc. 149, Nr. 708, Amer. Math. Soc., Providence, RI, 2001.

[19] C. R. Leedham-Green and E. A. O'Brien, Recognising tensor products of matrix groups, Internat. J. Algebra Comput. 7 (1997), 541–559.

[20] Peter M. Neumann and Cheryl E. Praeger, A recognition algorithm for special linear groups, Proc. London Math. Soc. (3) 65 (1992), 555–603.

[21] Peter M. Neumann and Cheryl E. Praeger, Derangements and eigenvalue-free elements in finite classical groups, J. London Math. Soc. (2), 58 (1998), 564–586.

[22] Alice C. Niemeyer and Cheryl E. Praeger, Implementing a recognition algorithm for classical groups, in: [13], 273–296.

[23] Cheryl E. Praeger, Primitive prime divisor elements in finite classical groups, in: Groups St Andrews 1997 in Bath, II (C. M. Campbell et al., ed.), London Math. Soc. Lect. Note Ser. 261, Cambridge Univ. Press, Cambridge, 1999, 605–623.

The Queen's College
Oxford OX1 4AW, U.K.
peter.neumann@queens.ox.ac.uk

Cyclic matrices and the MEATAXE

Peter M. Neumann and Cheryl E. Praeger

Abstract. We present variants of two of the procedures for analysing matrix representations of groups which form part of the so-called MEATAXE collection of algorithms. Our variants make use of cyclic matrices in both their design and analysis. This leads to simplifications in the procedures, more complete complexity analyses, and potential improvements in practical performance.

1991 Mathematics Subject Classification: primary 20C40; secondary 16-04, 20-04, 20H30, 68Q25.

1. Introduction

The MEATAXE was created by Richard Parker as a computationally effective tool for analysing matrix representations of groups. Although his original description in [7] was for representations of 2-generator finite simple groups over the field with 2 elements, he and others quickly developed it as a general-purpose algorithm for application to matrix representations of a wide range of groups and algebras. Over the years substantial effort has been invested in further development and sophisticated versions of the MEATAXE based on that of Holt and Rees [2], and suites of programs containing variants of it, have been incorporated into algebraic computing systems such as GAP (Aachen/St Andrews) and MAGMA (Sydney).

In modern practical implementations the MEATAXE has many forms and many functions. Already in Parker's original description [7] it emerges that an algorithm for certifying irreducibility of a module may (and indeed should) look different from a method for finding submodules. We focus in this note just on the former. The point of our work is to introduce variants of the MEATAXE that are based on the exploitation of cyclic matrices. They should be no less efficient in practice than the majority of currently available versions. But their main justification lies in their analysability, which is based upon the existence of accurate estimates for the probability that a randomly chosen matrix is cyclic [4, 5].

The body of this paper is structured as follows. Section 2 contains preparation. In Section 3 we review the Norton Irreducibility Test, which contains the basic idea underlying several of the procedures, and discuss the variant of it which underlies the

Groups and Computation III
Ohio State Univ. Math. Res. Inst. Publ. 8

MEATAXE procedures designed by Holt and Rees. In Section 4 we state the theorems about the frequency of occurrence of cyclic matrices in irreducible matrix algebras on which our analyses are based. Then in Sections 5 and 6 we present and comment on our variants of MEATAXE procedures for testing irreducibility of a matrix algebra, and for finding the centraliser algebra of an irreducible matrix algebra, respectively. We restrict ourselves here to just these two computational problems: variants of other MEATAXE procedures for computing with irreducible matrix algebras are proposed and discussed in **[6]**.

We are grateful to the referees of this paper for comments which led to improvements in the exposition and also in the analyses reported in Sections 5 and 6. We also thank Alice Niemeyer for her interest in this work, especially for her implementations of our algorithms and for explaining to us the fast spinning algorithm.

2. Preparation

As in our earlier papers **[4, 5]**, we establish the following notation and terminology as standard throughout this paper:

F is a finite field, and $q := |F|$;

d is a natural number;

$\mathrm{M}(d, F)$ or $\mathrm{M}(d, q)$ is the algebra of $d \times d$ matrices over F;

$V := F^d$, the vector space of $1 \times d$ row vectors over F, and $\mathrm{M}(d, F)$ acts by right multiplication;

V^* is the vector space of $d \times 1$ column vectors over F, and $\mathrm{M}(d, F)$ acts on V^* by left multiplication;

if $X \in \mathrm{M}(d, F)$ then $\mathrm{null}_V(X)$ and $\mathrm{null}_{V^*}(X)$ denote the null spaces of X in V and V^* respectively;

if $X \in \mathrm{M}(d, F)$ then $c_X(t) := \det(tI - X)$, the characteristic polynomial of X, and $m_X(t)$ is the minimal polynomial of X;

if $X \in \mathrm{M}(d, F)$ then $F\langle X\rangle$ is the algebra generated by X, that is, the subalgebra of $\mathrm{M}(d, F)$ spanned by $I, X, X^2, \ldots$;

if $v \in V$ and $X \in \mathrm{M}(d, F)$ then $\langle v\rangle_X$ is the cyclic $F\langle X\rangle$-module generated by v, that is, the subspace of V spanned by $v, vX, vX^2, \ldots$;

if $v \in V$ and A is a subalgebra of $\mathrm{M}(d, F)$ then $\langle v\rangle_A$ is the A-module generated by v, that is, the subspace $\{vX \mid X \in A\}$ of V; similarly, if $v^* \in V^*$, then $\langle v^*\rangle_A$ is the A-submodule of V^* generated by v^*;

if $v \in V$ and $X \in \mathrm{M}(d, F)$ then the order $\mathrm{ord}_X(v)$ (of $\langle v\rangle_X$, or of v with respect to X) is defined to be the monic polynomial $p(t)$ of least degree such that $v\,p(X) = 0$;

if $\langle v \rangle_X = V$ for some v, then X is said to be cyclic, v is called a cyclic vector for X, and (X, v) is called a cyclic pair.

We shall give complexity estimates for our procedures in terms of the number of 'arithmetic operations' required. What this means is that we disregard differences in the costs of performing the field operations $+$, $-$, $\times$ and $\div$, which might be exacerbated by the size of the field F. Although it would not be difficult to refine the analysis to recognise that $+$ and $-$ are cheaper operations than $\times$ and $\div$, we treat $+$, $-$, $\times$ and $\div$ as costing the same computationally.

3. Irreducibility tests

The Norton Irreducibility Test, as described by Parker in [7], proceeds as follows.

Procedure 3.1. (Norton Irreducibility Test)

Input: Generators $X_1, \ldots, X_k$ for a subalgebra A of $\mathrm{M}(d, q)$.

Aim: Decide whether A is irreducible on V.

Step 1. Find $X \in A$ and $v \in V$ such that $\mathrm{null}_V(X) = \langle v \rangle$ (1 dimensional).

Step 2. Test whether $\langle v \rangle_A = V$.

Step 3. Find $u^* \in V^*$ such that $\mathrm{null}_{V^*}(X) = \langle u^* \rangle$; test whether $\langle u^* \rangle_A = V^*$.

Conclusion: A is irreducible if and only if both $\langle v \rangle_A = V$ and $\langle u^* \rangle_A = V^*$.

Proof. Since 'only if' is trivial, to see that the conclusion is valid we suppose that $\langle v \rangle_A = V$ and $\langle u^* \rangle_A = V^*$ and prove that V is irreducible as A-module. Let U be an A-invariant proper subspace of V. Define $U^* := \{w^* \in V^* \mid y \cdot w^* = 0 \text{ for all } y \in U\}$. Then U^* is A-invariant; in particular, both U and U^* are X-invariant. Now $v \notin U$ since $\langle v \rangle_A = V$, so the restriction $X|_U$ of X to U is non-singular, and hence also X induces a non-singular action on V^*/U^*. It follows that $u^* \in U^*$, but then $\langle u^* \rangle_A \subseteq U^*$, so $U^* = V^*$, whence $U = \{0\}$. Thus V is irreducible. □

As we mentioned in the introduction, Parker's main application of the Norton Irreducibility Test concerned irreducible groups and algebras over a field of size 2. There are a number of issues that arise in its application to an arbitrary irreducible algebra of matrices. The main problem is that matrices with nullity 1 do not exist in every irreducible subalgebra A of $\mathrm{M}(d, q)$. For example, if $d = bc$ with $b > 1, c > 1$, then the space V may be regarded as a c-dimensional vector space over a field of order q^b, and if $A := \mathrm{M}(c, q^b)$ then A may be regarded as an irreducible subalgebra of $\mathrm{M}(d, q)$. The nullity of every element of A, considered as a matrix in $\mathrm{M}(d, q)$, is a multiple of b. Note that a subalgebra A of $\mathrm{M}(d, q)$ is irreducible if and only if

$A \cong \mathrm{M}(c, q^b)$ for some b, c with $d = bc$. Thus the following questions needed to be considered.

1. What matrices in an irreducible subalgebra A could play a similar role to that of X in the Norton Irreducibility Test described above?

2. How can we find such matrices efficiently?

3. How can we find v and u^* (or their equivalents) efficiently?

4. How can we find $\langle v \rangle_A$ and $\langle u^* \rangle_A$ efficiently?

Finding answers to these questions led to the versions of the MEATAXE based on that of Holt and Rees [2] which form part of the packages GAP and MAGMA. It was observed in [2] that, if a matrix $X \in \mathrm{M}(d, q)$ has the property that $c_X(t) = f(t)g(t)$, where $f(t)$ is irreducible and does not divide $g(t)$, then $V = \mathrm{null}_V\, f(X) \oplus \mathrm{null}_V\, g(X)$, and the restriction of X to $\mathrm{null}_V\, f(X)$ is irreducible with characteristic polynomial $f(t)$. Similarly $V^* = \mathrm{null}_{V^*}\, f(X) \oplus \mathrm{null}_{V^*}\, g(X)$, and the restriction of X to $\mathrm{null}_{V^*}\, f(X)$ is also irreducible with characteristic polynomial $f(t)$. Let v, u^* be non-zero vectors in $\mathrm{null}_V\, f(X)$ and $\mathrm{null}_{V^*}\, f(X)$ respectively. A similar argument to that above proves that a matrix algebra A containing X is irreducible on V if and only if both $\langle v \rangle_A = V$ and $\langle u^* \rangle_A = V^*$. This variant of Norton's Irreducibility Test is exploited in [2] to design general purpose MEATAXE procedures. Holt and Rees observe that, if $A \cong \mathrm{M}(c, q^b)$ with $d = bc$, then the proportion of matrices X in A for which $c_X(t)$ has a multiplicity one irreducible factor is usably large. For example, it is close to $1 - \mathrm{e}^{-1}$ if $b = 1$ and d is large. This ensures that a matrix with the requisite property may be readily obtained by a suitable number of random selections of elements from A.

4. Cyclic pairs

Recall that, for $X \in \mathrm{M}(d, q)$ and $w \in V = F^d$, the pair (X, w) is called a cyclic pair if the matrix X is cyclic and w is a cyclic vector for it. A crucial observation to make, regarding extensions of the Norton Irreducibility Test, is that if X is cyclic and $c_X(t) = f(t)g(t)$ with $f(t)$ irreducible, then the restriction of X to $\mathrm{null}_V\, f(X)$ is irreducible whether or not $f(t)$ divides $g(t)$. We propose therefore to exploit the class of cyclic matrices, rather than the class of matrices X for which $c_X(t)$ has an irreducible factor of multiplicity 1, in the design of our MEATAXE procedures.

It is very simple and efficient in practice to test whether (X, w) is a cyclic pair. The process is known as spinning the vector w under X. For each $i := 1, \ldots, d$, we test whether wX^i is linearly independent of $w, wX, \ldots, wX^{i-1}$. If (and only if) this is true, then (X, w) is a cyclic pair, and $w, wX, \ldots, wX^{d-1}$ is the corresponding cyclic basis. Also the characteristic polynomial $c_X(t)$ may be simply written down: if

$wX^d = \sum_{0\le i\le d-1} c_i wX^i$ then $c_X(t) = t^d - \sum_{0\le i\le d-1} c_i t^i$. Using echelonisation to test whether wX^i lies in the subspace spanned by the vectors $w, wX, \ldots, wX^{i-1}$, for each i, requires $O(d^3)$ arithmetic operations. However, we could instead use a "fast spinning" procedure (see p. 436 of **[1]**) to decide whether $w, wX, \ldots, wX^{d-1}$ is linearly dependent, and hence to decide whether (X, w) is a cyclic pair. The cost of this is $O(d^\omega \log d)$, where ω, the exponent of matrix multiplication, is known to be less than 2.38 (see p. 420 of **[1]**). However, the method leading to this good asymptotic estimate does not give practical improvements over simple-minded spinning using echelonisation until d is reasonably large.

The usefulness of cyclic pairs for an extension of the Norton Irreducibility Test therefore depends on there being sufficient such pairs in $A \times V$, where A is an irreducible subalgebra of $\mathrm{M}(d, q)$, to ensure that we have a good chance of finding one by random selection. In **[6]** we show that there is indeed an adequate supply of cyclic pairs. Recall that each irreducible subalgebra of $\mathrm{M}(d, q)$ is isomorphic to $\mathrm{M}(c, q^b)$ for some b, c such that $d = bc$. For an extension field K of F of degree b define

$$\mathcal{S}(c, K) := \{f(t) \in K[t] \mid f \text{ monic, } \deg f(t) = c, \text{ and } f^\sigma, f \text{ are co-prime} \\ \text{for every non-identity } \sigma \in \mathrm{Gal}(K/F)\}.$$

It is not hard to prove (see **[6]**) that if A is an irreducible subalgebra of $\mathrm{M}(d, q)$ isomorphic to $\mathrm{M}(c, K)$ then $(X, w) \mapsto (w, wX, \ldots, wX^c)$ defines a one-to-one map from the set of cyclic pairs $(X, w) \in A \times V$ to V^{c+1} having the property that a $(c+1)$-tuple $(w_0, w_1, \ldots, w_c)$ lies in the image if and only if $(w_0, w_1, \ldots, w_{c-1})$ is a K-basis for V and the monic polynomial $t^c - \sum_{0\le i<c} a_i t^i$, where $w_c = \sum_{0\le i<c} a_i w_i$, lies in $\mathcal{S}(c, K)$. The following theorem is also proved in **[6]**.

Theorem 4.1. *Consider $A = \mathrm{M}(c, K)$ embedded as an irreducible subalgebra of $\mathrm{M}(d, q)$, where $d = bc$ and K is an extension field of F of degree b.*

(a) *The proportion $\mathrm{Prob}_{\mathrm{cycpair}}(c, q; b)$ of cyclic pairs in $A \times V$ is given by*

$$\mathrm{Prob}_{\mathrm{cycpair}}(c, q; b) = \prod_{i=1}^{c}\left(1 - \frac{1}{q^{bi}}\right) \cdot \frac{|\mathcal{S}(c, K)|}{q^{bc}}.$$

In particular, for $b = 1$ we have

$$\mathrm{Prob}_{\mathrm{cycpair}}(d, q; 1) = \prod_{i=1}^{d}\left(1 - \frac{1}{q^{i}}\right) > 1 - \frac{1}{q} - \frac{1}{q^2} + \frac{1}{q^5},$$

and in all cases $\mathrm{Prob}_{\mathrm{cycpair}}(c, q; b) \ge \delta(q)$, where $\delta(2) > 33/128 > 0.2578$, and if $q \ge 3$ then $\delta(q) > 0.5$.

(b) *For a nonzero $w \in V$, the proportion of cyclic pairs in $A \times \{w\}$ is independent of the choice of w, and is greater than $\mathrm{Prob}_{\mathrm{cycpair}}(c, q; b)$.*

For the procedure designed to test absolute irreducibility and find the centraliser algebra of an irreducible subalgebra $A \cong \mathrm{M}(c, q^b)$ of $\mathrm{M}(d, q)$, we need cyclic pairs (X, w) for which $c_X(t)$ has an irreducible factor of degree b. There are also sufficient such pairs of this type in $A \times V$.

Theorem 4.2. [6] *Let $A := \mathrm{M}(c, K)$ embedded as an irreducible subalgebra of $\mathrm{M}(d, q)$, where $d = bc$ and K is an extension field of F of degree b. Let w be a non-zero vector of F^d. Then the proportion of cyclic pairs (X, w) in $A \times \{w\}$ such that $c_X(t)$ has an irreducible factor of degree b is equal to*

$$\prod_{i=1}^{c}(1 - \frac{1}{q^{bi}}) \cdot \frac{|\mathcal{S}^*(c, K)|}{q^{bc} - 1},$$

where $\mathcal{S}^*(c, K)$ is the subset of $\mathcal{S}(c, K)$ comprising those polynomials which have a linear factor of the form $t - \lambda$ with λ a generator for K over F. In all cases this proportion is greater than 0.1.

5. A cyclic irreducibility test

Our proposed test for irreducibility of a matrix algebra is a variant of that suggested by Holt and Rees. It is a Las Vegas algorithm, as is the analogous procedure in **[2]**. That is to say, there is a non-zero probability that the procedure will fail to recognise that an irreducible algebra is, in fact, irreducible. If the algebra is reported to be irreducible however, then this result is guaranteed to be correct. We present an outline of the procedure below, and then comment on the methods used and the places where the procedure may fail to return an answer.

Procedure 5.1. (Cyclic Irreducibility Test)

Input: Generators $X_1, \ldots, X_k$ for a subalgebra A of $\mathrm{M}(d, q)$.

Purpose: To determine whether A is irreducible on V.

Step 1. Choose nonzero vectors $w \in V$, and $w^* \in V^*$.

Step 2. Find (by random search) $X \in A$ such that (X, w) is a cyclic pair.

Step 3. Find the order $c^*(t)$ of $\langle w^* \rangle_X$.

Step 4. Find an irreducible divisor $f(t)$ of $c^*(t)$, and let $g(t) := c_X(t) \div f(t)$ and $g^*(t) := c^*(t) \div f(t)$.

Step 5. Compute $v := wg(X)$ and $u^* := g^*(X)w^*$; these are nonzero vectors in $\mathrm{null}_V\, f(X)$ and $\mathrm{null}_{V^*}\, f(X)$ respectively.

Step 6. Find (by random search) $Y_1, Y_2 \in A$ such that (Y_1, v) and (Y_2, u^*) are cyclic pairs.

Conclusion: If all steps succeed then A is irreducible.

Remark 5.2. (a) The proof that the conclusion is valid is similar to those for the Parker version (given in Section 3) and the version of Holt and Rees. Details are given in **[6]**.

(b) Steps 2 and 6 may fail. But if A is irreducible Theorem 4.1 provides a lower bound $1-x$, where $x = 1-\delta(q) < 0.74$, for the probability of finding an appropriate cyclic matrix by random selection, and if n independent random selections are made of matrices $X \in A$ (in Step 2) or $Y_1, Y_2 \in A$ (in Step 6), then the probability that none of these matrices is cyclic is less than x^n. By choosing n sufficiently large we may control the probability of failing to recognise irreducibility because of failure at either of these steps.

(c) Step 4 may fail for very large q because the polynomial factorisation algorithm is non-deterministic, see, for example, Chapter 1 of **[8]**. Again the probability of failure at this step may be controlled. Thus we may specify a threshold $\varepsilon = 2\varepsilon_1 + \varepsilon_2$ such that if A is irreducible then the probability of failure at either Step 2 or Step 6 is less than ε_1 and the probability of failure at Step 4 is less than ε_2. In this way we may be confident, with probability greater than $1-\varepsilon$, of certifying that an irreducible subalgebra *is* irreducible.

(d) At Step 5, we do not need an echelonising process to find the vectors v and u^*. If $g(t) = \sum a_i t^i$, then v is the linear combination $\sum a_i(wX^i)$ of the vectors wX^i which will have been stored after their computation at Step 2. A similar process gives us u^*. Thus finding v and u^* requires $O(d^2)$ arithmetic operations.

(e) By means of Step 6 we avoid the need to compute the A-submodules generated by v and u^*. For, since (Y_1, v) and (Y_2, u^*) are cyclic pairs, we have $\langle v\rangle_A = \langle v\rangle_{Y_1} = V$, and $\langle u^*\rangle_A = \langle u^*\rangle_{Y_2} = V^*$. We have been informed by John Cannon that the computation of these submodules in current MEATAXE implementations is a major problem affecting their performance. We expect that this new feature alone may lead to improved practical performance.

(f) The number of arithmetic operations required for the Cyclic Irreducibility Test is $O(\mu_X + d^\omega \log d)$ for small q, and $O(\mu_X + d^\omega \log d + \mathrm{PF}(d,q))$ for large q, where μ_X is the cost of selecting one random matrix $X \in A$, $O(d^\omega)$ is the cost of multiplying together two $d \times d$ matrices over F, and $\mathrm{PF}(d,q)$ is the cost of factorising a degree d polynomial over F. The constants implicit in the 'Oh' notation may be made explicit (see **[6]**). It is known that $\omega < 2.38$ (see p. 420 of **[1]**). However, if a standard echelonisation procedure is used for the cyclic pair test, then the term $d^\omega \log d$ should be replaced by d^3. Several probabilistic algorithms are available for factorising polynomials over finite fields. They have different performance characteristics for different relative values of the degree d and field size q. For example, there is an algorithm for which the cost is $\mathrm{PF}(d,q) = O(d^{1.815} \log q)$ arithmetic operations (see **[3]**; also see Chapter 1 of **[8]** for an overview of polynomial factorisation algorithms).

(g) A preliminary implementation of this algorithm by Alice Niemeyer in GAP indicates that it should perform very well in practice.

6. Finding centraliser algebras

Once a subalgebra A of $\mathrm{M}(d, q)$ is known to be irreducible the next task is to find its centraliser algebra $C(A) = \{Z \mid XZ = ZX \text{ for all } X \in A\}$. If $A \cong \mathrm{M}(c, q^b)$ then $\dim C(A) = b$ and $C(A)$ is isomorphic to the algebra of scalar matrices of $\mathrm{M}(c, q^b)$. In particular, A will be absolutely irreducible (that is, $b = 1$) if and only if $C(A)$ consists just of the scalar matrices over F. Again the procedure we suggest is a variant of one in **[2]** and the main innovation is the use of cyclic matrices to avoid submodule computations and, in this case, to avoid also some polynomial-ring computations.

Like the Cyclic Irreducibility Test, this procedure too is a Las Vegas algorithm. Theorem 4.2 gives a lower bound for the proportion of matrices X in $\mathrm{M}(c, q^b)$ such that (X, w) is a cyclic pair, where w is a fixed nonzero vector, and $c_X(t)$ has an irreducible factor of degree b. Write this probability as $1 - x'$ and define $N(d, q; \varepsilon) := \log(\varepsilon^{-1})/\log(x'^{-1})$. If we make N independent random selections of matrices X from $\mathrm{M}(c, q^b)$, where $N \geq N(d, q; \varepsilon)$, then the probability that (X, w) is a cyclic pair for at least one of them, will be greater than $1 - \varepsilon$.

Procedure 6.1. (Cyclic Centraliser Algebra)

Input: Generators $X_1, \ldots, X_k$ for an irreducible subalgebra A of $\mathrm{M}(d, q)$.

Purpose: To find $C(A)$, and in particular to determine the value of b such that $A \cong \mathrm{M}(c, q^b)$.

Step 1. Choose a nonzero vector $w \in V$.

Step 2. Make N independent random selections of matrices X from A, where $N := \lceil N(d, q; \varepsilon) \rceil$, and whenever (X, w) is cyclic, find the minimum $e(X)$ of the degrees of the irreducible factors of $c_X(t)$. Let e be the greatest common divisor of d and all of these numbers $e(X)$.

Step 3. If $e = 1$ then A is absolutely irreducible, that is $b = 1$ and $C(A)$ consists of the scalar matrices (and the procedure is finished).

Step 4. If $e > 1$ and we have not found a cyclic pair (X, w) such that $c_X(t)$ has an irreducible factor of degree e then the procedure has failed.

Step 5. So suppose that $e > 1$ and (X, w) is a cyclic pair such that $c_X(t)$ has an irreducible factor $f(t)$ of degree e; we verify that $e = b$ as follows. Set $g(t) := c_X(t) \div f(t)$ and $v := wg(X)$, so that $\langle v \rangle_X = \mathrm{null}_V\, f(X)$.

Step 6. Find (by random search) $Y \in A$ such that (Y, v) is cyclic.

Step 7. Define $Z : vY^i \mapsto vXY^i$ for $0 \leq i < d$, and check that Z commutes with $X_1, \ldots, X_k$.

Conclusion: If all steps succeed then $e = b$. If $b = 1$ then $C(A)$ consists of the scalar matrices, while if $b > 1$ then Z is an algebra generator for $C(A)$.

Remark 6.2. (a) Step 2 may fail either because we fail to find cyclic pairs (X, w), or (for very large q) because of a failure of polynomial factorisation. Similarly Step 4 may fail because we have failed to find a suitable matrix X. Once we have found X, however, the only remaining task is to verify that $e = b$. Note that e must be divisible by b. The vector v may be computed efficiently as described in the previous section. But then the procedure may fail at Step 6 if we fail to find a suitable matrix Y.

(b) In Step 7, the fact that (Y, v) is cyclic means that we have a straightforward definition for the map Z. In particular, there is no need to spin a vector under the generators of A in order to define Z. Once we have verified that Z commutes with the generators $X_1, \ldots, X_k$ of A, we know that Z lies in $C(A)$. Since $vZ = vX$, and since X acts irreducibly on $\langle v \rangle_X$, it follows that Z acts irreducibly on $\langle v \rangle_X$, and hence $b = \dim C(A) \geq \dim \langle v \rangle_X = e$. Thus, since e is divisible by b, we conclude that $e = b$, and that Z generates $C(A)$.

(c) The number of arithmetic operations required for the Cyclic Centraliser Algebra procedure is $O(\mu_X + (k + \log d)d^\omega)$ for small q, and $O(\mu_X + (k + \log d)d^\omega + \mathrm{PF}(d, q))$ for large q, where, recall, k is the number of generators of A and $\mathrm{PF}(d, q)$ is the cost of factorising a polynomial of degree d over F.

References

[1] Peter Bürgisser, Michael Clausen and M. Amin Shokrollahi, Algebraic complexity theory, Springer-Verlag, Berlin and Heidelberg, 1997.

[2] D. F. Holt and S. Rees, Testing modules for irreducibility, J. Austral. Math. Soc. Ser. A 57 (1994), 1–16.

[3] Erich Kaltofen and Victor Shoup, Subquadratic-time factoring of polynomials over finite fields, Math. Comp. 67 (1998), 1179–1197.

[4] Peter M. Neumann and Cheryl E. Praeger, Cyclic matrices over finite fields, J. London Math. Soc. (2) 52 (1995), 263–284.

[5] Peter M. Neumann and Cheryl E. Praeger, Cyclic matrices in classical groups over finite fields, J. Algebra, to appear.

[6] Peter M. Neumann and Cheryl E. Praeger, Exploiting cyclic matrices in computer algebra: Sharpening the MEATAXE, in preparation.

[7] R. A. Parker, The computer calculation of modular characters (the Meat-Axe), Computational group theory, Procs London Math. Soc. Symposium on Computational Group Theory (Michael Atkinson, ed.), Academic Press, London, 1984, 267–274.

[8] Igor E. Shparlinski, Finite Fields: Theory and Computation, Kluwer, Dordrecht, 1999.

The Queen's College
Oxford OX1 4AW, U.K.
peter.neumann@queens.ox.ac.uk

Department of Mathematics and Statistics
University of Western Australia
Nedlands WA 6907, Australia
praeger@maths.uwa.edu.au

What do we know about the product replacement algorithm?

Igor Pak

Abstract. The *product replacement algorithm* is a commonly used heuristic to generate random group elements in a finite group G, by running a random walk on generating k-tuples of G. While experiments showed outstanding performance, until recently there was little theoretical explanation. We give an extensive review of both positive and negative theoretical results in the analysis of the algorithm.

1991 Mathematics Subject Classification: primary 68W40; secondary 20F69.

Introduction

In the past few decades the study of groups by means of computations has become a wonderful success story. The whole new field, Computational Group Theory, was developed out of needs to discover and prove new results on finite groups. More recently, the probabilistic method became an important tool for creating faster and better algorithms. A number of applications were developed which assume a fast access to (nearly) uniform group elements. This led to a development of the so called "product replacement algorithm", which is a commonly used heuristic to generate random group elements in finite groups. The main object of this paper is a rigorous study of this algorithm.

The story behind the product replacement algorithm is the following. Initially research in Computational Group Theory was mostly focused on working with permutation groups, where the fundamental algorithms of Sims (see **[Si]**) led the way to current advances. Sims's method is based on the computation of a base and strong generating set for a permutation group; after that computation, uniformly distributed, independent random elements of the group are available at a small cost.

Currently, matrix groups are in the focus of research in Computational Group Theory. In this setting, there is no efficient (neither in the theoretical nor in the practical sense of efficiency) analogue of the strong generating set constructions and so the method of random element generation in permutation groups cannot be duplicated. However, random element generation seems to be crucial for matrix group computations. It appeared that advancement in the "recognition project" (understanding the structure of a matrix group given by a set of generators) became possible only with a fast access to (nearly) uniform group elements (see e.g. **[BrP, Kn2, KS, NP1]**). The simplest idea does not work: it was observed on several occasions that getting random group elements by running a simple random walk on a group takes a very long time, or otherwise gives inadequate results (see **[Bb5, HR, DS3]**).

The problem of generating random group elements has two solutions: one practical and one theoretical. On a theoretical side, Babai in **[Bb3]** (see also Section 4.1) found a general black box algorithm (see **[Bb5]**) which produces (nearly) uniform group elements at a cost of $O(\log^5|G|)$ group multiplications. Being provably polynomial, albeit practically slow, this algorithm became a fundamental result on which subsequent theoretical work could be built. It did not resolve, however, the practical need for an efficient random group element generator.

The practical "product replacement algorithm" was discovered by Leedham-Green and Soicher **[LG]**. It was later tested **[CLMNO]** and proved to have remarkably good performance in several practically interesting cases (cf. **[HR]**). As success of the algorithm became widely acknowledged, it was included as a standard routine in two major group algebra packages GAP (see **[GAP]**) and MAGMA (see **[BSM]**).

Unfortunately, the reasoning why algorithm has such a good performance remained a mystery. Until recently, all attempts to prove theoretical results on performance of the algorithm either failed or produced incremental results (see **[Bb5, CG1, DG, DS2]**). A major work of Diaconis and Saloff-Coste **[DS3]** and several (joint) results of the author **[BbP, LP, P4, P5, P6, PB]** gave a new life to the hopes of fully understanding the algorithm.

The aim of this review article is to give an up to date review of the state of our knowledge on the algorithm. We tried to make this review accessible to nonspecialists, so on occasion we give proofs of standard or elementary results, as well as spend much time giving background in certain areas. The attempt was to have a standard point of reference to most (often basic) results that come up in the analysis of the algorithm.

The product replacement algorithm is defined as follows (**[CLMNO]**). Given a finite group G, let $\mathcal{N}_k(G)$ be the set of k-tuples $(g) = (g_1, \ldots, g_k)$ of elements of G such that $\langle g_1, \ldots, g_k\rangle = G$. We call elements of $\mathcal{N}_k(G)$ the *generating k-tuples*. Given a generating k-tuple $(g_1, \ldots, g_k)$, define a *move* on it in the following way. Choose uniformly a pair (i, j), such that $1 \le i \ne j \le k$, then apply one of the following four operations with equal probability:

$$R^{\pm}_{i,j} : (g_1, \ldots, g_i, \ldots, g_k) \to (g_1, \ldots, g_i \cdot g_j^{\pm 1}, \ldots, g_k),$$

$$L^{\pm}_{i,j} : (g_1, \ldots, g_i, \ldots, g_k) \to (g_1, \ldots, g_j^{\pm 1} \cdot g_i, \ldots, g_k).$$

Note that these moves map a generating k-tuple into a generating k-tuple. Now apply these moves t times (the choice of the move must be uniform and independent at each step), and return a random component of the resulting generating k-tuple. This is the desired "random" element of the group G.

Another way to describe the algorithm, is to define on $\mathcal{N}_k(G)$ a structure of a graph induced by maps $R^{\pm}_{i,j}$ and $L^{\pm}_{i,j}$. This makes $\mathcal{N}_k(G)$ into a $4k(k-1)$-regular graph with no orientation on edges, but with loops when $k > d(G)$, where $d(G)$ is the minimal number of generators of G. Now the algorithm consists of running a nearest neighbor random walk on this graph (for t steps) and returning a random component of the stopping state. We refer to this random walk as the *product replacement random walk* $\mathcal{W} = \mathcal{W}_k(G)$.

About the presentation of a group. We assume the group is given as a black box group, which means that there is an oracle which can multiply elements, invert them, and compare them with identity (see **[Bb5]**). The group is then defined by a set of generators $(g_1, \dots, g_l)$. Now, in the algorithm one should take $k \geq l$ and set $g_{l+1} = \dots = g_k = \text{id}$ (see **[CLMNO]**).

A few words about the parameters k and t. In the original paper **[CLMNO]** the authors showed that when $k > 2\log|G|$ and t is large enough, the algorithm will provably work. Practical computations suggested that small values work as well. Much of our work is concentrated on a proper choice of k and t.

Our analysis of the algorithm breaks into three mostly separate parts:

1) *Bias.*

2) *Connectivity.*

3) *Mixing time.*

Each of these is the subject of a separate section of this review. We sum up things at the end (Section 4) where we allow ourselves some speculations on the possible future developments. Let us now give a brief overview of the sections.

First, observe that it is unclear whether $\Gamma_k(G)$ is connected. When it isn't, little can be said about the connected component $\Gamma'_k(G)$ which contains (g). In Section 2 we review what is known about this problem, especially in connection with an algebraic notion of (T) systems. Following **[P4, P5]** we apply results in probabilistic group theory to show that, for reasonably small k, graphs $\Gamma_k(G)$ already have "large" connected component.

In a different direction, Section 1 is dedicated to bias of the output. Formally, if $(g) = (g_1, \dots, g_k)$ is chosen uniformly from $\mathcal{N}_k(G)$, consider the bias in the distribution Q of a random component g_i. While mentioned in **[CLMNO]**, this problem was largely ignored until recently, when the first examples of strong bias have been found (**[PB, BbP]**). In our treatment we follow the recent paper **[BbP]** of Babai and the author.

Finally, is Section 3, we review the results on the mixing time of the product replacement random walk $\mathcal{W}_k(G)$. In the past two decades there has been a spectacular progress in the study of discrete Markov chains, in particular, random walks on groups. Several attempts have been made to proceed with the analysis in the case of $\mathcal{W}_k(G)$. We will concentrate on the paper **[DS3]** of Diaconis and Saloff-Coste, where the authors use a state of the art analytic approach (largely developed in their previous papers) to obtain subexponential bounds on the mixing time for general groups.

A different approach was used by Lubotzky and the author in **[LP]**, based on the Kazhdan's property (T) from the representation theory of Lie groups. We show that the positive solution of an important open problem (whether a group of automorphisms of a free group $\mathrm{Aut}(F_k)$ has property (T)) implies that the graphs $\Gamma_k(G)$ are expanders, with an expansion constant $\varepsilon = \varepsilon(k)$ depending only on k. In special cases this gives a rigorous proof of the rapid mixing. We follow **[LP]** to give what seem to be the first explanation of the rapid mixing phenomenon.

A last warning before we conclude the introduction. This paper is purely theoretical and we do not claim (and do not review) any practical results on performance of the algorithm. We review a number of theoretical results, and try to give proper acknowledgements when appropriate. The remaining results are either folklore or due to the author.

1. The bias in the output

Much of this section is dedicated to probabilistic group theory. We do not attempt to give a review of this interesting and rapidly developing area for several reasons. First, there are already several such reviews which cover different parts of the subject (see **[Kn2, Sh]**). Second, this issue is rather tangential to the main subject of our study and is used mostly as a powerful tool in analysis of the algorithm. Still, for the sake of completeness we present here a somewhat biased review of results we will use later on.

1.1. Probability of generating a finite group. Let G be a finite group, and let $d(G)$ be the minimal number of generators of G. By $\varphi_k(G)$ denote the probability that k uniformly and independently chosen elements in G generate the whole group:

$$\varphi_k(G) = \frac{|\mathcal{N}_k(G)|}{|G|^k}$$

The subject of this section is to give some estimates on these probabilities.

Theorem 1.1.1. *For every sequence of nonisomorphic simple groups $\{G_i\}$ we have*

$$\varphi_2(G_i) \to 1, \quad as \ |G_i| \to \infty$$

This theorem is a combination of several results. First, it was proved for alternating groups A_n by Dixon **[Dx]**. Then, for classical groups (and few more series), it was established by Kantor and Lubotzky **[KL]** (see also **[Kn1]**). They used Aschbacher's classification of maximal subgroups of linear groups **[As]**. Finally, Liebeck and Shalev **[LS1, LS2]** completed the proof for the remaining series. We should add that these papers (except for **[Dx]**) use extensively the classification of finite simple groups.

The asymptotic behavior of the probabilities φ_2 is also known and the results can be summarized as follows.

Theorem 1.1.2. *Let A_n be alternating groups, $G_n(q)$ be a simple group of Lie type of (untwisted) rank n. Then:*

(i) $\varphi_k(A_n) = 1 - \frac{1}{n^{k-1}} + O\left(\frac{1}{n^{2k-1}}\right)$, *where* $k \geq 2$

(ii) $\varphi_2(G_n(q)) = 1 - O\left(\frac{n^3 \log^2 q}{q^n}\right)$.

Part (i) follows from the work of Babai **[Bb2]**, part (ii) was conjectured and partly proved by Kantor and Lubotzky **[KL]**, and proved in full by Liebeck and Shalev **[LS1, LS2]**.

Proposition 1.1.3. *Let G be a finite p-group, $d = d(G)$. Then:*

(i) $\varphi_d(G) > 1 - \frac{1}{p} - \frac{1}{p^2}$,

(ii) $\varphi_{d+1}(G) > 1 - \frac{1}{p(p-1)}$,

(iii) $\varphi_{d+r-1}(G) > 1 - \frac{1}{p^r - p^{r-1} - p^{r-2} - 1}$, *where* $r \geq 1$.

The result is obtained by direct calculation due to the fact that the quotient $G/\Phi(G) \simeq \mathbb{Z}_p^m$, and $\varphi_k(G) = \varphi_k(G/\Phi(G))$, where $\Phi(G)$ is a Frattini group of G. Different versions of this result were obtained in **[DS3, P4]**. For nilpotent groups we also obtain:

Proposition 1.1.4. *Let G be a finite nilpotent group, $d = d(G)$. Then:*

(i) $\varphi_d(G) > \frac{1}{5 \log\log |G|}$,

(ii) $\varphi_{d+1}(G) > \frac{1}{e}$,

(iii) $\varphi_{d+r-1}(G) > 1 - \frac{8}{12^{r/4}}$, *where* $r \geq 1$.

Again, see **[DS3]**, Remark after Lemma 6.3; and **[P4]**.

Theorem 1.1.5. *There exists a universal constant $C < 10^7$ such that for any solvable group G, $d = d(G)$, and $r > 0$ we have*

$$\varphi_{(\beta+1)\cdot d+C+r} > 1 - \alpha^r,$$

where $\beta = (3 \cdot \ln 48 + \ln 24)/(3 \cdot \ln 9) \approx 2.243991050$ *is the Pálfy–Wolf constant, and* $\alpha = 1/\sqrt[4]{12} < 1$.

This result is due to the author **[P4]** and is obtained by using the product formula of Gachütz **[Ga]**. A weaker version is due to A. Mann **[Mn]** who used a somewhat different method to show that $\varphi_{(\beta+1)d+C}(G)$ is greater that some universal constant $\epsilon > 0$. In the other direction, Mann in **[Mn]** showed that $\varphi_{\beta d-C}(G) \to 0$ for certain series of groups. This implies that the constant $(\beta + 1)$ in Theorem 1.1.5 cannot be improved to a constant smaller than β.

Question 1.1.6. Can one improve the constant $(\beta + 1)$ in Theorem 1.1.5 ? Can one find a "reasonable" bound on C (compared to 10^7 in the theorem) ?

Finally, for general groups we have the following weak result:

Theorem 1.1.7. *For any finite group G, $1 > \epsilon > 0$, $m = \lceil \log_2 |G| \rceil$, we have*

(i) $\varphi_k(G) \geq \varphi_k(\mathbb{Z}_2^m)$,

(ii) $\varphi_k(G) > 1 - \epsilon$, *where* $k > m + 2 + \log_2 1/\epsilon$,

(iii) $\varphi_m(G) > \frac{1}{4}$ *and* $\varphi_{m+1}(G) \geq \frac{1}{2}$.

Part (ii) of the theorem is a slight improvement over a more general classical result of Erdős and Rényi in **[ER]** (see also **[P3]**). The proof of part (i) is based on "random group process" defined in **[P4]** (cf. **[Ac, PV]**). Parts (ii), (iii) then follow from (i) and Proposition 1.1.3.

We conclude with a somewhat related recent result of Guralnick and Kantor. While technically different, we find it amazingly useful in several application.

Theorem 1.1.8. *Let G be a nonabelian finite simple group. Define*

$$\mathrm{PC}(G) = \max_{C} \min_{g \in G, g \neq \mathrm{id}} \mathbf{P}\big(\langle g, h\rangle = G \mid h \in C\big),$$

where max *is over all conjugacy classes $C \subset G$, and the probability is taken over uniform $h \in C$. Then:*

(i) $\mathrm{PC}(G) > 1/10$ *for every G,*

(ii) $\mathrm{PC}(G) \to 1$ *as $|G| \to \infty$, for G not isomorphic to A_n or $\Omega(2m + 1, q)$ with a bounded q.*

1.2. Generating k-tuples in products of simple groups. A famous result of Philip Hall **[H1]** (see also **[KL]**) can be formulated as follows:

Theorem 1.2.1. *Let G be a nonabelian simple group. Then the maximal number N, such that group $H = G^N$ (direct product of N copies of a G) is generated by k elements,* *is equal to* $|\mathcal{N}_k(G)|/|\mathrm{Aut}(G)|$. *Further, an element* $(g) = (g_1, \dots, g_k) \in H^k$, *where* $g_i = \left(g_i^{(1)}, \dots, g_i^{(m)}\right) \in H$, *is a generating* k*-tuple of* $H = G^N$ *if and only if all* k*-tuples* $\left(g_1^{(j)}, \dots, g_k^{(j)}\right)$, $1 \le j \le m$ *generate the group* G *and lie in different orbits of the diagonal action of* $\mathrm{Aut}(G)$ *on* $\mathcal{N}_k(G)$.

In a celebrated special case of $G = A_5$ and $k = 2$, Hall showed **[H1]** that $N = 19$, i.e. that $d(A_5^{19}) = 2$, while $d(A_5^{20}) = 3$.

Recall that by Theorem 1.1.1, for any fixed $k > 1$ the probabilities of generation $\varphi_k(A_n) \to 1$. Since $\mathrm{Aut}(A_n) = S_n$ for $n \ge 5$, $n \ne 6$, we obtain that the N from the theorem satisfies:

$$N = \varphi_k(A_n)\frac{|A_n|^k}{|S_n|} > \frac{n!^{k-1}}{2^{k+2}},$$

where n is large enough. In particular, if $G = A_n$ and n is large enough, the group $A_n^{n!/8}$ can be generated by two elements, while $A_n^{n!/4}$ cannot.

Now let $N = n!/8$, $k > 5$. Consider a set of k-tuples in $H = A_n^N$. From the above and the birthday paradox (see **[F, KL, BbP]**) one can easily see that the probability that two k-tuples $\left(g_1^{(j)}, \dots, g_k^{(j)}\right)$ and $\left(g_1^{(j')}, \dots, g_k^{(j')}\right)$ lie in the same orbit of $\mathrm{Aut}(G)$ goes to zero superexponentially. Thus one can ignore this probability and conclude that N generating k-tuples almost surely correspond to a generating k-tuple of A_n^N.

First, observe that $\varphi_k(A_n^N) \to 0$ as $n \to \infty$, given $k = o(n)$ (**[KL]**). Indeed, the probability that any given k-tuple generates A_n is at most $(1 - n^{-k})$, which is the probability that none of the k permutations fixes 1. Therefore the probability that *all* of the N k-tuples generate A_n is at most $(1 - 1/n^k)^N \to 0$ as $n \to \infty$.

From here it is immediate that in order to have $\varphi_k > \epsilon > 0$, one needs $k = \Omega(n)$. Now, in the reverse direction one has the following general result:

Theorem 1.2.2. *If* G *is a direct product of simple nonabelian groups (possibly, with repetitions), then* $\varphi_k(G) > 1/2$ *for* $k \le C \log m$, *where* m *is the maximal number of isomorphic copies of each group and* C *is a universal constant.*

The result is due to the author **[P4]**. The proof uses a number of technical calculations involving the generation probabilities in Theorem 1.1.2, as well as the classification of finite simple groups. The constant C is expected to be reasonable, but probably should be hard to calculate due to difficulties with calculation on large sporadic groups.

1.3. Proving the bias. Now we are ready to present the first version of the bias in the output of the product replacement algorithm. We shall start with preliminary observation.

Let G be a finite group, and let Q_k, Q_k^* be the probability distributions of the first and of the random component in a uniform generating k-tuple $(g) \in \mathcal{N}_k(G)$. Then symmetry gives:

Lemma 1.3.1. *For all G, $k \geq d(G)$ we have* $\mathrm{Q}_k = \mathrm{Q}_k^*$.

The lemma reduces the problem of bias of the output to the problem of bias of the first component in generating k-tuples of G.

Let $H = A_n^N$, $N = n!/8$, $k > 5$. Consider a random generating k-tuple $(h) = (h_1, \dots, h_k) \in \mathcal{N}_k(H)$. We write $h_1 = (\sigma_1, \dots, \sigma_N)$, where $\sigma_j \in A_n$. Now, each σ_j is the first component of a random generating k-tuple of A_n. While they may seem close to being nearly uniform in A_n, a small bias remains:

Lemma 1.3.2 ([BbP]). *If σ is the first element in a uniform generating k-tuple of A_n, then*

$$\mathbf{P}\big(\sigma(1) = 1\big) = \frac{1}{n} - \frac{1}{n^k} + O\left(\frac{1}{n^{2k-1}}\right)$$

The proof is immediate from part (i) of Theorem 1.1.2. See **[BbP]** for a complete proof and other versions of the lemma.

Now, in an unbiased sample we have $\mathbf{P}(\sigma_j = 1) = 1/n$. Since the number of independent samples is N, we must have the bias in each of them less than N, or otherwise the bias becomes statistically significant. In other words, if $k = o(n)$, then $n^k = o(n!/8)$, and the bias can be detected.

Theorem 1.3.3. *Let $H_n = A_n^N$, where $N = n!/8$. Let $k = o(n)$. Then there exists a sequence of subsets $B_n \subset \mathcal{N}_k(H_n)$ such that $\mathrm{Q}_k(B_n) \to 0$, and $|B_n|/|\mathcal{N}_k(G)| \to 1$ as $n \to \infty$.*

Proof. Take B_n to be the set of all $(h) = (\sigma_1, \dots, \sigma_N) \in H_n$ such that

$$\#\big\{j \mid \sigma_j(1) = 1\big\} > N\left(\frac{1}{n} - \frac{1}{2\,n^k}\right).$$

The Chernoff bound (see e.g. **[ASE]**) gives that the N random permutations $\sigma_j \in A_n$ are in B_n with high probability.

On the other hand, for σ_j chosen from generating k-tuple of A_n we have

$$\mathbf{E}\left(\#\big\{j \mid \sigma_j(1) = 1\big\}\right) = N \cdot \left(1 - \frac{1}{n^k}\right),$$

so the Chernoff bound gives that such N independent σ_j do not lie in B_n with high probability. □

Remark 1.3.4. The first results on bias was announced in **[PB]**, and rigorously proved in **[BbP]**. One can generalize the result to other sets B_n as well as to other series of simple groups of Lie type. We refer to **[BbP]** for a more detailed treatment.

Before we finish this section, let us elaborate on the findings. Basically, we showed that even if the graph $\Gamma_k(H)$ is connected, even if the product replacement random walk mixes rapidly, the resulting distribution of the output can still be biased. Of course, for that to happen we need $\varphi_k(H_n) \to 0$ as $n \to \infty$, which is indeed true when $H_n = A_n^{n!/8}$. While this is only a necessary condition, it seems likely that bias exists in all natural cases when $\varphi_k(H_n) \to 0$. However proving bias in such cases can be a delicate task.

Open Problem 1.3.5. Can one exhibit the bias for a sequence of solvable groups?

Remark 1.3.6. We believe a natural candidate for H_n in 1.3.5 to be the group $H_n = R_n \ltimes \left(\mathbb{Z}_2^n\right)^N$, where R_n is an n-th iterated wreath product of S_4:

$$R_n = S_4 \wr S_4 \wr \ldots \wr S_4 \;\; (n \text{ times}),$$

where the action of R_n is independent on each copy of $\mathbb{Z}_2^n$, and N is the largest number such that H_n still can be generated by three (or any fixed constant number of k) elements. Note that it is not even obvious that R_n is two-generated. On the other hand, using Gaschütz formulas A. Mann showed in **[Mn]** that $\varphi_2(R_n) > 1/2$ for any n. Still, we challenge the reader to find a single "nice" generating 2-tuple.

1.4. Detecting the bias. While from general mathematical point of view the previous section completely answers the question about existence of the bias, from computational point of view the answer is somewhat weak. The idea is that generating random group elements is not a goal, but rather a routine used in various randomized algorithms (see e.g. **[Bb4, Bb5, BP, Kn2, P2]**). Thus the "real" threshold for bias is much higher: one must show that these "random elements" can actually "mess up" some calculations.

In the original paper **[CLMNO]** the authors tested the algorithm by using χ^2 on the distribution of the orders of the elements. The order statistics is of great importance, especially in various recognition algorithms (see e.g. **[BrP, Kn2, KS, NP1]**). Let us show that they are useless in case of powers of A_n.

Proposition 1.4.1. *The proportion of elements of* $H = A_n^N$ *which have the same order is at least* $1 - e^{-N/(n \log n)}$, *for any integers* n, N.

Proof. First, let us determine this order. Denote by $m_p(n) = p^{\lfloor \log_p n \rfloor}$ the maximal $m \le n$ such that $m = p^r$. Let $M = \prod_p m_p(n)$, where the product is over all primes $p \le n$. Since M is divisible by every possible length i of the cycle, $1 \le i \le n$, for every $h \in H$ we have $h^M = \mathrm{id}$.

Let us now calculate the probability that $h \in H$ has order M. Observe that the length of the cycle containing 1 is uniform in $\{1, \ldots, n\}$. Thus with probability at least $1/n$ an element $\sigma \in A_n$ contains a cycle of length $m_p(n)$. The probability that no component of a random $h \in H$ contains a cycle of length $m_p(n)$ is at most $(1 - 1/n)^N$. Finally, the probability that components h contain cycles of length $m_p(n)$ for every prime p is at least $1 - n \cdot (1 - 1/n)^N > 1 - \exp(-N/n \log n)$. But then h has order M which completes the proof. □

The proposition implies that the bias found in the previous section cannot be detected by the order statistics. This begs the following question, possibly related to Problem 1.3.5 :

Question 1.4.2. Can one find a sequence of groups H_n which exhibit bias in the order statistics ?

Now let us show that the bias can nevertheless be detected by a short straight line program. First we need several definitions.

Let $\mathbf{w}$ be a word over the alphabet $\{x_i^{\pm 1}, i = 1, 2, \dots\}$. Substituting elements of G for the x_i assigns $\mathbf{w}$ a value in G. Assume that the x_i are chosen independently from the probability distribution P over G. We denote by $\mathbf{w}[P]$ the probability distribution of the value of $\mathbf{w}$.

Let G be a finite group. As in the previous section, by $\mathrm{Q}_k = \mathrm{Q}_k^*$ denote the probability distribution on G of the random component in a generating k-tuple $(g) \in \mathcal{N}_k(G)$. By U denote the uniform distribution on G.

Theorem 1.4.3. *Let $k = k(n) \geq 4$ and $k = o(n)$, $n \to \infty$. Then there exists a sequence of words $\mathbf{w}_{n,k}$ with the following properties. The length of $\mathbf{w}_{n,k}$ is $n^{O(k)}$. Let $\omega(n) \to \infty$, $\omega(n) = o(n)$. Set $m = n^{k\omega(n)}$, and $G = A_n^m$. Then $\mathbf{w}[\mathrm{Q}_k] = \mathrm{id}$ has probability $1 - O(n^{-cn})$, while $\mathbf{w}[\mathrm{U}] = \mathrm{id}$ has probability $O(n^{-cn})$, where c is a universal constant.*

Let us sketch the idea of the proof. First, let n be a prime. Rather than look at the number of σ_j in $g = (\sigma_1, \dots, \sigma_m)$, such that $\sigma_j(1) = 1$ (see the proof of Theorem 1.3.3) we look at the number of those σ_j which are n-cycles. But since n is prime, we have $\sigma^n = \mathrm{id}$ implies σ is a long cycle or $\sigma = \mathrm{id}$. The analog of Lemma 1.3.2 is also straightforward. Therefore we have a (positive) bias in the number of trivial components in random element $g^n \in G$.

Now, a general theory (the Ajtai–Komlós–Szemerédi sorting network in [**AKS**]) shows how to construct a short monotone Boolean circuit which "detects" the bias. The problem is to construct a short word $\mathbf{w}$ from the circuit. Basically we need to find probabilistic simulation of **AND** and **OR** by group operations.

Formally, let H be a group and $g \in H$. Consider the predicate $\mathcal{E}(g)$ meaning "$g = \mathrm{id}$". We wish to construct words w_1 and w_2 corresponding to the predicates $\mathcal{E}_1(g, h) = \mathcal{E}(g) \wedge \mathcal{E}(h)$ and $\mathcal{E}_2(g, h) = \mathcal{E}(g) \vee \mathcal{E}(h)$, respectively. Clearly, there is no word which would be id exactly if $\mathcal{E}_1$ holds, nor is there one for $\mathcal{E}_2$. But the product $w_1 = gh$ and the commutator $w_2 = [g, h] = g^{-1}h^{-1}gh$ go part of the way: $\mathcal{E}_1(g, h)$ implies $w_1 = \mathrm{id}$ and $\mathcal{E}_2$ implies $w_2 = \mathrm{id}$; and the converse holds often enough in each case.

Now, what we do is consider $w_1 = g'h'$ and $w_2 = [g', h']$ where $g' = g^{a_1} \cdot \ldots \cdot g^{a_l}$, where $g^a = a^{-1}ga$, and a_i are independent and chosen from the same distribution as g; h' is defined analogously. Using rapid mixing of random walks on $H = A_n$ we conclude w_1, w_2 are the desired probabilistic simulations of the monotone Boolean operations. We refer to **[BbP]** for details and references.

Remark 1.4.4. A priori it may seem unlikely that the graph $\Gamma = \Gamma_k(A_n^{n!/8})$ is connected when $k \geq 4$. In this case the result of Theorem 1.4.3 is only a theoretical exercise as the algorithm does not produce uniform generating k-tuples. Nevertheless the graph Γ is probably connected indeed for all $k \geq 4$. In particular, we know that $\Gamma_4(A_5^N)$, $N \leq 1140$ is connected. We refer to Corollary 2.4.5, Remark 2.4.7 and Conjecture 2.5.5 below.

1.5. Avoiding the bias. To avoid having a bias of Q on G one should take k to be sufficiently large. How large? That depends on your definition of the bias and structural properties of G.

Let P be a probability distribution on a finite set X, and U be a uniform distribution on X. Define the *total variation* between P and U as follows:

$$\| \mathrm{P} - \mathrm{U} \|_{\mathrm{tv}} = \max_{B \subset X} \left| \mathrm{P}(B) - \frac{|B|}{|X|} \right| = \frac{1}{2} \sum_{x \in X} \left| \mathrm{P}(x) - \frac{1}{|X|} \right|,$$

where $\mathrm{P}(B) = \sum_{x \in B} \mathrm{P}(x)$. It is easy to see that $0 \leq \| \mathrm{P} - \mathrm{U} \|_{\mathrm{tv}} \leq 1$. We postpone discussion of the total variation and other distances until Section 3.1. Instead, let us make the following elementary observation.

Proposition 1.5.1. *Let* P *be a probability distribution on* $X \subset G^k$ *and let* Q *be a probability distribution on* G *defined as a projection of* P *onto the first component. Then*

$$\| \mathrm{Q} - \mathrm{U} \|_{\mathrm{tv}} \leq \| \mathrm{P} - \mathrm{U} \|_{\mathrm{tv}} + \left(1 - \frac{|X|}{|G|^k} \right)$$

In particular, when $X = \mathcal{N}_k(G)$ *we have* $\| \mathrm{Q} - \mathrm{U} \|_{\mathrm{tv}} \leq 1 - \varphi_k(G) + \| \mathrm{P} - \mathrm{U} \|_{\mathrm{tv}}$.

Proof. Denote by P′ the probability distribution on $G^k \supset X$, which is P on $(g) \in X$ and zero otherwise. Clearly,

$$\begin{aligned} \| \mathrm{P}' - \mathrm{U} \|_{\mathrm{tv}} &= \frac{1}{2} \sum_{(g) \in X} \left| \mathrm{P}'(g) - \frac{1}{|G|^k} \right| + \frac{1}{2} \sum_{(g) \in G^k \setminus X} \frac{1}{|G|} \\ &\leq \frac{1}{2} \sum_{(g) \in X} \left| \mathrm{P}'(g) - \frac{1}{|X|} \right| + \frac{1}{2} \sum_{(g) \in X} \left| \frac{1}{|X|} - \frac{1}{|G|^k} \right| \\ &\quad + \frac{1}{2} \left(1 - \frac{|X|}{|G|^k} \right) = \| \mathrm{P} - \mathrm{U} \|_{\mathrm{tv}} + \left(1 - \frac{|X|}{|G|^k} \right). \end{aligned}$$

Now, since a uniform distribution on G is a projection of the uniform distribution on G^k, we immediately obtain:

$$\|Q - U\|_{tv} \le \|P' - U\|_{tv} \le \|P - U\|_{tv} + \left(1 - |X|/|G|^k\right). \qquad \square$$

Now, fix a sequence $\{G_n\}$ of groups. Let us assume that $\varphi_k(G_n) \to 1$ as $n \to \infty$. Then Proposition 1.5.2 implies that the bias of Q is bounded by that of the random walk. In the remainder of this section we will be concerned with the problem of finding an appropriate function $k = k(G_n)$ for a given sequence of groups to satisfy the assumption above.

First, let us translate the previous results into this language. By $\omega(n)$ denote a function $\omega : \mathbb{N} \to \mathbb{N}$ such that $\omega(n) \to \infty$ as $n \to \infty$.

Corollary 1.5.2. *Let $\{G_n\}$ be any sequence of groups. Then $\varphi_k(G_n) \to 1$ given $k = k(n)$ satisfies either of the following:*

(i) $k = d(G_n) + \omega(n)$, *if G_n is nilpotent,*

(ii) $k = (\beta + 1) \cdot d(G_n) + \omega(n)$, *if G_n is solvable,*

(iii) $k = (\log m) \cdot \omega(n)$, *if G_n is a direct product of finite nonabelian simple groups, each copy at most m times,*

(iv) $k = \log |G_n| + \omega(n)$, *for any G.*

While parts (i)–(iii) seem to be satisfactory, the general case (iv) remains quite weak. The following conjecture if true would be a significant advancement:

Conjecture 1.5.3. *Let $\{G_n\}$ be a sequence of groups. Then $\varphi_k(G_n) \to 1$ given $k = \omega(n) \cdot d(G_n) \cdot \log\log |G_n|$.*

Remark 1.5.4. Note that the results (i)–(iii) in Proposition 1.5.2 are stronger than what Conjecture predicts. Indeed, for the only nontrivial case (iii), if $G = F_1^m \times F_2^l \times \cdots$ we have $\log\log |G| \ge \log\log |F_1|^m \ge \log m$.

We remark that a large body of work on positive generation of profinite groups can be translated into this language. Let us especially note a paper **[BPS]** where the authors show that if a profinite group G has no sections isomorphic to A_n for large n (i.e. for all $n > N$), then this group is positively generated, i.e. $\varphi_k(G) \ge \epsilon$ for a universal k and $\epsilon > 0$ independent of the group G.

Question 1.5.5. Can one quantify the results in **[BPS]** (i.e. find explicit expressions of k and ϵ in terms of N)?

2. Connectivity of $\Gamma_k(G)$

2.1. Preliminaries. Let G be a finite group. Recall that $\mathcal{N}_k(G)$ denotes the set of generating k-tuples $(g) = (g_1, \dots, g_k)$, where $\langle g_1, \dots, g_k \rangle = G$. Let $\Gamma_k(G)$ be as in the Introduction, the graph on $\mathcal{N}_k(G)$ with edges corresponding to moves $R^{\pm}_{i,j}$, $L^{\pm}_{i,j}$:

$$R^{\pm}_{i,j} : (g_1, \dots, g_i, \dots, g_k) \to (g_1, \dots, g_i \cdot g_j^{\pm 1}, \dots, g_k),$$

$$L^{\pm}_{i,j} : (g_1, \dots, g_i, \dots, g_k) \to (g_1, \dots, g_j^{\pm 1} \cdot g_i, \dots, g_k).$$

We call $\Gamma_k(G)$ the *product replacement graph*, and its edges the *Nielsen moves*.

Note that while $\Gamma_k(G)$ is defined as an oriented graph, it is symmetric, so orientation can be disregarded. As before, let $d(G)$ be the minimal number of generators which generate G. In this section we are concerned with the following problem.

Problem 2.1.1. For a given G, $k \geq d(G)$, is $\Gamma_k(G)$ connected?

It is quite embarrassing how little is known about this problem. For example, we do not know the answer to the following innocent looking question:

Question 2.1.2. Is it true that if $\Gamma_k(G)$, $k \geq d(G)$ is connected, then $\Gamma_m(G)$ is connected for every $m > k$?

In the next sections we shall the examples when $\Gamma_{d((G)}$ is disconnected. However we do not know the answer to the following "easy" question:

Question 2.1.3. Is there a finite group G, $d = d(G)$ such that $\Gamma_{d+1}(G)$ is disconnected?

Later in this section we will show that the answer to this question is negative when G is solvable, and it is probably negative for finite simple groups and their direct products. We reserve our judgement with regard to the question in full generality.

By $\varkappa_k(G)$ denote the number of connected components of $\Gamma_k(G)$. Now let us prove some general results on $\varkappa_k(G)$ before turning to examples and special cases.

First, note that one cannot expect that $H \subset G$ implies that $\varkappa_k(H) \leq \varkappa_k(G)$ since we already do not have $d(H) \leq d(G)$. For example, $H = \mathbb{Z}_2^n \subset G = \mathbb{Z}_2 \wr \mathbb{Z}_n$, but $d(H) = n > d(G) = 2$. On the other hand, if H is a quotient of G, then $d(H) \leq d(G)$. This result can be extended to show that $\varkappa_k(H) \leq \varkappa_k(G)$ for any $k \geq d(G)$.

Theorem 2.1.4. *If $\alpha : G \to H$ is an epimorphism between finite groups, then for every $k \geq d(H)$ the map $\alpha_k : \Gamma_k(G) \to \Gamma_k(H)$, defined by*

$$\alpha_k(g_1, \dots, g_k) = (\alpha(g_1), \dots, \alpha(g_k))$$

is a surjective graph projection. In particular, the number of connected components of $\Gamma_k(H)$ is bounded by that of $\Gamma_k(G)$.

The result easily follows from the following important lemma:

Lemma 2.1.5 (Gaschütz). *Let $\psi : G \to H$ be an epimorphism between finite groups, $k \geq d(G)$, and let $(h_1, \dots, h_k)$ be a generating k-tuple of H. Then there exists a generating k-tuple $(g_1, \dots, g_k)$ of G with $\psi(g_i) = h_i$ for $i = 1, \dots, k$.*

Proof. Denote by $N \lhd G$ the kernel of the epimorphism ψ: $H \simeq G/N$. Consider any $(g_1, \dots, g_k)$ such that $\langle g_1 N, \dots, g_k N \rangle = H$. We need to find $(u_1, \dots, u_k) \in N^k$ such that $\langle g_1 u_1, \dots, g_k u_k \rangle = G$.

Let us fix a k-tuple $(g) = (g_1, \dots, g_k)$ as above. Consider the set $R(g)$ of all k-tuples $(g_1 u_1, \dots, g_k u_k)$. Clearly, $|R(g)| = |N|^k$. For every subgroup $S \subset G$ we have $R(g) \cap S^k \neq \emptyset$ if and only if $SN = G$. Further, in this case $|R(g) \cap S^k| = |S \cap N|^k$. This number is therefore independent of the choice of (g).

Now, the number of k-tuples in $R(g)$ which generate G is equal to $|N|^k$ minus the number of k-tuples in $R(g)$ and proper subgroups. But the latter can be calculated by the Möbius summation formula of the numbers $|R(g) \cap S^k|$ (see **[H1]**). Therefore the number N of k-tuples in $R(g)$ which generate G is *independent* of the choice of (g).

By hypothesis, there does exist a generating k-tuple $a = (a_1, \dots, a_k)$ of G. Then for $\psi_k(a) = (\psi(a_1), \dots, \psi(a_k))$ the number N is positive. But since it is the same for all (g), we obtain the result. □

In our proof we closely follow the original proof (see **[Gr]**, Proposition 6.14.) We refer to **[Gr]** for a number of other related results.

2.2. Extended product replacement graph. Define an *extended product replacement graph* $\widetilde{\Gamma}_k(G)$ to be a graph on $\mathcal{N}_k(G)$ with edges corresponding to $R^{\pm}_{i,j}$, $L^{\pm}_{i,j}$, and $\pi_{i,j}$, ι_m, $1 \leq i, j, m \leq k$, where

$$\pi_{i,j} : (g_1, \dots, g_i, \dots, g_j, \dots, g_k) \to (g_1, \dots, g_j, \dots, g_i, \dots, g_k),$$

$$\iota_m : (g_1, \dots, g_m, \dots, g_k) \to (g_1, \dots, g_m^{-1}, \dots, g_k).$$

Proposition 2.2.1. *Let $\varkappa_k(G)$, $\widetilde{\varkappa}_k(G)$ be the number of connected components in $\Gamma_k(G)$ and $\widetilde{\Gamma}_k(G)$ respectively. Then*

$$\widetilde{\varkappa}_k(G) \leq \varkappa_k(G) \leq 2\,\widetilde{\varkappa}_k(G).$$

Moreover, if $\widetilde{\Gamma}_k(G)$ is connected and $k \geq d(G) + 1$, then $\Gamma_k(G)$ is also connected.

Proof. Let $(g) = (g_1, \dots, g_k)$. We easily have:

$$(g) = (\dots, g_i, \dots, g_j, \dots) \longrightarrow^{R^-_{i,j}} (\dots, g_i \cdot g_j^{-1}, \dots, g_j, \dots)$$

$$\longrightarrow^{L^+_{j,i}} (\dots, g_i \cdot g_j^{-1}, \dots, g_i, \dots) \longrightarrow^{L^-_{i,j}} (\dots, g_j^{-1}, \dots, g_i, \dots).$$

Denote this move by $P_{i,j}$. Repeating $P_{i,j}$ twice we obtain:

$$(g) \longrightarrow^{P_{i,j}} (\ldots, g_j^{-1}, \ldots, g_i, \ldots) \longrightarrow^{P_{i,j}} (\ldots, g_i^{-1}, \ldots, g_j^{-1}, \ldots).$$

Denote this move by $I_{i,j}$. Observe that $P_{i,j}$, $I_{i,j}$ generate a subgroup of index two of rotations in a hyperoctahedral group H_n (the Weyl group of root system D_n).

Let us assume that $(g) = (g_1, \ldots, g_k)$ is connected to $(h) = (h_1, \ldots, h_k)$ by a certain sequence of moves in graph $\widetilde{\Gamma}_k(G)$. Now, in the graph $\Gamma_k(G)$ apply the same moves as in $\widetilde{\Gamma}_k(G)$ (up to $\pm$) and use $I_{i,j}$ and $P_{i,j}$ in place of inversions ι_j and transpositions $\pi_{i,j}$. This gives $(h_1^{\pm 1}, \ldots, h_k^{\pm 1})$. Let us apply a few more $I_{i,j}$ to obtain $(h_1, \ldots, h_{k-1}, h_k^{\pm 1})$. If $\Gamma' \subset \widetilde{\Gamma}_k(G)$ is a connected component, then every $(g) \in \widetilde{\Gamma}'$ is connected to a fixed (h). This implies that such (g) is connected to either $(h_1, \ldots, h_k)$ or $(h_1, \ldots, h_{k-1}, h_k^{\pm 1})$. Thus $\Gamma_k(G)$ has at most two times the number of connected components in $\widetilde{\Gamma}_k(G)$.

For the second part, if $k \geq d(G) + 1$, and $\widetilde{\Gamma}_k(G)$ is connected, then every k-tuple $(g) \in \Gamma_k(G)$ is connected to some $(h_1, \ldots, h_{k-1}, \mathrm{id})$. This proves the result. □

Let us show that for k large enough, the graphs $\Gamma_k(G)$ do become connected. Let $\bar{d} = \bar{d}(G)$ be the maximal size of the minimal generating set, which is a set such that no generator can be omitted. By $\ell = \ell(G)$ denote the length of the longest increasing sequence of subgroups:

$$G_0 = \{\mathrm{id}\} \subset G_1 \subset \ldots \subset G_\ell = G,$$

where $G_i \neq G_{i+1}$. Note that

$$d(G) \leq \bar{d}(G) \leq \ell(G) \leq \log_2 |G|,$$

where the latter inequality follows from $|G_{i+1}|/|G_i| \geq 2$.

Proposition 2.2.2. *If $k \geq d(G) + \bar{d}(G)$, then $\Gamma_k(G)$ is connected.*

Proof. By Proposition 2.2.1, it suffice to prove that $\widetilde{\Gamma}_k(G)$ is connected.

Let $d = d(G)$, $\bar{d} = \bar{d}(G)$, $k \geq d + \bar{d}$. Fix a generating set $\langle h_1, \ldots, h_d\rangle = G$ and consider a generating k-tuple $(h) = (h_1, \ldots, h_d, \mathrm{id}, \ldots, \mathrm{id})$. Observe that every generating k-tuple $(g_1, \ldots, g_k) \in \mathcal{N}_k(G)$ is connected (in graph $\widetilde{\Gamma}_k(G)$) to (h). Indeed, the set of generators is redundant, so up to a permutation we have $\langle g_{d+1}, \ldots, g_{d+\bar{d}}\rangle = G$. Then (g) is connected to $(h_1, \ldots, h_d, g_{d+1}, \ldots, g_{d+\bar{d}})$ and finally to (h). □

Theorem 2.2.3 (Babai). *Let G be a finite group, $k = 2\lceil \log_2 |G| \rceil$. Then the diameter $\Delta = \Delta(G, k)$ of the product replacement graph $\Gamma_k(G)$ is at most $C \cdot \log^2 |G|$, where C is a universal constant.*

The proof of Babai **[Bb5]** is an extension of the proof of Proposition 2.2.2. It uses an important "cube doubling" idea (see **[ER, Bb3]**) which has other important applications and is worth noting.

Proof. Let $r = \lceil \log_2 |G| \rceil$. We say that $(g) = (g_1, \dots, g_r) \in \mathcal{N}_r(G)$ is *good*[1] if $C^{-1}C = G$, where

$$C = C(g) = \left\{ g_1^{\epsilon_1} \cdots g_r^{\epsilon_r} | \epsilon_1, \dots, \epsilon_r \in \{0, 1\} \right\}.$$

Observe that by definition of one can move from $(g, h) = (g_1, \dots, g_r, h_1, \dots, h_r)$ to $(g, h') = (g_1, \dots, g_r, h'_1, \dots, h'_r)$ in at most $2r^2$ Nielsen moves.

We claim that for every k-tuple $(g) \in \mathcal{N}_k(G)$ one can obtain a k-tuple of the form (u, v), where $(u), (v) \in \mathcal{N}_r$, and (u) is good. Indeed, permute elements of (g) so that the last r of them generate G. That will be our (v). Now let us construct a *good* r-tuple (u) by induction. Set $C_0 = \emptyset$. Assume that at step i we determined $u_1, \dots, u_i$ such that $|C_i| = 2^i$, where $C_{i-1} = C(u_1, \dots, u_i)$. If $C_i^{-1}C_i = G$, the this is already a good set. Otherwise there exists $w \in \{\mathrm{id}, v_1, \dots, v_r\}$ such that $g_{i+1}D_i w \neq D_i$, where $D_i = C_i^{-1}C_i$. Indeed, if $g_{i+1}D_i = D_i$ for $w = \mathrm{id}$, then for $w = v_i$ we have $g_{i+1}D_i v_i = D_i v_i$. But since (v) is a generating r-tuple, and $D_i G \neq D_i$, there exists v_i such that $g_{i+1}C_i v_i \neq C_i$. Now move g_{i+1} to an element $u_{i+1} = g_{i+1}z$, where $z = D_i^{-1}w$, $u_{i+1} \notin D_i$. Then $|C_{i+1}| = 2\,|C_i|$, which completes the induction step in at most $(2i + 1)$ Nielsen moves.

The total number of induction steps is at most $\lceil \log_2 |G| \rceil = r$ since we are "doubling the cube" C_i at every step. We need therefore at most r^2 Nielsen moves and $O(r)$ transpositions and inversions to obtain a good k-tuple. From the previous observation, two good k tuples are connected by at most $2r^2$ Nielsen moves and $O(r)$ transpositions and inversions. Recall now that by the proof of Proposition 2.2.2 each transpositions and inversions can be substituted by at most c Nielsen moves, where c is a universal constant. This completes the proof. □

Remark 2.2.4. Proposition 2.2.2 is an observation in **[DS3]**. A weaker version with $k \geq 2\bar{d}(G)$ was given in the original paper **[CLMNO]** (see also **[Gi]**). Theorem 2.2.3 and the proof closely follow the paper **[Bb5]**.

Let us also note that computing $\bar{d}(G)$ is not an easy task. A recent result $\bar{d}(S_n) = n - 1$ of Whiston **[Wh]** is worth mentioning, especially in comparison with another hard result $\ell(S_n) = \lfloor \frac{3n-1}{2} \rfloor - b_n$, where b_n is the number of ones in the binary expansion of n (see **[CST, Bb1]**).

2.3. Abelian, nilpotent and solvable groups. The main result of this section is due to Dunwoody **[Du2]**, who showed that $\Gamma_{d(G)+1}(G)$ is connected for any finite solvable group G. The case $k = d(G)$ is more delicate, and is also considered.

Example 2.3.1. Let $G = (\mathbb{Z}_p)^m$, where p is a prime. Then $\mathcal{N}_k(G)$ is in one to one correspondence with the set $\mathrm{GL}(m, p)$ of nonsingular $m \times m$ matrices over $\mathbb{F}_p$. Note that Nielsen moves $R_{i,j}^{\pm}$, $L_{i,j}^{\pm}$ become elementary transformations (adding or subtracting a row), and therefore do not change the determinant. Thus $\Gamma_m(G)$ has at

[1] In **[Bb5]** such (g) are called "cube-semigenerators"

least $(p-1)$ components. Further, $\Gamma_m(\mathbb{Z}_p^m)$ is a Cayley graph of $\mathrm{GL}(m,p)$ with a generating set $E_{i,j}^{\pm}$.

In the other direction, the elementary matrices $E_{i,j}^{\pm}$ with ones on the diagonal, ± 1 at (i,j) and zero elsewhere, generate $\mathrm{SL}(m,p)$. Therefore $\Gamma_m(\mathbb{Z}_p^m)$ has exactly m isomorphic connected components.

Note also that $\widetilde{\Gamma}(\mathbb{Z}_p^m)$ is a Cayley graph of $\Gamma_m(\mathbb{Z}_p^m)$ with a generating set consisting of elementary matrices, permutation matrices, and matrices which have (-1) at (i,i), 1 elsewhere on diagonal, and zero outside of the main diagonal. For $p=2$ graph $\widetilde{\Gamma}(\mathbb{Z}_p^m)$ is also connected, but for $p \geq 3$, p a prime, it has $(p-1)/2$ connected components.

Theorem 2.3.2 (**[NN, DG]**). *Let G be a finite abelian group, given as*

$$G \simeq \mathbb{Z}_{m_1} \times \mathbb{Z}_{m_2} \times \cdots \times \mathbb{Z}_{m_r},$$

where $m_1 \mid m_2 \mid \ldots \mid m_r$. Then $d(G)=r$ and

(i) *$\Gamma_k(G)$ is connected for all $k \geq r+1$,*

(ii) *$\Gamma_r(G)$ has $\phi(m_1)$ components of equal size,*

where $\phi(m)$ is the Euler function (the number of integers less than m which are relatively prime with m).

The first part of the theorem [2] is due to B. H. Neumann and H. Neumann **[NN]** (see also **[DG, Ev2]**). The second part is due to Diaconis and Graham **[DG]**, who also rediscovered the first part.

The proof is an elementary, though nontrivial generalization of the observation in Example 2.3.1. One proves in part (ii) that when an element $g_i \in G$ is written as $(a_{i,1},\ldots,a_{i,r})$, then $\det(a_{i,j})$ is the only invariant of generating k-tuples under Nielsen moves (see **[DG]** for details).

For nilpotent groups the situation with the case $k=d(G)$ is more complicated. First, using Gaschütz' Lemma, one can lift "invariants" from abelian quotients to the whole group. This idea was successfully used in **[Du1]** by Dunwoody. A somewhat different approach was proposed in **[Ne]**:

Lemma 2.3.3 (Higman). *Let Γ' be a connected component of $\Gamma_2(G)$ which contains $(g_1,g_2) \in \mathcal{N}_2(G)$. Then for every $(h_1,h_2) \in \Gamma'$ we have $[h_1,h_2]$ is a conjugate of $[g_1,g_2]$.* [3]

In other words, the conjugacy class of the commutator $[g_1,g_2]$ is invariant under Nielsen moves. This can be easily checked directly. Note that in $\widetilde{\Gamma}_k(G)$, the transposition of generators inverts the commutator: $[g_1,g_2] \to [g_2,g_1] = [g_1,g_2]^{-1}$, so

[2] The result in **[NN]** was formulated in a slightly different language of T-systems, see below.

[3] Higman's Lemma has appeared in **[NN]** in a somewhat different form.

one has to add an inverse of the conjugacy class. In **[Ne]** B. H. Neumann used Higman's Lemma to find distinct connected components in a particular nilpotent group $G \simeq \mathbb{Z}_8 \ltimes D^4$, where D is a dihedral group of order 8.

Question 2.3.4 (B.H. Neumann). Can one find a generalization of Higman's Lemma for more than two generators, that is for $k > 2$?

Open Problem 2.3.5. Describe all connected components of $\mathcal{N}_k(G)$, where G is a finite nilpotent group and $k = d(G)$.

Now let us prove the main result of this section:

Theorem 2.3.6 (Dunwoody). *Let G be a finite solvable group, $k \geq d(G) + 1$. Then $\Gamma_k(G)$ is connected.*

Proof. The proof is by induction on the length ℓ of a chief series of G:

$$\{\text{id}\} = G_0 \subset G_1 \subset \ldots \subset G_\ell = G\,,$$

where G_{i-1} is a minimal normal G-invariant subgroup of G_i, and $G_i \lhd G$, $i = 1, \ldots, \ell$.

When $\ell = 0$ there is nothing to prove. Assume $\ell \geq 1$. Fix any $(k-1) \geq d(G)$ generators $h_1, \ldots, h_{k-1}$ of G. We will show that any generating k-tuple $(g) = (g_1, \ldots, g_k) \in \Gamma_k(G)$ is connected to $(h) = (\text{id}, h_1, \ldots, h_{k-1}) \in \Gamma_k(G)$ by Nielsen moves. This clearly implies connectivity of the whole graph.

Denote G_1 by M. By the induction hypothesis on G/G_1, (g) is connected to a k-tuple $(m, m_1h_1, \ldots, m_{k-1}h_{k-1})$, where $m, m_1, \ldots, m_{k-1} \in M$. If $m = \text{id}$, the elements m_ih_i generate G and we can use $R^{\pm}_{1,i}$ to get $m \neq \text{id}$. Therefore we can assume that $m \neq \text{id}$.

Now note that $h_i^{-1}mh_i = (m_ih_i)^{-1}m\,(m_ih_i)$. For any given $g \in G$ write $g = \mathbf{w}(h_1, \ldots, h_{k-1})$ as a word in generators of G. Then

$$\big(\mathbf{w}(m_1h_1, \ldots, m_{k-1}h_{k-1})\big)^{-1}\, m\, \mathbf{w}(m_1h_1, \ldots, m_{k-1}h_{k-1}),$$

$$= \big(\mathbf{w}(h_1, \ldots, h_{k-1})\big)^{-1}\, m\, \mathbf{w}(h_1, \ldots, h_{k-1}) = g^{-1}\, m\, g = m^g.$$

Therefore by successive application of $R^{+}_{1,i}$ and $L^{-}_{1,i}$ one can connect the generating k-tuple $(m, m_1h_1, \ldots, m_{k-1}h_{k-1})$ with $(m^g, m_1h_1, \ldots, m_{k-1}h_{k-1})$ for any g.

Since M is a minimal normal subgroup of G, every $m' \in M$ is a product of conjugates of m. Therefore by iterating the previous procedure, one can use $L^{\pm}_{i,1}$ to get rid of all the m_i and obtain $(m, h_1, \ldots, h_{k-1})$. But then, since $h_1, \ldots, h_{k-1}$ generate G, we easily obtain $(h) = (\text{id}, h_1, \ldots, h_{k-1})$ by the Nielsen moves. This completes the proof. □

Remark 2.3.7. As shown by Evans **[Ev2]**, the graphs $\Gamma_k(G)$ with $k \geq d(G) + 1$ are connected for all (even infinite) nilpotent groups. This fails badly for $k = d(G)$ (see **[Du1]**), even when the so called "swap moves" are allowed :

$$(g_1, \dots, g_i, \dots, g_k) \to (g_1, \dots, g_i', \dots, g_k),$$

where $g_i' \in G$ is any element such that $(g_1, \dots, g_i', \dots, g_k)$ is a generating k-tuple. We refer to **[TT, Ro]** for reasoning behind this definition and various examples. As we shall see later, all but a finite number of simple groups are connected by these swap moves for $k \geq 3$ (see proof of Proposition 2.5.13).

We challenge the reader to generalize (or disprove) the above mentioned Evans' and Dunwoody theorems to all (even infinite) solvable groups. We refer to a related result of Myasnikov **[My]** who proved the Andrews–Curtis Conjecture for certain solvable groups.

Let us mention that in a different context Question 2.1.3 has a negative answer for infinite groups. This is related to the following Waldhausen's question: "If $G = \langle x_1, \dots, x_n, r_1, \dots, r_m \rangle$ can be generated by fewer than n elements, then does the normal closure of $\langle r_1, \dots, r_n \rangle$ in F_n contain a primitive element of F_n? " (see **[LyS]**, p. 92). The answer for this question is negative (see **[No, Ev1]**). On the other hand, it was noted in **[Du2]** (in fact, motivated his work) that this is equivalent to the question of whether a generating set for G with n elements can always be changed by Nielsen moves to a set containing the identity element, given $d(G) = n - 1$. But this is a weaker condition than the connectivity of $\widetilde{\Gamma}_{n-1}(G)$. Thus $\widetilde{\Gamma}_{n-1}(G)$ and therefore $\Gamma_{n-1}(G)$ is disconnected in this case.

2.4. *T*-systems. Let G be a finitely generated group, and $d(G)$ be the minimal number of generators of G. *Systems of transitivity* (or T-systems) are defined as follows.

We say that N is a G-defining subgroup of a free group F_k if N is normal in F_k and $F_k/N \simeq G$. Denote by $\Sigma_k(G)$ the set of G-defining subgroups of F_k, and consider an action of $\mathrm{Aut}(F_k)$ on $\Sigma_k(G)$. The orbits of this action are called T_k-systems. They were introduced in **[NN]** by B.H. Neumann and H. Neumann, and studied on and off since then.

Recall also that $\mathrm{Aut}(F_k)$ is generated by the set $\overline{\Upsilon}_k$ of Nielsen moves, transpositions and inversions. These moves define a graph structure on $\Sigma_k(G)$. Now T_k-systems correspond to connected components of $\Sigma_k(G)$.

Recall that $\widetilde{\Gamma}_k(G)$ is also defined by the set of moves $\overline{\Upsilon}_k$. Consider the natural action of $\mathrm{Aut}(G)$ on the set $\mathcal{N}_k(G)$ of generating k-tuples of G. Identifying under this action defines a projection of graph $\widetilde{\Gamma}_k(G)$ onto $\Sigma_k(G)$. We obtain:

Proposition 2.4.1. *The number of connected components in $\Sigma_k(G)$ (the number of T_k-systems) is less or equal to the number of connected components in $\widetilde{\Gamma}_k(G)$. In the other direction, if $\Sigma_k(G)$ is connected and $k \geq 2\,d(G)$, then $\widetilde{\Gamma}_k(G)$ is also connected.*

Proof. The first part follows from the previous observation. For the second part, connectivity of $\Sigma_k(G)$ implies that every k-tuple $(g) \in \mathcal{N}_k(G)$ is connected in $\widetilde{\Gamma}_k(G)$ to some $\varphi(h)$, where $\varphi \in \mathrm{Aut}(G)$ and $(h) \in \mathcal{N}_k(G)$ is fixed. Now take $(h) = (h_1, \dots, h_d, \mathrm{id}, \dots, \mathrm{id})$, where $d = d(G)$. Then $\varphi(h) = (h'_1, \dots, h'_d, \mathrm{id}, \dots, \mathrm{id}) \in \mathcal{N}_k(G)$ is a generating k-tuple which is connected to (h) in $\widetilde{\Gamma}_k(G)$. Indeed, permute elements in $\varphi(h)$ to make the first d of them to be id. Now generate $(h_1, \dots, h_d)$ in the first d positions, and then make the remaining $(k-d)$ elements become id. We conclude that any $(g) \in \mathcal{N}_k(G)$ is connected in $\widetilde{\Gamma}_k(G)$ to a fixed (h) as above. □

Corollary 2.4.2. *If $k \geq 2d(G)$, then $\Gamma_k(G)$ is connected if and only if G has exactly one T_k-system.*

Proof. The corollary follows immediately from Proposition 2.4.1 and Proposition 2.2.1. □

Theorem 2.4.3 (Gilman). *Let G be a finite simple group and $k \geq 3$. Then* $\mathrm{Aut}(F_k)$ *acts on at least one T_k-system of G as the alternating or symmetric group.*

This result is quite miraculous since it basically says that for simple groups on some orbit the group action of $A = \mathrm{Aut}(F_k)$ is not only simply transitive, but highly transitive. The proof is also quite interesting, but goes beyond the scope of this survey.

In fact, Theorem 2.4.3 is a compilation of several results. The main part, which is the case $k \geq 4$, is due to Gilman **[Gi]**. It was subsequently extended to the case $k = 3$ by Evans **[Ev3]** (see Lemma 5.2) under assumption that a simple group is generated by two elements, one of which is an involution. The proof in **[Ev3]** follows closely **[Gi]**, so we will refer to Theorem 2.4.3 as Gilman's Theorem.

Now, the assumption that a simple group is generated by two elements one of which is an involution, holds in fact for every simple group (see Remark 2.4.8). The easiest way to see this is to apply Theorem 1.1.8 of Guralnick and Kantor. Recall that by the Feit–Thompson Theorem (see e.g. **[Go]**) every finite simple group contains an involution. Therefore by 1.1.8 (i), it generates the whole group with some elements of a certain conjugacy class.

Corollary 2.4.4. *If G is simple and has exactly one T_k-system, $k \geq 3$, then G^m has also exactly one T_k-system, given $m < N = |\Sigma_k(G)|$.*

Proof. By Hall's Theorem 1.2.1, the generating k-tuples of G^m correspond to those m-tuples of generating k-tuples of G which all lie in different orbits of the diagonal action of $\mathrm{Aut}(G)$ on $\mathcal{N}_k$. Assume now that $\Sigma_k(G)$ is connected. By Theorem 2.4.3, $\mathrm{Aut}(F_k)$ acts $(N-1)$-transitively on $\Sigma_k(G)$. But that means that for any $m \leq N$ one can use extended Nielsen moves to get from m elements in different orbits (of $\mathrm{Aut}(G)$ on $\mathcal{N}_k(G)$) to m elements in any other different m orbits. But this is more than enough to show that $\Sigma_k(G^m)$ is connected, as the action of the group $\mathrm{Aut}(G^m)$ on $\mathcal{N}_k(G^m)$ contains separate actions of m copies of $\mathrm{Aut}(G)$ on different $\mathcal{N}_k(G)$. □

Corollary 2.4.5. *If G is simple and has exactly one T_k-system, $k \geq 4$, then $\Gamma_k(G^m)$ is connected, given $m < |\Sigma_k(G)|$.*

Proof. Recall that for all finite simple groups $d(G) = 2$. Now the result follows from Corollary 2.4.4 and Corollary 2.4.2. □

Unfortunately we cannot improve the condition $k \geq 4$ to $k \geq 3$. Indeed, the group A_n having only one T_3-system does not imply that $\Gamma_3(A_n)$ is connected since in Corollary 2.3.2 we need k to be at least two times the minimal number of generators.

Let us conclude by formulating a stronger version of Question 2.1.3:

Question 2.4.6. Is there a finite group G, $d = d(G)$ such that G has more than one T_{d+1}-system (equivalently: $\Sigma_{d+1}(G)$ is disconnected) ?

Remark 2.4.7. Corollary 2.4.4 is a refined version of the observation made by Dunwoody in a review of the paper **[Gi]** (Math. Reviews 55#8186, see also **[Du2]**). The story is that in **[Du2]** Dunwoody proposed P. Hall's example A_5^{19} as a possible candidate for a positive answer to the question 2.4.6.[4]

In a review Dunwoody asserts that Gilman's results imply that A_5^{19} cannot satisfy 2.4.6. This does not seem clear to us in view of the condition $k \geq 4$ in **[Gi]**, but certainly Gilman was very close to resolving the A_5^{19} problem. Only after a paper **[Ev3]** of Evans about 15 years later, one can extend the result to the case $k = 3$, so to prove that $G = A_5^{19}$ has only one T_3-system. Indeed, since $\varphi_3(A_5) > \varphi_2(A_5) = 19/30$ (this is Hall's result) by Corollary 2.4.4 we have

$$|\Sigma_3(A_5)| = \frac{\varphi_3(A_5)|A_5^3|}{|\mathrm{Aut}(A_5)|} > \frac{\frac{19}{30} \cdot 60^3}{120} = 1140.$$

From here we obtain a remarkable connectivity of the graphs $\Sigma_3(A_n^N)$ for $N \leq 1140 < |\Sigma_3(A_5)|$. It is unclear, however, whether $\mathrm{Aut}(F_3)$ acts as a symmetric (not as an alternating) group on $\Sigma_3(A_5)$, i.e. whether $A_5^{|\Sigma_3(A_5)|}$ has only one T_3-system as suggested in Question 2.4.6.

Remark 2.4.8. A few words about simple groups being generated by two elements, one of which is an involution. For sporadic groups this follows from their construction, and for several series (such as alternating groups, $\mathrm{PSL}(n, q)$, etc.) it was known for decades (see e.g. **[CM, Go]**). In full generality this was first proved by Malle, Saxl and Weigel in **[MSW]**.

[4]Actually, Dunwoody proposes A_5^{19} as a possible challenge problem for computational group theory. Note that the order of the group $|A_5^{19}| \approx .61 \cdot 10^{34}$ is much too large for any brute force computation. Imagine now how much challenge it was in 1967 when Dunwoody proposed the problem!

2.5. Simple groups. In this section we elaborate on the connectivity of the product replacement graph in the special case of simple groups. As we shall see, this case seems to be of particular importance.

Example 2.5.1. Let $G = A_5$. In a pioneering paper **[NN]** the authors show that A_5 has at least two T_2-systems. To see that, simply use Higman's Lemma and observe that the commutator of two generating elements must be either a 5-cycle or a 3-cycle. Since both can occur, there are at least two T_2-systems.

Note that when $G = S_5$, we already have at least three T_2-systems: elements with two 2-cycles can occur as commutators of a generating pair of element.

Proposition 2.5.2. *The number of connected components of* $\Gamma_2(A_n)$ *increases as* $n \to \infty$.

Proof. Let n be odd. Simply observe that when $(m, p) = 1$ the elements of the form $a = (1, 2, \ldots, n)$ and $b = (1, 2, \ldots, p)$, (in cycle notation) generate the whole A_n. Their commutators belong to different conjugacy classes (for $p < m/2$). Now Higman's Lemma implies the result. □

Question 2.5.3. What is the number of conjugacy classes of the commutator $[a, b]$ of two generating elements $(a, b) \in \mathcal{N}_2(A_n)$?

Note, of course, that $[A_n, A_n] = A_n$, $n \geq 5$, and two elements in A_n generate A_n with probability $\mathbf{P} = \left(1 - 1/n + O(1/n^2)\right)$. Also, if $[a, b] = c$, then $[\sigma a, \sigma b] = \sigma c$ for all $\sigma \in S_n$. Therefore elements with the same cycle structure either can be all obtained, or can't be all obtained as commutators of two generating elements in A_n. Of course, some conjugacy classes, e.g. id, can't be obtained in this way.

Open Problem 2.5.4. Describe all connected components of $\Gamma_2(A_n)$, $n \geq 5$.

For $k \geq 3$ very few results are known, so let us start with the general conjecture of obvious importance.

Conjecture 2.5.5 (Wiegold). *For every* $k \geq 3$ *and every nonabelian group* G*, there exists only one* T_k*-system (equivalently: the graph* $\Sigma_k(G)$ *is connected).*

While never published, this conjecture was attributed to Wiegold on several occasions (see **[Da, Ev2, Ev3]**). It seems inspired by Gilman's paper **[Gi]**. Let us state here a somewhat stronger companion conjecture.

Conjecture 2.5.6. *For every nonabelian simple group* G *and every* $k \geq 3$*, the product replacement graph* $\Gamma_k(G)$ *is connected.*

Note that since $d(G) = 2$ for all finite simple groups, for $k \geq 4$ Conjectures 2.5.5 and 2.5.6 are equivalent. A similar conjecture for $G = S_n$ was made by Diaconis and Graham in **[DG]**. Let us briefly formulate what is known in this direction:

Theorem 2.5.7. *Wiegold's Conjecture holds in the following cases:*

(i) $G = \mathrm{PSL}(2, p)$, *where* $p \geq 5$ *is a prime, and* $k \geq 3$ (**[Gi]**),

(ii) $G = \mathrm{PSL}(2, 2^m)$ *or* $G = \mathrm{Sz}(2^{2m-1})$, *where* $m \geq 2$, *and* $k \geq 3$ (**[Ev3]**),

(iii) $G = A_6, A_7$, *and* $k = 3$ (**[Da]**),

(iv) $G = A_8, A_9, A_{10}$, *and* $k = 3$ (**[CP]**).

Note that the case $G = A_5$ is covered in (i): $A_5 \simeq \mathrm{PSL}(2, 5)$. In fact, **[CP]** proves that the stronger Conjecture 2.5.6 holds for $k = 3$ and $G = A_n$, $6 \leq n \leq 10$.

Proposition 2.5.8. *Let* $G = \mathrm{PSL}(2, p)$, $H = G^m$, *where* $m \geq p(p+1)$. *Then* $d(H) \geq 3$ *and for every* $k \geq d\big(G^{(m+1)}\big)$, *the group* H *has only one* T_k*-system. Further, if* $m \geq p^2(p-1)(p+1)^2$, *then* $d(H) \geq 4$ *and the product replacement graph* $\Gamma_k(G)$ *is connected.*

Proof. Recall that $|\mathrm{PSL}(2, p)| = p(p-1)(p+1)$, $\mathrm{Aut}(\mathrm{PSL}(2, p)) = \mathrm{GL}(2, p)$, $|\mathrm{GL}(2, p)| = p(p-1)^2(p+1)$. By Hall's Theorem 1.2.1, the largest power N such that $d(G^N) \leq 2$ is given by

$$N = \varphi_2(G) \cdot \frac{|G|^2}{|\mathrm{Aut}(G)|} < \frac{\big[p(p-1)(p+1)\big]^2}{p(p-1)^2(p+1)} = p(p+1).$$

Therefore for all $m \geq p(p+1)$ we have $d(H) \geq 3$. Analogously, whenever $m \geq p^2(p-1)(p+1)^2$, we have $d(H) \geq 4$, which proves the first part of both claims.

The second parts follows easily from Corollaries 2.4.4 and 2.4.5. Indeed, uniqueness of T_k system condition in 2.4.4,5 is satisfied by part (i) of Theorem 2.5.7. By Hall's Theorem, the condition $m \leq |\Sigma_k(G)|$ is equivalent to the condition $k \geq d(H)$, so $m < |\Sigma_k(G)|$ is equivalent to $k \geq d\big(G^{(m+1)}\big)$. This completes the proof. □

There is one more result in favor of Conjecture 2.5.6. To state it we need the notion of *spread* (see **[BW1]**).

Definition 2.5.9. A 2-generator group G is said to have spread m if for every m group elements $g_1, \ldots, g_m \neq \mathrm{id}$ there exists an element $h \in G$ such that $\langle g_1, h\rangle = \cdots = \langle g_m, h\rangle = G$.

Much is known about the spread of simple and quasisimple groups. We refer to papers **[BW1, BW2, GK, GS]** for various extensions and references. Let us just mention that all symmetric groups S_n and alternating groups A_n, $n \geq 5$, have spread 2 (see **[Bi, BW1]**).

Conjecture 2.5.10 ([BW1, GS]). *All finite nonabelian simple groups have spread* 2.

Theorem 2.5.11 ([GS]). *All but a finite number of finite nonabelian simple groups have spread* 2.

Proof. By Theorem 1.1.8 (ii), $\mathrm{PC}(G) \to 1$ as $|G| \to \infty$ unless G is isomorphic to alternating groups A_n or odd-dimensional orthogonal groups $\Omega(n, q)$ with a bounded size q of the field.

Now let G be a simple group, C be a conjugacy class on which $\mathrm{PC}(G)$ is maximized. For any $g_1, g_2 \in G$ we have

$$\mathbf{P}\big(\langle g_1, h\rangle = \langle g_2, h\rangle = G \mid h \in C\big) \geq 1 - 2\big(1 - \mathrm{PC}(G)\big).$$

Therefore $\mathrm{PC}(G) > 1/2$ implies that $\mathbf{P}(\cdot) > 0$, and hence G has spread 2.

We conclude that for all but the two exceptional series there is only a finite number of simple groups which have $\mathrm{PC}(G) \leq 1/2$. As we mentioned before, all alternating groups have spread 2. The case of odd orthogonal groups is considered separately (see **[GK, GS]**). □

Actually, one can describe explicitly the set of possible exceptions in Theorem 2.5.11. It would be interesting to check Conjecture 2.5.10 by computational means, or otherwise.

Definition 2.5.12. A generating k-tuple $(g_1, \dots, g_k) \in \mathcal{N}_k(G)$ is called *redundant*[5] if it is not a minimal generating set, which means that one generators g_i can be omitted so that the remaining elements generate the whole group G.

Proposition 2.5.13 ([Ev3]). *Let G be a 2-generated group of spread* 2 *and let $k \geq 3$. Then all redundant generating k-tuples belong to the same connected component in $\Gamma_k(G)$.*

Proof. Clearly, every redundant generating k-tuple is connected in $\Gamma_k(G)$ to a k-tuple with some entry id. It follows from the proof of Proposition 2.2.1 that the connected component in $\widetilde{\Gamma}_k(G)$ which contains such a k-tuple is also connected in $\Gamma_k(G)$. Thus it suffices to prove that all redundant generating k-tuples belong to the same connected component in $\widetilde{\Gamma}_k(G)$.

Let (g), (h) be two redundant generating k-tuples. One can always permute their elements and make the last element to be identity. Thus we can assume that $(g) = (g_1, \dots, g_{k-1}, \mathrm{id})$, $(h) = (h_1, \dots, h_{k-1}, \mathrm{id})$. Without loss of generality we can also assume that $g_1, h_2 \neq \mathrm{id}$. Since G has spread 2, there exists an element $z \in G$ such that $\langle g_1, z\rangle = \langle h_2, z\rangle = G$. Clearly, (g) is connected to $(g_1, g_2, \dots, g_{k-1}, z)$, and therefore to $(g_1, h_2, \dots, h_{k-1}, z)$ and to $(h_1, h_2, \dots, h_{k-1}, z)$ and $(h_1, h_2, \dots, h_{k-1}, \mathrm{id})$. This completes the proof. □

Observe that by Theorem 1.1.1 a random generating k-tuple in G must be redundant with high probability (given $k \geq 3$). Thus in a sense which will be made precise in

[5]Note that our use of the term is different from that in **[Ev3, Da]**

the next section, Proposition 2.5.13 and Theorem 2.5.11 imply that Conjecture 2.5.6 holds "for most" generating k-tuples of a large enough simple group G.

We conclude with the following important observation:

Proposition 2.5.14. *Let G be a 2-generated group of spread 2 and let $k \geq 3$. Then the connectivity of $\Sigma_k(G)$ implies the connectivity of $\Gamma_k(G)$.*

Proof. Proposition 2.4.1 covers the case $k \geq 2 \cdot d(G) = 4$. It remains to be proved that if G has only one T_3-system, then $\Gamma_3(G)$ is connected. But this is a simple combination of the proof of Proposition 2.4.1 with Proposition 2.5.13. In the notation of the proof of 2.4.1, we need to show that $\varphi(h) = (h'_1, h'_2, \mathrm{id})$ is connected to $(h) = (h_1, h_2, \mathrm{id})$. But that's clear since G has spread 2 and both 3-tuples are redundant. Proceeding as in the proof of 2.4.1, we obtain the result. □

Corollary 2.5.15. *If G is a finite nonabelian simple group of spread 2, then Conjecture* 2.5.6 *is equivalent to Wiegold's Conjecture* 2.5.5. □

Remark 2.5.16. As noted in **[BW1]**, if G has spread r, so does any quotient H. Also, if G_1, G_2 have spread 2, then $G_1 \times G_2$ has spread 1. It would be interesting to investigate when G^m has spread 2, when G is simple. Let us note that the real power of the bounds in **[GK]** combined with proof of Theorem 2.5.11 give large spread for various series of simple groups (see **[GS]**). The latter can be used to show that certain large powers of simple groups can have spread 2.

Remark 2.5.17. Parts (i), (ii) of Theorem 2.5.7 are obtained by an elaborate calculation using subgroup structure and conjugacy classes. It seems that generalizations of the brute force technique to other series of groups (especially of rank ≥ 2) must be difficult.

The proof of David **[Da]** of the case (iii) is based on a tedious calculation. In **[CP]** we present a "proof by computer" of part (iv). We describe the algorithm in Section 2.7.

Let us note that symmetric groups S_n, $n \geq 3$, also have spread 2. This supports the conjecture by Diaconis and Graham **[DG]** on the connectivity of $\Gamma_3(S_n)$ (see above).

Clearly, Proposition 2.5.8 is slightly weaker than what one would like, which is a condition $k \geq d(G^m)$ rather than $k \geq d(G^{(m+1)})$. Nevertheless for powers of $\mathrm{PSL}(2, p)$ as in 2.5.8, the answer to the Question 2.4.6 is positive. Still, it would be interesting to check whether $\mathrm{Aut}(F_k)$ acts as a symmetric or an alternating group on $\Sigma_k(G)$ (cf. Remark 2.4.7.)

Our proof of Theorem 2.5.11 by Guralnick and Shalev **[GS]**, follows closely the proof in their paper. As we mentioned above, it is a direct application of the previous result of Guralnick and Kantor **[GK]**.

2.6. The "large" connected component. We give here a probabilistic approach to the connectivity problem, by showing existence of the large connected component for k moderately large.

Theorem 2.6.1 ([P5]). *Let $\{G_n\}$ be a sequence of finite nonabelian simple groups such that $|G_n| \to \infty$ as $n \to \infty$. Fix $k \geq 3$. Then graphs $\Gamma_k(G_n)$ have connected components $\Gamma'_k(G_n) \subset \Gamma_k(G_n)$, such that*

$$\frac{|\Gamma'_k(G_n)|}{|\Gamma_k(G_n)|} \to 1 \quad as \ n \to \infty.$$

Furthermore, for n large enough $\Gamma'_k(G_n)$ contain all the redundant generating k-tuples.

Proof. By Theorem 2.5.11, every large enough simple group has spread 2. Further, by Proposition 2.5.13, the set of redundant generating k-tuples is connected in $\Gamma_k(G)$. Finally, Theorem 1.1.1 implies that, given $l \geq 2$, a random l-tuple of elements in G is a generating l-tuple, with probability $\to 1$ as $n \to \infty$. Taking $l = k - 1$ and $l = k$, we obtain that a random k-tuple, $k \geq 3$ is a redundant generating k-tuple with the first $(k - 1)$ elements generating the whole group G. This completes the proof. □

In particular, the graphs $\Gamma_3(A_n)$ have a "large" connected component, i.e. a component $\Gamma' = \Gamma'_3(A_n)$ such that $|\Gamma'|/|\Gamma| \to 1$, as $N \to \infty$. Compare this with Proposition 2.2.2, which proves connectivity for $k \geq d(A_n) + \overline{d}(A_n) = 2 + (n-1) = n+1$ (see **[Wh]**).

The following generalization is straightforward:

Theorem 2.6.2 ([P5]). *Let $\{G_n\}$ be a sequence of finite groups, and let k_n be an integer sequence such that*

$$\varphi_{k_n}(G_n) \to 1, \quad as \ n \to \infty.$$

Then the graphs $\Gamma_n = \Gamma_{k_n + d(G_n)}(G_n)$ have connected components Γ'_n, such that

$$\frac{|\Gamma'_n|}{|\Gamma_n|} \to 1 \quad as \ n \to \infty.$$

Furthermore, graphs Γ'_n contain all the generating $\big(k_n + d(G_n)\big)$-tuples of the form $(g_1, \dots, g_{k_n}, \mathrm{id}, \dots, \mathrm{id})$.

Proof. The proof is an easy modification of the proof of Theorem 2.6.1. Indeed, every generating $(k + d)$-tuple of the form $(g) = (g_1, \dots, g_k, \mathrm{id}, \dots, \mathrm{id})$, where $k = k_n$, $d = d(G_n)$, is connected to a fixed $(\mathrm{id}, \dots, \mathrm{id}, h_1, \dots, h_d)$, where $\langle h_1, \dots, h_d \rangle = G$. Now, by hypothesis, the fraction of elements (g) among *all* k-tuples goes to 1 as $n \to \infty$. Thus fraction of (g) among all *generating* k-tuples goes to 1 as $n \to \infty$. By the observation above, they all lie in the same connected component, which implies the result. □

Note now that Corollary 1.5.2 proves the conditions of Theorem 2.6.2 in various special cases. The first two cases which deal with nilpotent and solvable groups were completely resolved by Dunwoody's Theorem 2.3.6. In the last two cases we have:

Corollary 2.6.3. *The product replacement graphs $\Gamma_n = \Gamma_{k_n}(G_n)$ contain connected components Γ'_n such that $|\Gamma'_n|/|\Gamma_n| \to 1$ as $n \to \infty$ if one of the following holds:*

(i) $k_n = \omega(n) \cdot \log(m_n)$, *and each given G_n is a product of finite nonabelian simple groups, with at most m_n copies of each,*

(ii) $k_n = \log_2 |G_n| + d(G_n) + \omega(n)$, *for any $\{G_n\}$,*

where $\omega(n)$ is any function such that $\omega(n) \to \infty$ as $n \to \infty$. □

Note that part (ii) implies that to have a large component in general product replacement graphs it suffices to have $d(G) + \log_2 |G| + C$ generators. This is somewhat weaker than the result of Proposition 2.2.2 which proves that $k = d(G) + \overline{d}(G) \leq d(G) + \log_2 |G|$ is enough to imply connectivity. On the other hand, as we have seen before, part (i) does not follow from 2.2.2.

We conclude with another unexpected bonus from Gilman's Theorem 2.4.3, which gives a generalization of Theorem 2.6.1 to any fixed power of a simple group.

Theorem 2.6.4. *Let $\{G_n\}$ be a sequence of finite nonabelian simple groups such that $|G_n| \to \infty$ as $n \to \infty$, and let m be a fixed integer. Fix $k \geq 3$. Then the graphs $\Gamma_k(G_n^m)$ have connected components $\Gamma'_k(G_n^m) \subset \Gamma_k(G_n^m)$, such that*

$$\frac{|\Gamma'_k(G_n^m)|}{|\Gamma_k(G_n^m)|} \to 1 \quad as \ n \to \infty.$$

Furthermore, for n large enough $\Gamma'_k(G_n)$ contain all the redundant generating k-tuples.

Proof. The orbit in Gilman's Theorem 2.4.3 contains redundant elements by construction (see **[Gi]**). This implies that $\Gamma_k(G_n^m)$ contains a connected component Γ', where each of the m generating k-tuples of G is redundant. This component is "large" since as $|G_n| \to \infty$, the probability that the k-tuple is redundant goes to 1. But, then the probability that m independently chosen k-tuples lie in different orbits of $\mathrm{Aut}(G_n)$ and are all redundant goes to 1 as well. This proves the claim that $|\Gamma'|/|\Gamma| \to 1$. The second part is clear by construction. □

Remark 2.6.5. The results in this section are in **[P5]**. The concept of a "large" connected component is not new as it is used heavily in various versions in the random graphs community (see e.g. **[ASE]**).

Let us note that from the proof of Theorem 2.6.4 one cannot conclude that $\Gamma_k(A_n^{n!/8})$ has a large connected component. Indeed, by 1.1.2 (i), the probability that all of the $n!/8$ k-tuples are redundant is about $(1 - 1/n^{k-1})^{n!/8} \to 0$ as $n \to \infty$. Thus for all $k = o(n)$ the fraction of the k-tuples considered in the proof of Theorem 2.6.4, is "small". On the other hand, for $k = \Omega(n)$ we already know that $\Gamma_k(A_n^{n!/8})$ has a large connected component by Theorem 2.6.3 (i).

2.7. Checking connectivity. Given the importance of the Problem 2.1.1 and Conjecture 2.5.5, one might ask whether the claims are supported by the computational evidence. We describe here a simple algorithm developed in **[CP]** (based on the "large connected component concept" in Section 2.6) for checking connectivity of product replacement graphs.

Let us remark first that beside the paper **[CP]**, the only computational evidence that we know, was reported in **[DG]**: John Laffretty and Dan Rockmore showed in 1997 that $\Gamma_3(S_4)$ and $\Gamma_3(S_5)$ are connected. While they probably could have checked $\Gamma_3(S_6)$ by "naïve" methods, further progress is complicated by a superexponential increase in the complexity: $|\mathcal{N}_3(S_n)| = \frac{7}{8}(n!)^3(1 - o(1))$. In fact, $|\mathcal{N}_3(S_5) = 1,401,120$ while $\frac{7}{8}(5!)^3 = 1,512,000$ (see **[DG]**).

Problem 2.7.1. Prove or disprove computationally that $\Gamma_k(G)$ is connected, when G is simple of reasonable size, and k is small (3 or 4).

We will illustrate the algorithm of **[CP]** in the case $G = A_n$ and $k = 3$. We start with the following key observation: *it suffice to show that all the generating k-tuples lie in the "large" connected component.* As we already know that all redundant triples lie in the "large" component, so checking connectivity to redundant triples gives an efficient check for inclusion in the "large" connected component. Once formulated this way, one can also utilize the symmetry of the situation and obtain a great speed up as compared to the "brute force" method.

By $\mathcal{R}_n$ denote the set of all redundant triples in $\Gamma_3(A_n)$. Recall that by Proposition 2.5.13. they lie in the same connected component. The algorithm checks connectivity to $\mathcal{R}_n$ of all generating triples $(\sigma_1, \sigma_2, \sigma_3)$ of A_n such that $\langle \sigma_1, \sigma_2 \rangle \neq A_n$. To achieve this, we search over all tuples $(M, \sigma_1, \sigma_2, \sigma_3)$, where M is a maximal subgroup of A_n and $\sigma_1, \sigma_2 \in M$. The maximal subgroups M in A_n for small n are known and can be precomputed. Of course some generating triples will be considered several times, but their fraction is small while speed up we achieve by pruning our search based on symmetry is quite large.

Here is how the symmetry can be utilized. Fix an appropriate subgroup H of the group of automorphisms of G. In our case we take $G = A_n$ and $H = S_n$. If $(\sigma_1, \sigma_2, \sigma_3)$ is not connected to $\mathcal{R}_n$ in $\Gamma_3(A_n)$, then $(\sigma_1^h, \sigma_2^h, \sigma_3^h)$ is also not connected to $\mathcal{R}_n$ for every $h \in H$.

By $\mathcal{O}_n$ denote the set of orbits of the natural action of H on $\mathcal{N}_3(A_n)$. For every orbit $O \in \mathcal{O}_n$, denote by $\eta(O)$ the lexically first element in O. In our situation, orbits of $H = S_n$ on A_n correspond to partitions of n into an even or odd number of parts (depending on the parity of n). Take $\sigma_1 \in A_n$ to be lexically first in the conjugacy class corresponding to such a partition. The centralizer $C_H(\sigma_1)$ fixes σ_1, so we can now take σ_2 to be lexically first under action of $C_H(\sigma_1)$, etc.

Now, for every such element $\eta(O)$ we check whether it's connected to $\mathcal{R}_n$. To do that we run the product replacement random walk until we find a redundant triple. Note that of the three subgroups $\langle\sigma_1, \sigma_2\rangle$, $\langle\sigma_1, \sigma_3\rangle$ and $\langle\sigma_2, \sigma_3\rangle$, after either of the Nielsen moves $R_{i,j}^{\pm}$, $L_{i,j}^{\pm}$, two subgroups will always remain. Thus at every step it suffices to check whether the third subgroup is G. We simply check whether both permutations in a pair lie in one of the maximal subgroups. Once the product replacement random walk hits a redundant triple, we proceed to the next orbit representative. The algorithm stops when all triples $\eta(O)$ are checked.

We discovered in **[CP]** that not only is $\Gamma_3(A_n)$ connected, when $n = 5, \ldots, 10$, but also that it usually takes very few steps of the product replacement walk to reach a redundant triple. This proves part (iv) of Theorem 2.5.7, and leads to the following question.

Question 2.7.2. Is there a universal constant C such that every generating triple $(g) \in \mathcal{N}_3(A_n)$ is connected in the graph $\Gamma_3(A_n)$ to a redundant triple by at most C steps ?

A negative answer to this question will shine a new light at the structure of generating sets of alternating groups.

3. Mixing time

3.1. Preliminaries. Let Γ be an r-regular connected unoriented graph, and let $v \in \Gamma$ be a fixed vertex in Γ. Consider a *nearest neighbor random walk* $\mathcal{W}_v = \mathcal{W}_v(\Gamma)$ defined as follows: start at a vertex v; at every step choose an adjacent edge with equal probability and move along that edge; repeat this ad infinitum. For technical reasons it is often useful to study *lazy random walk* $\mathcal{W}_v^\circ$, which is defined similarly, but now the walk stays at the same vertex with probability $1/2$ and moves along every adjacent edge with probability $1/2r$.

Denote by Q_v^t the probability distribution of the lazy random walk $\mathcal{W}_v^\circ$ after t steps. Since the graph is connected and regular, the stationary distribution is uniform:

$$\mathrm{Q}_v^t \to \frac{1}{|\Gamma|}, \quad \text{as } t \to \infty.$$

Recall the definition of the total variation distance of distribution P on Γ:

$$\|\mathrm{P} - \mathrm{U}\|_{\mathrm{tv}} = \max_{B \subset X} \left| \mathrm{P}(B) - \frac{|B|}{|X|} \right| = \frac{1}{2} \sum_{x \in X} \left| \mathrm{P}(x) - \frac{1}{|X|} \right|,$$

where $P(B) = \sum_{x \in B} P(x)$. To simplify the notation, denote $\mathbf{d}_v(t) = \|Q_v^t - U\|_{tv}$. It is easy to see that $0 \le \mathbf{d}_v(t) \le 1$, $\mathbf{d}_v(t) \to 0$ as $t \to \infty$, and that $\mathbf{d}_v(t+1) \le \mathbf{d}_v(t)$ for every t (see **[AF]**). Also, if $\mathbf{d}(t) = \min_{v \in \Gamma} \mathbf{d}_v(t)$, then $\mathbf{d}(t+s) \le 2\,\mathbf{d}(t) \cdot \mathbf{d}(s)$, so once the total variation distance becomes small, it decreases exponentially (see **[AF]**).

Definition 3.1.1. Define *mixing time* mix_v of the lazy random walk $\mathcal{W}_v^\circ(\Gamma)$ as follows:

$$\text{mix}_v = \min\left\{t \mid \mathbf{d}_v(t) \le \frac{1}{4}\right\}$$

Let $\mathcal{A}$ be the $|\Gamma| \times |\Gamma|$ adjacency matrix of graph Γ. The transition matrix of the random walk $\mathcal{W}(\Gamma)$ is given by $\mathcal{P} = \mathcal{A}/r$. Let $1 = \lambda_0 > \lambda_1 \ge \cdots \ge \lambda_{|\Gamma|-1|} \ge -1$ be the eigenvalues of $\mathcal{P}$ (all eigenvalues are real since the matrix is symmetric, the first inequality comes from the Perron–Frobenius Theorem). Now, for the lazy random walk $\mathcal{W}^\circ$ the transition matrix is given by $\mathcal{P}' = (Id + \mathcal{P})/2$, and for the eigenvalues of $\mathcal{P}$ we have $\lambda_i' = (1+\lambda_i)/2$, $1 = \lambda_0' > \lambda_1' \ge \cdots \ge \lambda_{|\Gamma|-1}' \ge 0$.

Proposition 3.1.2. *Let $\mathcal{W}_v^\circ(\Gamma)$ be a lazy random walk on a regular graph Γ starting at vertex $v \in \Gamma$. Then for the mixing time we have:*

$$\text{mix}_v \le \frac{4}{1-\lambda_1} \log|\Gamma|$$

Proof. We have

$$(\mathcal{P}')^n \cdot f = f_0 + (\lambda_1')^n f_1 + (\lambda_2')^n f_2 + \cdots,$$

where f is a function on vertices of Γ and f_i are projections of f on the eigenvectors of $\mathcal{P}'$. Now take f to be a function concentrated at $v \in \Gamma$. Then $Q_v^t = (\mathcal{P}')^t \cdot f$, $f_0 = U$, and

$$\left|Q_v^t(w) - \frac{1}{|\Gamma|}\right| \le (\lambda_1')^t \cdot |\Gamma|$$

for every $w \in \Gamma$. The proposition follows immediately. □

Remark 3.1.3. There are a number of distances on Γ one can study. While total variation can be thought as rescaled ℓ_1-norm, one can also define ℓ_p-norms for any $1 \le p < \infty$, as well as ℓ_∞-norm:

$$\|P - U\|_p = \left(\sum_{v \in \Gamma}\left(P(v) - \frac{1}{|\Gamma|}\right)^p\right)^{1/p}, \quad \|P - U\|_\infty = \max_{v \in \Gamma}\left|P(v) - \frac{1}{|\Gamma|}\right|$$

It also useful to define a *separation distance*, which can be thought as a one-sided ℓ_∞-norm:

$$\text{sep}(P) = |\Gamma| \cdot \max_{v \in \Gamma}\left(\frac{1}{|\Gamma|} - P(v)\right)$$

Denote $\mathbf{s}_v(t) = \text{sep}(Q_v^t)$. It is known that the separation distance satisfies similar monotonicity and submultiplicativity properties as the total variation distance (see **[AF, Di]**). Also, $\mathbf{d}_v(t) \leq \mathbf{s}_v(t)$ for any $t < \infty$.

Finally, let us note a straightforward generalization of Proposition 3.1.2 to all these norms, including ℓ_∞-norm. We leave the details to the reader.

3.2. The product replacement random walk. Let $\Gamma_k(G)$ be as before, the graph of generating k-tuples of the finite group G; let (g) be a fixed starting k-tuple. By Γ' denote the connected component of $\Gamma_k(G)$ which contains (g). By $\mathcal{PR} = \mathcal{PR}(G, k; (g))$ denote the lazy random walk on Γ' starting at (g). By $\text{mix}_{(g)}$ denote the mixing time of $\mathcal{PR}(G, k; (g))$. For the rest of this section, let mix stand for the maximum $\text{mix}_{(g)}$ over all $(g) \in \mathcal{N}_k(G)$. We call mix the *mixing time* of the product replacement random walk.

As far as theoretical computer science is concerned, the following conjecture is central:

Conjecture 3.2.1. *For every $k \geq d(G)$, the mixing time* mix *is polynomial in k and* $\log|G|$.

In this section we will cover the known "general results" on the subject. Let us note, however, that conjecture is wide open even for such special cases as $k = 3$ and $G = S_n$. But first a stronger version of the conjecture:

Question 3.2.2. Is it true that for every fixed $k \geq d(G)$ the relaxation time $\tau = 1/(1 - \lambda_1(\Gamma'))$ is bounded by a polynomial of k ?

Note that by Proposition 3.1.2 this would imply that the mixing time is *linear* in $\log|G|$.

Theorem 3.2.3 ([CG1, DS3]). *Let G be fixed and let $k \to \infty$. Then the mixing time* mix *of the lazy product replacement random walk $\mathcal{PR}$ satisfies:*

$$\text{mix} = O(k^2 \log k).$$

In other words, for k sufficiently large, the product replacement random walk mixes rapidly (compared to the size of the graph $|\Gamma| \sim |G|^k$). In fact, even more rapid mixing has been conjectured (see **[DS2, DS3]**):

Conjecture 3.2.4. *Let G be fixed and let $k \to \infty$. Then*

$$\text{mix} = O(k \log k).$$

The reasoning behind this conjecture is the following: for large k there seems little interaction between components of the k-tuple. As $|G|$ is bounded, we need just a bounded number of times to "hit" each component to make it "random" in G. Thus the problem becomes similar to the coupon collector's problem with k coupons, where each we need to collect a bounded number of times (cf. **[Di, F]**) This is where $O(k \log k)$ come from.

For $G = \mathbb{Z}_2$ an ad hoc method was used by Chung and Graham **[CG2]** to prove Conjecture 3.2.4 in this case.

Remark 3.2.5. Recently the author announced **[P6]** a positive solution of Conjecture 3.2.1 in case when G is nilpotent. This itself is much stronger than the partial solution when k and the nilpotency class i are bounded. It nevertheless gives quite a large degree polynomial. The solution is yet to be checked and fully written.

3.3. The Diaconis and Saloff-Coste bound. In an important paper **[DS3]** the authors found the first general bounds for the mixing time of the product replacement random walk. Since the authors do not use any special properties of the walk as compared to other random walks, the results are quite weak from the theory of computation point of view (cf. Conjecture 3.1.1.) On the other hand, a refined analytic technique used by the authors seem to be so advanced that we expect the result to be difficult (if at all possible) to improve by means of the better estimates.

Rather than give here the full power of the result by Diaconis and Saloff-Coste (which may be difficult to quickly digest), we choose to sketch the approach and state slightly weaker version of the main result.

The idea is that given any connected graph Γ one can bound the relaxation time $\tau = 1/(1 - \lambda_1)$ via geometric properties of the graph (we refer to **[AF, DSt]**) for references and details). Formally, for every two vertices $v, v' \in \Gamma$ fix a path $\gamma(v, v')$. By Δ denote the maximum length of such paths, and by N denote the maximum congestion through edge e in Γ (the maximum of the number of paths which go through e, divided by $|\Gamma|$). Roughly, the relaxation time τ is bounded by a polynomial of Δ, N. The problem is that if the paths are chosen in a "stupid" way, the congestion can be as large as size of the graph.

In **[DS3]** the authors consider an interesting set of paths (see p. 268) which split into "big steps": change any coordinate by any element (as long as k-tuple still generates G). Now, these "big steps" can be easily used to connect any $(h) \in \Gamma_k(G)$ to any $(g) \in \Gamma_k(G)$ (given that k is large enough, see **[DS3]**). Now big steps can be decomposed into "small" steps of the walk on $\Gamma_k(G)$ by use of the geometry of generating sets, not unlike the one used in Proposition 2.2.2.

Now, once the path are constructed, "the fight" is over the final shape of the estimates. The authors involve an advanced ℓ_2-technique and largely win this fight. Still, they "lose the battle" since their paths are just too long and this problem is inherent in the construction. One would suggest using short paths of Babai (see the proof of Theorem 2.2.3), but these paths are of a special type and hard to analyze for congestion. Thus in the final estimate Diaconis and Saloff-Coste have the following parameter:

$\Delta(G) =$ *maximal diameter of the Cayley graph* Cayley(G, S) *over all generating sets* $S \subset G$

This $\Delta(G)$ is the main ingredient of the length of the small steps of nice paths of Diaconis and Saloff-Coste. While $\Delta(G)$ is conjectured to be nice for various families of groups (such as simple groups) it is provably large for a number of other families (such as abelian and nilpotent groups). This is the key problem with the argument which prevents the final bound on the mixing time from being polynomial (even in the cases when it has been confirmed by other methods)

Now we are ready to formulate a "toy version" of the main result in **[DS3]**:

Theorem 3.3.1 (Diaconis, Saloff-Coste). *Let G be a finite group, $d = d(G)$, $\overline{d} = \overline{d}(G)$, $k \geq \overline{d} + 2d$, $\varphi_k(G) = |\Gamma_k(G)|/|G^k|$, and let $\Delta(G)$ be as above. Let* mix *be the mixing time of the lazy product replacement random walk on $\Gamma_k(G)$ (the maximum is taken over all starting points $(g) \in \Gamma_k(G)$). Then*

$$\text{mix} \leq O\left(A\, k^2 \log|G| \left(\log k + \log\log|G|\right) \frac{\Delta^2(G)}{\varphi_k^2(G)}\right),$$

where A is a parameter defined as follows:

$$A = \frac{d^2 \cdot k(k-1) \cdot \ldots \cdot (k - \overline{d} + 1)}{(k-2d)(k-2d-1) \cdot \ldots \cdot (k - 2d - \overline{d} + 1)}.$$

The actual result is much more refined, to the extend of being difficult to comprehend. Among others, the constant implies by $O(\cdot)$ notation is calculated (it is about 80) and several little improvements are made. We present here just "the first order" bound to give the reader an easy access to the result.

A few words about the quality of this result for general groups: when $k = \theta(d \cdot \overline{d})$ we have $A = O(d^2)$. Now using bounds $d, \overline{d} = O(\log|G|)$, $\varphi_k(G) \sim 1$, we obtain a bound on the mixing time which is polynomial in k, $\log|G|$ and $\Delta(G)$.

Perhaps comparison with the mixing of random walks on groups is in order. Let G be a finite group, S be a symmetric ($S = S^{-1}$) generating set, $|S| = k$, and let $\Delta(G, S)$ be the diameter of the corresponding Cayley graph $\Gamma =$ Cayley(G, S). Then for the mixing time mix of the lazy nearest neighbor random walk on Γ (starting at id $\in G$) we have:

$$\text{mix} = O(\Delta^2(G, S) \cdot k)$$

This result was proved in **[DS1]** and is an improvement over a more traditional bound mix $= O(\Delta^2(G, S) \cdot k \cdot \log|G|)$. Note now the similarity with the main result in Theorem 3.2.1. Observe that the bound obtained there is worse that the bound on the mixing time of *any random walk on the same group with the same number k of generators*. On the other hand, the main reason the product replacement random walk is used, is because it mixes *faster* than the ordinary random walk. We conclude that the bound in Theorem 3.3.1 by no means explains the rapid mixing phenomenon of the product replacement random walk.

Example 3.3.2. Let G be abelian. Then $\Delta(G)$ can be as large as $|G|^2$ (for $G \simeq \mathbb{Z}_n$). Thus the mixing time bound in Theorem 3.3.1 becomes exponential in this case (in $\log|G|$) while as we show below (see Theorem 3.5.9) the mixing time is in fact polynomial.

Example 3.3.3. Let $G = S_n$. It is conjectured that $\Delta(S_n) = O(n^2)$, while the best known bound is $\Delta(S_n) = \exp(O(\sqrt{n}))$ (see **[Bb4]**). Now observe that the conjecture combined with Theorem 3.3.1 gives a polynomial bound on the mixing time in this case. Analogously, for every simple group G it is conjectured that $\Delta(G) = (\log|G|)^{O(1)}$. See **[Bb4, DS3]** for references and details.

3.4. $\Gamma_k(G)$ as Schreier graphs. Denote by F_k the free group on k generators $x_1, \dots, x_k$ and let $\mathrm{Aut}(F_k)$ be a group of automorphisms of F_k. Consider the set Υ_k of the following automorphisms:

$$R^{\pm}_{i,j}(x_i) = x_i x_j^{\pm 1}, \text{ and } R^{\pm}_{i,j}(x_l) = x_l \text{ if } l \neq i$$
$$L^{\pm}_{i,j}(x_i) = x_j^{\pm 1} x_i, \text{ and } R^{\pm}_{i,j}(x_l) = x_l \text{ if } l \neq i$$

These are exactly *Nielsen moves* when $G = F_k$. A classical result of Nielsen **[Ni]** (see **[LyS, MKS]**) shows that $\mathrm{Aut}(F_k)$ is generated by the Nielsen moves and elementary automorphisms of permutation and inversion of generators.

Proposition 3.4.1. *Let $A^+ = A^+(F_k)$ be the subgroup of* $\mathrm{Aut}(F_k)$ *generated by Nielsen moves. Then $A^+(F_k)$ is a normal subgroup of index two in* $\mathrm{Aut}(F_k)$.

The group $A^+(F_k)$ is called *special group of automorphisms* of a free group F_k.

Proof. The proposition seems to be well known (cf. **[Ge]**). An easy proof that A^+ has index at most two follows along the lines of the proof of Proposition 2.2.1. As to why it's exactly two, consider the natural projection $\pi : \mathrm{Aut}(F_k) \to \mathrm{Aut}(F_k/[F_k, F_k]) \simeq GL_k(\mathbb{Z})$. Observe that $\pi(R^{\pm}_{i,j}) = \pi(L^{\pm}_{i,j}) \in \mathrm{SL}_k(\mathbb{Z})$ and gives elementary transvections $E^{\pm}_{i,j}$, for all $i \neq j$. Therefore $\pi(A^+) = \mathrm{SL}_k(\mathbb{Z})$ and A^+ has index at least two in $\mathrm{Aut}(F_k)$. □

Proposition 3.4.2. *The graph $\Gamma_k(F_k)$ has two connected components, each of them is isomorphic to the right Cayley graph of the special automorphism group $A^+(F_k)$ with respect to the Nielsen moves.*

Proof. Indeed, $\mathrm{Aut}(F_k)$ acts simply transitively on the vertices of $\Gamma_k(F_k)$. This gives a one to one correspondence ϱ between $\mathrm{Aut}(F_k)$ and $\Gamma_k(F_k)$ defined as

$$\varrho : \alpha \in \mathrm{Aut}(F_k) \to \big(\alpha(x_1), \dots, \alpha(x_k)\big).$$

It is easy to see that $\alpha R^{\pm}_{i,j}$ and $\alpha L^{\pm}_{i,j}$ correspond to the neighbors of $\varrho(\alpha) \in \Gamma_k(G)$. This shows that $\Gamma_k(F_k)$ is the Cayley graph of $\mathrm{Aut}(F_k)$ with respect to the Nielsen moves. As they do not generate the group but rather a subgroup of index two, the graph $\Gamma_k(F_k)$ has two connected components, each one isomorphic to $\mathrm{Cayley}\big(A^+(F_k); \{R^{\pm}_{i,j}, L^{\pm}_{i,j}\}\big)$. □

Let H and G be two (possibly infinite) groups and ψ be an epimorphism from H onto G. Then ψ induces a map $\psi_k : \Gamma_k(H) \to \Gamma_k(G)$, where

$$\psi_k : (h_1, \dots, h_k) \to \big(\psi(h_1), \dots, \psi(h_k)\big),$$

provided $k \geq d(H)$. It is easy to see that ψ_k is a projection of graphs which preserves adjacency relations. For $H = F_k$ with a little work one can deduce:

Proposition 3.4.3. *Let $X = \mathrm{Cayley}\left(A^+(F_k); \Upsilon_k\right)$ be the Cayley graph of $A^+(F_k)$ with respect to the Nielsen generators. Then for every G and every connected component Y of $\Gamma_k(G)$, there exists a surjective graph projection $\psi : X \to Y$.*

Proof. Identify $\Gamma_k(G)$ with the set $\mathrm{Epi}(F_k, G)$ of all epimorphisms $\psi : F_k \twoheadrightarrow G$, where an epimorphism ψ is identified with $(\psi(x_1), \dots, \psi(x_k))$. Then $A = \mathrm{Aut}(F_k)$ acts on $E = \mathrm{Epi}(F_k, G)$ by: $\alpha(\psi) = \psi \circ \alpha^{-1}$ for $\alpha \in \mathrm{Aut}(F_k)$ and $\psi \in E$. Fix $\psi : F_k \to G$ and let B be the subgroup of $A^+ = A^+(F_k)$ defined as $B = \{\alpha \in A^+ \mid \alpha(\psi) = \psi\}$. Then A^+/B is naturally identified with connected component of $(g) = (\psi(x_1), \dots, \psi(x_k))$ in $\Gamma_k(G)$. Indeed, if $\alpha_1, \alpha_2 \in A^+$, then $\alpha_1 B = \alpha_2 B$ if and only if $\alpha_1(\varphi) = \alpha_2(\psi)$. Moreover, $\big[R^{\pm}_{i,j} \circ \alpha\big](\psi)$ and $\big[L^{\pm}_{i,j} \circ \alpha\big](\psi)$ are exactly the neighbors of $\alpha(\psi)$. □

The proposition implies that every connected component of $\Gamma_k(G)$ is a Schreier graph of $A^+(F_k)$ with respect to the Nielsen moves Υ_k, modulo finite index subgroup. This is a very strong property which will be used in the next section.

Note that we do not claim that $\psi_k : X \to \Gamma_k(G)$ is surjective. In fact this is not true as graphs $\Gamma_k(G)$ can have any number of connected components (see Section 2.3). On the other hand, for finite groups the map this is indeed surjective, as follows from Theorem 2.1.4.

Now let us generalize our results from free groups to "relatively free groups". Let W be a characteristic subgroup of F_k, i.e. $\alpha(W) = W$ for every $\alpha \in \mathrm{Aut}(F_k)$. There is a natural homomorphism $\pi : \mathrm{Aut}(F_k) \to \mathrm{Aut}(F_k/W)$. Denote $\pi(A^+(F_k))$ by $A^+(F_k/W)$, and call it the special automorphism group of F_k/W. Note that in general π is not necessarily an epimorphism, so $A^+(F_k/W)$ can be of large (even infinite) index in $\mathrm{Aut}(F_k/W)$. Still, the Nielsen moves generate $A^+(F_k/W)$ and we have the following general result:

Proposition 3.4.4. *Let W be a characteristic subgroup of F_k and G a finite quotient of F_k/W. Then every component of $\Gamma_k(G)$ is a Schreier graph of $A^+(F_k/W)$ with respect to the Nielsen moves and modulo some finite index subgroup of $A^+(F_k/W)$.* □

Example 3.4.5. Let $W = [F_k, F_k]$ be the commutator subgroup of F_k. Then $\mathrm{Aut}(F_k/W) = GL_k(\mathbb{Z})$ and $A^+(F_k/W) = \mathrm{SL}_k(\mathbb{Z})$. The Nielsen moves $R^{\pm}_{i,j}$ and $L^{\pm}_{i,j}$ correspond to the elementary matrices $E^{\pm}_{i,j}$ with 1's along diagonal, ± 1 at the (i, j) entry, and 0 elsewhere. We therefore conclude:

Proposition 3.4.6. *Let G be a finite abelian group. Then any connected component of $\Gamma_k(G)$ is a Schreier graph of $\mathrm{SL}_k(\mathbb{Z})$ with respect to the elementary matrices $E^{\pm}_{i,j}$ modulo a finite index subgroup of $\mathrm{SL}_k(\mathbb{Z})$.* [6] □

Remark 3.4.7. Let G be as in Theorem 2.3.2. It is not difficult to see that a finite index subgroup of $\mathrm{SL}_k(\mathbb{Z})$ in Proposition 3.3.6 is a congruence subgroup containing $\mathrm{Ker}\big(\mathrm{SL}_k(\mathbb{Z}) \to \mathrm{SL}_k(\mathbb{Z}/m_1\mathbb{Z})\big)$, where m_1 is as in Theorem 2.3.2. This explains part (ii) of the Theorem.

Example 3.4.8. Define the lower central series of F_k by $\gamma_1(F_k) = F_k$ and $\gamma_{i+1}(F_k) = \big[F_k, \gamma_i(F_k)\big]$. Let $W = \gamma_{i+1}(F_k)$, so F_k/W is the "free nilpotent group of class i". Let $A^+(F_k/W)$ and $\mathrm{Aut}(F_k/W)$ be as above. It is known that when $k \geq 2$ and $i \geq 4$, $A^+(F_k/W)$ is of infinite index in $\mathrm{Aut}(F_k/W)$ (see **[An, Bh]**). Nevertheless, we still have that for every nilpotent group G of class i, every connected component of $\Gamma_k(G)$ is a quotient of the Cayley graph of $A^+(F_k/W)$ with respect to the Nielsen moves.

3.5. Kazhdan's property (T) and mixing time. The main result in this section is a connection between Kazhdan's property (T) and the mixing time of the product replacement random walk. The exposition follows the paper **[LP]** of Lubotzky and the author.

Definition 3.5.1. A topological group G is said to have (Kazhdan) *property* (T) if there exists a compact subset Q of G such that $\mathcal{K} = \mathcal{K}(G, Q) > 0$, where

$$(*) \qquad \mathcal{K}(G, Q) = \inf_{\rho} \inf_{v} \max_{q \in Q} \frac{\|\rho(q)\, v - v\|}{\|v\|},$$

where ρ runs over all unitary representations $(\mathcal{H}, \rho)$ of G which do not contain the trivial representation (i.e., no non-zero G-fixed vector), and v runs over all vectors $v \neq 0$ in $\mathcal{H}$.

[6] Formally, one should include each elementary matrix $E^{\pm}_{i,j}$ twice, which is due to the fact that $L^{\pm}_{i,j} = R^{\pm}_{i,j}$ for abelian groups.

We say that (Q, ε) is a *Kazhdan constant* for G if $\varepsilon \leq \mathcal{K}(G, Q)$ (see **[Lu1]**).

Let now Γ be a discrete group. It is well known and not difficult to prove that if Γ has (T), then Γ is finitely generated, and if (Q, ε) is a Kazhdan constant for Γ, then Q generate Γ (see **[Kz, Lu1]**).

Open Problem 3.5.2. Does $\mathrm{Aut}(F_k)$ $\big($or equivalently $A^+(F_k)\big)$ has property (T) with respect to Nielsen generators, given $k > 3$?

Remark 3.5.3. For $k = 2, 3$ the answer is negative. This follows from **[Lu1, Mc]**. Let note also that if $\mathrm{Aut}(F_k)$ has (T), then in view of Gilman's Theorem 2.4.3 and his part (i) of Theorem 2.5.7, one can obtain constructions of expanders from Cayley graphs of the alternating groups (cf. **[Mr, Lu1]**). This would solve positively open problem 10.3.4 and negatively Open Problem 10.3.2 in **[Lu1]**.

Proposition 3.5.4. *Let Γ be a discrete group generated by a finite set S. Assume Γ has property (T) with Kazhdan constant $(S, \mathcal{K})$. Then for every finite index subgroup N of Γ, the Schreier graph X on Γ/N with respect to S satisfies $\beta \geq \mathcal{K}^2/2|S|$, where $\beta(X) = 1 - \lambda_1(X)$ is an eigenvalue gap of X.*

The proof can be found in **[HRV]**, corollary to Proposition III, p. 89.

Proposition 3.5.5 ([PZ]). *In the condition of Proposition 3.5.4, assume that there is a finite group $H < \mathrm{Aut}(\Gamma)$ such that $H(S) = S$ and action of H on S has m equal size orbits. Then $\beta \geq \mathcal{K}^2/(2m)$.*

The proof will appear in a forthcoming paper **[PZ]**. Let us note that one can always take $H = \{1\}$. Then $m = |S|$ and the bound in Proposition 3.5.5 coincides with a bound in Proposition 3.5.4.

As a combination of Proposition 3.5.5 and 3.4.3 we obtain:

Theorem 3.5.6 ([LP]). *If $\mathrm{Aut}(F_k)$ (or equivalently $A^+(F_k)$) has Kazhdan's property (T), then for every finite group G generated by k elements, the mixing time $\mathrm{mix}_{(g)}$ of the lazy product replacement random walk on a connected component $\Gamma' \subset \Gamma_k(G)$, $(g) \in \Gamma'$, is bounded as*

$$\mathrm{mix}_{(g)} = C(k) \log |G|,$$

where $C(k)$ depends only on k.

Analogously, for "relatively free groups" we have:

Theorem 3.5.7 ([LP]). *Let W be a characteristic subgroup of F_k. If the special automorphism group $A^+(F_k/W)$ has (T), then the conclusion of Theorem 3.4.5 is satisfied for every finite quotient G of F_k/W.*

Example 3.5.8. Let $W = [F_k, F_k]$ be as in Example 3.4.5. Then $A^+(F_k/W) \simeq \mathrm{SL}(k, \mathbb{Z})$, which indeed has property (T) (see **[Lu1]**). Further, recently Shalom **[Sh2]** gave bounds on the Kazhdan constant exactly for the elementary matrices as a generating set. Combining all these results we get:

Theorem 3.5.9 ([LP]). *Let G be an abelian group, $(g) = (g_1, \dots, g_k)$ be the initial generating k-tuple, and let $\Gamma' \subset \Gamma_k(G)$ be a connected component containing (g). Then for the mixing time $\mathrm{mix}_{(g)}$ of the lazy product replacement random walk $\mathcal{PR}$ starting at (g) we have*

$$\mathrm{mix}_{(g)} \le C \cdot k^5 \cdot \log |G|,$$

where C is a universal constant.

Remark 3.5.10. Note that we write $\mathrm{mix}_{(g)}$ as we prove rapid mixing for random walks on *every* connected component. If the connected component Γ' is small compared to Γ, one can use $\mathrm{mix}_{(g)} \le C \cdot k^4 \log |\Gamma'|$ bound instead.

A very special case of groups $\mathbb{Z}_p^m$, p-prime, was consider earlier in **[DS2]**. The authors obtained a bound on the mixing time by using just diameter bound for $\mathrm{SL}(2, p)$, supplementing it with random walk techniques. The resulting bound for $G \simeq \mathbb{Z}_p$ is $\mathrm{mix} = O(k^4 \log^3 p)$. As in situations of interest $k = O(\log p)$, this bound is weaker than that in Theorem 3.5.9. Still, this suggests possible improvement of the power k^5 in the latter bound.

Example 3.5.11. Let $W = \gamma_{i+1}(F_k)$ be as in Example 3.4.8. Then, as showed in **[LP]**, the quotient F_k/W has property (T). The problem is that we have yet to estimate the Kazhdan constant in this case. Still, this implies the following weak generalization of Theorem 3.5.9:

Theorem 3.5.12 ([LP]). *Let k, i be fixed, and let G be a nilpotent group of class at most i, $d(G) \le k$. Let $(g) = (g_1, \dots, g_k)$ be the initial generating k-tuple, and let $\Gamma' \subset \Gamma_k(G)$ be a connected component containing (g). Then for the mixing time $\mathrm{mix}_{(g)}$ of the lazy product replacement random walk $\mathcal{PR}$ starting at (g) we have*

$$\mathrm{mix}_{(g)} \le C(k, i) \cdot \log |G|,$$

where $C(k, i)$ is a constant which depends on k, i, and is independent of the group G.

Conjecture 3.5.13. *The constant $C(k, i)$ is bounded by a universal constant $C(k)$, independent of i and polynomial in k.*

Remark 3.5.14. The proof of the Theorem 3.5.12 is based on first proving that the corresponding Lie group has property (T), which is based on the Mautner phenomenon (see **[Wa]**). Unfortunately it is often difficult to obtain good bounds for Kazhdan constants of lattices (cf. **[Bu, Sh1, Sh2]**). Still, we have a high confidence that *some*, perhaps an exponential bound, can be obtained for $C(k,i)$. The idea is based on applying the technique of Shalom **[Sh1]** combined with the bounds in the work **[KB]** on the Smith normal form. We challenge the reader to obtain any bounds on $C(k,i)$ in this direction.

4. Putting it all together

4.1. What one would like to have. Let G be a black box group with the cost μ of multiplication and inversion (see **[Bb5, KS, P2]** for definitions). For convenience, one can simply assume that G is given as a subgroup of a large permutation or matrix group (over the finite field). We also assume that an upper bound N is given for $\log|G|$ (say, the bit length of the encoding of groups elements).

One would like to design a routine which when given generators $g_1, \dots, g_k$ of G, outputs independent elements of G from a certain distribution $\mathcal{P}$ on G such that $\|\mathcal{P} - \mathrm{U}\|_{\mathrm{tv}} \le 1/4$. The cost ρ of such a routine should be polynomial in both N and k:

$$\rho = \mathrm{Poly}(N,k) \cdot \mu$$

Note that successive iterations of the routine give a number of independent "random" elements of G. Taking their product gives better approximations of the uniform distribution U, as good as one likes (in any of the distances mentioned in 3.1.3) This is what we wanted at the first place. We call elements sampled from the distribution $\mathcal{P}$ as above the *(nearly) uniform groups elements*. Also, from now on we will replace N by $O(\log|G|)$ hoping this does not lead to confusion.

Now, the product replacement algorithm is yet to be proves a legitimate candidate for such a routine. In the meantime one should ask whether there is *any* algorithm which satisfies the above requirements. The answer is yes: there is one provably correct known example of such an algorithm; it was discovered by Babai in his pioneer paper **[Bb3]**. We will sketch the algorithm below for the reader's convenience.

Algorithm 4.1.1 (Babai).

0) Use "random subproducts" routine to reduce the number of generators in a generating set S from k to $O(\log|G|)$.
1) For $C_1 \cdot \log|G|$ rounds repeat the following procedure. Run a simple random walk on the Cayley graph $\mathrm{Cayley}(G,S)$ starting at $\mathrm{id} \in G$, for $C_2 \log^4|G|$ steps. Then add the endpoint of the walk to S and repeat.
2) Use the "Erdős–Rényi machine" to generate the output.

Theorem 4.1.2 (**[Bb3]**). *The above algorithm produces m independent (nearly) uniform elements of G at a cost* $O\big(k\log|G| + \log^5|G| + m\log|G|\big) \cdot \mu$.

We will always assume that $k = O(\log |G|)$. Then the preprocessing part 0) can be skipped. The "Erdős-Rényi machine" will be touched upon later in this section, but it is also unnecessary for our current purposes: one can simply replace it by additional rounds of a simple random walk of length $O(\log^2 |G|)$ with the stopping points output as "random elements"[7]. Thus we will concentrate on the main part 1) of the algorithm.

Several remarks before we analyze part 1). First, constants C_1, C_2 are universal and have been explicitly computed in **[Bb3]**. Second, for technical reason in the original paper the stopping time of the simple walks was chosen from a certain distribution. Recent advancements in the random walk technique proved this to be unnecessary so we use here a simplified version so as not to alarm the reader. Finally, for a reason of saving a small constant factor the author made the lengths of the walk to be increasing. We ignore this difference.

Now, the main result of Babai is based on the following lemma:

Lemma 4.1.3. *After round $C_1 \cdot i$, for a generating set $S = \{g_1, \dots, g_r\}$, $r = k + C_1 \cdot i$, the number of distinct elements in $C(S) = \{g_1^{\varepsilon_1} \cdot \dots \cdot g_r^{\varepsilon_r}\}$ is w.h.p.*[8] *at least 2^i, where $i \leq \log_2 |G| - 1$.*

Basically, the length $C_2 \log^4 |G|$ is chosen so that the size of the "cube" $C(S)$ is doubled with positive probability after a constant number of steps. At the end, when the diameter of the Cayley graph $\Delta(G, S)$ becomes of the order $O(\log |G|)$ while $|S| = O(\log |G|)$, the simple random walks on G mix rapidly (after a careful analysis, one can show that $\text{mix} = O(\log^2 |G|)$).

When compared to the product replacement random walk, few similarities are clear. Both walks use constantly changing generating sets to "reach" to more distant elements in the original Cayley graph. Both walks use the result of the previous walk as its "generating set" to make these steps.

The main difference is that the size of the generating set in the algorithm of Babai is increasing, absorbing all the new generators. As a result, in Babai's algorithm, after $\log |G|$ rounds the size of the generating set becomes $\geq \log |G|$. Since each short walk must be of length at least $\Omega(\log |G|)$, the lower bound of $\Omega(\log^2 |G| \cdot \mu)$ is in order[9].

On the other hand, in the product replacement random walk, the size k of the generating set remains fixed. Thus, theoretically the cost of the algorithm can be as small as $O(k^c \cdot \log |G| \cdot \mu)$, i.e. linear in $\log |G|$. In fact this would be the case if Open Problem 3.5.2 had a positive solution. While formally k can be as large as $\log_2 |G|$ (say, for $G \simeq \mathbb{Z}_2^n$), and the constant c can be larger than 5, in the practically important

[7] Of course, the use of the "Erdős-Rényi machine" has an important theoretical and perhaps practical advantage (see **[Bb2, Bb4]**). We skip it here for simplicity.

[8] w.h.p. = with high probability. Let us be somewhat vague here as one can always refer to **[Bb3]** for technical details.

[9] An improved lower bound $\Omega(\log^3 |G| \cdot \mu)$ for any version of Babai's algorithm seems also reasonable.

case when k is small this gives a desired speed up and seems to explain the rapid mixing phenomenon.

We realize that at present the above "explanation" is nothing but a speculation. However a fine distinction between the algorithms seem important to preserve and use small k in the product replacement algorithm, rather than take $k = \theta(\log |G|)$ as done in Theorem 2.2.3 and Theorem 3.3.1.

4.2. Running the product replacement algorithm. As we describe earlier, to run the product replacement algorithm, even before analyzing the mixing time, one needs to remove two main obstacles: connectivity and bias. Rather than repeat much said in Sections 1, 2 about these two, let us summarize what could be done in a general case.

As noted in Sections 1.5 and 2.6, both obstacles can be removed by taking k large enough so that $\varphi_k(G) \to 1$. Then bias is avoided automatically and when $d(G)$ is added to k, the graphs contain a large component containing the initial generating set (see Theorem 2.6.2.) Together the result is quantified in Proposition 1.5.1, while Corollary 1.5.2 summarizes what's known in special cases. We believe that a good rule of thumb is given by Conjecture 1.5.3: take $k = Cd(G) \log\log |G|$. This probably removes both obstacles, and the factor $O(\log\log |G|)$ is still a small price to pay for that (as compared to $\theta(\log |G|)$).

How long should one run the product replacement random walk? Well, for the mixing time *mix* steps, after which one obtains a (nearly) uniform generating k-tuple (with all components equally useful for random generation). Of course, we do not have good theoretical bounds on *mix* barring few special cases (see Section 3.3, 3.5.) Determining *mix* in practice is not an easy task. Thus from the theoretical point of view the algorithm should not be use other than in Las Vegas applications, which is to say that any Monte Carlo estimates are largely unreliable.

4.3. Modifying the product replacement algorithm. It was proposed recently **[LG]** by Leedham-Green et al. that while running the product replacement random walk one should consider a simple random walk on a group with a generating k-tuple as a generating set. Thus the generating set of this new random walk on will change with time. The stopping state of the walk is assigned to be the output. This modification[10] is proposed to remove the bias obstacle (**[LG]**).

We see several problems with such an approach. First, while it formally removes the bias obstacle, it doesn't remove the connectivity obstacle and tells nothing about the mixing. Second, it is seemingly worse than running a simple random walk on the generating k-tuple obtained at the end of the product replacement random walk. Indeed, heuristically one "wastes time" by running a simple random walk with the initial "bad" generators, rather than wait till the end of the product replacement walk

[10]The authors gave colorful British names to each version of the algorithm. We decided to avoid this terminology for the sake of clarity.

and use these "good" generators. The reason being that initial generating set may correspond to the "worst case" of a random walk, while the k-tuple obtained after running product replacement gives an "average case" of a random walk. Note that random walks on random generating sets is a known concept in the theory, and are called sometimes "random random walks" (see **[P3]** for the references).

Another cheap method to avoid bias is to use the "Erdős-Rényi machine" of Babai mentioned above. Simply take the resulting generating set of k elements and output random elements of the "cube" (these elements are also known as "random subproducts"). This method provably works for $k \geq 2 \log_2 |G|$, but has yet to be analyzed for smaller values of k (see **[ER, Bb3]**). In fact, this approach is again intimately related to taking simple random walks on G with this generating set: we refer to **[P3]** for making this claim formal.

If the author was to modify the algorithm, it would be the following construction. Simply run the product replacement random walk, but from time to time multiply random elements of the k-tuple by one of the randomly chosen *original* generators. Then one preserves the structure of the walk as a random walk on a Schreier graph of $\text{Aut}(F_k)$, while ensuring that the stationary distribution is uniform. Basically the walk will be moving from one connected component of $\Gamma_k(G)$ to another when we multiply by the original generators. As a bonus, in the limit *all* the components in an obtained k-tuple are uniform and independent in G, which is an improvement over a single such element in the previous versions of the algorithm.

Finally, let us conclude by saying that the product replacement algorithm is still largely mysterious and is perhaps a detour from the "right" solution. The problem of generating random group elements remains a challenge, and we hope that some radically new provably correct algorithms will appear in the future.

Acknowledgements. I would like to start by expressing my extreme gratitude to Persi Diaconis, Bill Kantor and Alex Lubotzky who supported this project and kindly encouraged writing of the paper. This review article is a compilation of several papers, including joint works with László Babai, Sergey Bratus, Gene Cooperman, Alex Lubotzky, and Andrzej Zuk. I am grateful for the wonderful opportunity to work with them.

I would like to thank Eamonn O'Brien, Derek Holt, Charles Leedham-Green, Scott Murray, Alice Niemeyer, and Leonard Soicher for the valuable conversations on the practical importance of the algorithm. Remarks of Alexander Borovik, Ken Brown, Keith Dennis, Martin Dunwoody, Robert Guralnick, László Lovász, Avinoam Mann, Gregory Margulis, László Pyber, Laurent Saloff-Coste, Jan Saxl, Yehuda Shalom, and James Wiegold were also very helpful. Special thanks to Bill Kantor for reading the preliminary version of the paper.

The author was supported by the NSF Postdoctoral Research Fellowship in Mathematical Sciences.

Added in proof. After the paper was written, the author proved that the product replacement algorithm is indeed polynomial for all $k = \Omega(\log|G| \log\log|G|)$. The proof was published in the proceedings of FOCS 2000 and is available from the authors's web page (http://www-math.mit.edu/~pak/pak.html).

References

[Ac] V. Acciaro, The probability of generating some common families of finite groups, Utililitas Math. 49 (1996), 243–254.

[AKS] M. Ajtai, J. Komlós, E. Szemerédi, Sorting in $c \log n$ parallel steps, Combinatorica 3 (1983), 1–19.

[AF] D. Aldous, J. Fill, Reversible Markov Chains and Random Walks on Graphs, monograph, in preparation, 1996.

[ASE] N. Alon, J. H. Spencer, P. Erdős, The Probabilistic Method, Wiley, New York, 1992.

[An] S. Andreadakis, On the automorphisms of free groups and free nilpotent groups, Proc. London Math. Soc. (3) 15 (1965), 239–268.

[As] M. Aschbacher, On the maximal subgroups of the finite classical groups, Invent. Math. 76 (1984), 469–514.

[Bb1] L. Babai, On the length of subgroup chains in the symmetric group, Comm. Algebra 14 (1986), 1729–1736.

[Bb2] L. Babai, The probability of generating the symmetric group, J. Combin. Theory Ser. A 52 (1989), 148–153.

[Bb3] L. Babai, Local expansion of vertex-transitive graphs and random generation in finite groups, in: Proc. 23rd ACM STOC, 1991, 164–174.

[Bb4] L. Babai, Automorphism groups, isomorphism, reconstruction, in: Handbook of Combinatorics (R. L. Graham, M. Groetschel, and L. Lovasz, eds.), Elsevier, 1996.

[Bb5] L. Babai, Randomization in group algorithms: Conceptual questions, in: Groups and Computation II (L. Finkelstein and W. M. Kantor, eds.), DIMACS Ser. Discrete Math. Theoret. Comput. Sci. 28, Amer. Math. Soc., Providence, RI, 1997, 1–17.

[BCFLS] L. Babai, G. Cooperman, L. Finkelstein, E. Luks, A. Seress, Fast Monte Carlo algorithms for permutation groups, J. Comput. System Sci. 50 (1995), 296–308.

[BbP] L. Babai, I. Pak, Strong bias of group generators: an obstacle to the "product replacement algorithm", Proc. Eleventh Annual ACM-SIAM Symposium on Discrete Algorithms, 2000, 627–635.

[Bh] S. Bachmuth, Automorphisms of free metabelian groups, Trans. Amer. Math. Soc. 118 (1965), 93–104.

[Bi] G. Ja. Binder, Bases of the symmetric group, Izv. Vuzov. Matem. 11 (1968), 19–25.

[BPS] A.V. Borovik, L. Pyber, A. Shalev, Maximal subgroups in finite and profinite groups, Trans. Amer. Math. Soc. 348 (1996), 3745–3761.

[BCP] W. Bosma, J. Cannon, C. Playoust, The Magma algebra system, in: "Computational algebra and number theory" (London, 1993), J. Symbolic Comput. 24 (1997), 235–265.

[BrP] S. Bratus, I. Pak, Fast constructive recognition of a gray box group isomorphic to S_n or A_n using Goldbach's Conjecture, J. Symbolic Comput. 29 (2000), 33–57.

[BW1] J. L. Brenner, J. Wiegold, Two-generator groups. I., Michigan Math. J. 22 (1975), 53–64.

[BW2] J. L. Brenner, J. Wiegold, Two-generator groups. II., Bull. Austral. Math. Soc. 22 (1980), 113–124.

[Bu] M. Burger, Kazhdan constants for SL(3, $\mathbb{Z}$), J. Reine Angew. Math. 413 (1991), 36–67.

[CST] P. J. Cameron, R. Solomon, A. Turull, Chains of subgroups in symmetric groups, J. Algebra 127 (1989), 340–352.

[CLMNO] F. Celler, C. R. Leedham-Green, S. Murray, A. Niemeyer, E.A. O'Brien, Generating random elements of a finite group, Comm. Algebra 23 (1995), 4931–4948.

[Ch] F. R. K. Chung, Spectral Graph Theory, CBMS Regional Conference Ser. Math 92, Amer. Math. Soc., Providence, RI, 1994.

[CG1] F. R. K. Chung, R. L. Graham, Random walks on generating sets for finite groups, Electron. J. of Combin. 4 (1997), #R7.

[CG2] F. R. K. Chung, R.L. Graham, Stratified random walk on the n-cube, Random Structures Algorithms 1 (1997), 199–222.

[CCNPW] J. H. Conway, R. T. Curtis, S. P. Norton, R. A. Parker, R. A. Wilson, Atlas of Finite Simple Groups, Clarendon Press, Oxford, 1985.

[CM] H. S. M. Coxeter, W. O. J. Moser, Generators and relations for discrete groups (third edition), Springer-Verlag, Berlin, 1972.

[Da] C. David, T_3-systems of finite simple groups, Rend. Sem. Mat. Univ. Padova, 89 (1993), 19–27.

[Di] P. Diaconis, Group Representations in Probability and Statistics, IMS, Hayward, California, 1988.

[DG] P. Diaconis, R. Graham, The graph of generating sets of an abelian group, Colloq. Math. 80 (1999), 31–38.

[DS1] P. Diaconis, L. Saloff-Coste, Comparison techniques for random walk on finite groups, Ann. Probab. 21 (1993), 2131–2156.

[DS2] P. Diaconis, L. Saloff-Coste, Walks on generating sets of abelian groups, Probab. Theory Related Fields 105 (1996), 393–421.

[DS3] P. Diaconis, L. Saloff-Coste, Walks on generating sets of groups, Invent. Math. 134 (1998), 251–199.

[DSt] P. Diaconis, D. Stroock, Geometric bounds for eigenvalues of Markov chains, Ann. Appl. Probab. 1 (1991), 36–61.

[Dx] J. D. Dixon, The probability of generating the symmetric group, Math. Z. 110 (1969), 199–205.

[Du1] M. J. Dunwoody, On T-systems of groups, J. Austral. Math. Soc. 3 (1963), 172–179.

[Du2] M. J. Dunwoody, Nielsen Transformations, in: Computational Problems in Abstract Algebra, 45–46, Pergamon, Oxford, 1970.

[ER] P. Erdős, A. Rényi, Probabilistic methods in group theory, J. Anal. Math. 14 (1965), 127–138.

[Ev1] M. Evans, Primitive elements in free groups, Proc. Amer. Math. Soc. 106 (1989), 313–316.

[Ev2] M. Evans, Presentations of groups involving more generators than are necessary, Proc. London Math. Soc. (3) 67 (1993), 106–126.

[Ev3] M. Evans, T-systems of certain finite simple groups, Math. Proc. Cambridge Philos. Soc. 113 (1993), 9–22.

[F] W. Feller, An introduction to Probability theory and its applications, Vol. 1 (third edition) John Wiley, New York, 1968.

[GAP] The GAP Group, GAP — Groups, Algorithms, and Programming, Version 4.2; Aachen, St Andrews, 2000 (`http://www-gap.dcs.st-and.ac.uk/~gap`).

[Ga] W. Gaschütz, Die Eulersche Funktion auflösbarer Gruppen, Illinois J. Math. 3 (1959), 469–476.

[Ge] S. M. Gersten, A presentation for the special automorphism group of a free group, J. Pure Appl. Algebra 33 (1984), 269–279.

[Gi] R. Gilman, Finite quotients of the automorphism group of a free group, Canad. J. Math. 29 (1977), 541–551.

[Go] D. Gorenstein, Finite Simple Groups, Plenum, New York, 1982.

[Gr] K. W. Gruenberg, Relation modules of finite groups, CBMS Regional Conference Ser. Math. 25, Amer. Math. Soc., Providence, R.I., 1976.

[GK] R. M. Guralnick, W. M. Kantor, Probabilistic generation of finite simple groups, J. Algebra, to appear.

[GS] R. M. Guralnick, A. Shalev, On the spread of finite simple groups, preprint, 1999.

[H1] P. Hall, The Eulerian functions of a group, Quart. J. Math. 7 (1936), 134–151.

[H2] M. Hall, The Theory of Groups, Chelsea, New York, 1976.

[HR] D. F. Holt, S. Rees, An implementation of the Neumann-Praeger algorithm for the recognition of special linear groups, Experiment. Math. 1 (1992), 237–242.

[HGV] P. de la Harpe, A. G. Robertson, A. Valette, On the spectrum of the sum of generators for a finitely generated group, Israel J. Math. 81 (1993), 65–96.

[KB] R. Kannan, A. Bachem, Polynomial algorithms for computing the Smith and Hermite normal forms of an integer matrix, SIAM J. Comput. 8 (1979), 499–507.

[Kn1] W. M. Kantor, Some topics in asymptotic group theory, in: Groups, combinatorics & geometry (Durham, 1990), Cambridge Univ. Press, Cambridge, 1992, 403–421.

[Kn2] W. M. Kantor, Simple groups in computational group theory, in: Proc. ICM Berlin, Vol. II, 1998, 77–86.

[KL] W. M. Kantor, A. Lubotzky, The probability of generating a finite classical group, Geom. Dedicata 36 (1990), 67–87.

[KS] W. M. Kantor and Á. Seress, Black box classical groups, Mem. Amer. Math. Soc. 149, Nr. 708, Amer. Math. Soc., Providence, RI, 2001.

[Kz] D. A. Kazhdan, On the connection of the dual space of a group with the structure of its closed subgroups (Russian), Funkcional. Anal. i Prilozhen. 1 (1967), 71–74.

[LG] C. R. Leedham-Green, personal communication.

[LS1] M. W. Liebeck, A. Shalev, The probability of generating a finite simple group, Geom. Dedicata 56 (1995), 103–113.

[LS2] M. W. Liebeck, A. Shalev, Simple groups, probabilistic methods, and a conjecture of Kantor and Lubotzky, J. Algebra 184 (1996), 31–57.

[Lu1] A. Lubotzky, Discrete Groups, Expanding Graphs and Invariant Measures, Birkhäuser, Boston, 1994.

[Lu2] A. Lubotzky, Eigenvalues of the Laplacian, the first Betti number and the congruence subgroup problem, Ann. of Math. (2) 144 (1996), 441–452.

[LP] A. Lubotzky, I. Pak, The product replacement algorithm and Kazhdan's property (T), J. Amer. Math. Soc. 52 (2000), 5525–5561.

[LyS] R. C. Lyndon, P. E. Schupp, Combinatorial group theory, Springer-Verlag, Berlin, 1977.

[MKS] W. Magnus, A. Karrass, D. Solitar, Combinatorial group theory. Presentations of groups in terms of generators and relations (Second edition), Dover, New York, 1976.

[MSW] G. Malle, J. Saxl, Th. Weigel, Generation of classical groups, Geom. Dedicata 49 (1994), 85–116.

[Mn] A. Mann, Positively finitely generated groups, Forum Math. 8 (1996), 429–459.

[Mr] G. A. Margulis, Explicit constructions of expanders, Problems Inform. Transmission 9 (1973), 325–332.

[Mc] J. McCool, A faithful polynomial representation of OutF_3, Math. Proc. Cambridge Philos. Soc. 106 (1989), 207–213.

[My] A. G. Myasnikov, Extended Nielsen transformations and the trivial group, Mat. Zametki 35 (1984), 491–495.

[Ne] B. H. Neumann, On a question of Gaschütz, Arch. Math. 7. (1956), 87–90

[NN] B. H. Neumann, H. Neumann, Zwei Klassen charakteristischer Untergruppen und ihre Faktorgruppen, Math. Nachr. 4 (1951), 106–125.

[NP1] P. Neumann, C. Praeger, A recognition algorithms for special linear groups, Proc. London Math. Soc. 65 (1992), 555–603.

[NP2] P. Neumann, C. Praeger, Cyclic matrices over finite fields, J. London Math. Soc. (2) 52 (1995), 263–284.

[Ni] J. Nielsen, Über die Isomorphismen unendlicher Gruppen ohne Relation, Math. Ann. 79 (1918), 269–272.

[No] G. A. Noskov, Primitive elements in a free group, Math. Notes 30 (1981), 739–740.

[P1] I. Pak, Random walks on groups: strong uniform time approach, Ph.D. Thesis, Harvard Univ., 1997.

[P2] I. Pak, When and how n choose k, in: Randomization methods in algorithm design (P. Pardalos et al., eds.), DIMACS Ser. Discrete Math. Theoret. Comput. Sci. 43, Amer. Math. Soc., Providence, RI, 1998, 191–238.

[P3] I. Pak, Random walks on finite groups with few random generators, Electron. J. Probab. 4 (1999), 1–11.

[P4] I. Pak, On probability of generating a finite group, preprint, 1999.

[P5] I. Pak, On the graph of generating sets of a simple group, preprint, 1999.

[P6] I. Pak, Generating random elements in solvable groups, in preparation, 1999.

[PB] I. Pak, S. Bratus, On sampling generating sets of finite groups and the product replacement algorithm, Proceedings of ISSAC'99, 1999, 91–96.

[PV] I. Pak, V. H. Vu, On finite geometric random walks, Discrete Math. Appl., to appear.

[PZ] I. Pak, A. Zuk, in preparation, 2000.

[Ro] V.A. Roman'kov, The Tennant-Turner swap conjecture, Algebra and Logic 34 (1995), 249–257.

[Si] C. C. Sims, Group-theoretic algorithms, a survey, in: Proc. ICM, Helsinki, 1978, 979–985.

[Sh] A. Shalev, Probabilistic group theory, St. Andrews Lectures, Bath, 1997.

[Sh1] Y. Shalom, Explicit Kazhdan constants for representations of semisimple and arithmetic groups, Annales de L'Institut Fourier, to appear.

[Sh2] Y. Shalom, Bounded generation and Kazhdan's property (T), Publ. Math. IHES, to appear.

[St] A. Stein, $1\frac{1}{2}$-generation of finite simple groups, Beiträge Algebra Geom. 39 (1998), 349–358.

[TT] R. F. Tennant, E.C. Turner, The swap conjecture, Rocky Mountain J. Math. 22 (1992), 1083–1095.

[Wa] S. P. Wang, On the Mautner phenomenon and groups with property (T), Amer. J. Math. 104 (1982), 1191–1210.

[Wh] J. Whiston, Maximal independent generating sets of the symmetric group, J. Algebra 232 (2000), 255–268.

Current address:
Department of Mathematics
MIT
Cambridge, MA 02139-4307, U.S.A.
pak@math.mit.edu

On the complexity of the endomorphism problem for free groups

Charles C. Sims

Abstract. This paper formalizes algorithmic ideas of Edmunds and demonstrates that for most free group elements u of length at most 6 determining whether another free group element v is a homomorphic image of u can be decided in time polynomial in the length of v.

1991 Mathematics Subject Classification: primary 20E05, 20F10; secondary 20-04, 68Q25.

1. Introduction

The endomorphism problem for free groups can be stated most simply as follows:

> Let F be a countably generated free group. Given elements u and v of F, decide whether there is an endomorphism of F that takes u to v.

An alternative formulation is:

> Let F and G be finitely generated free groups. Given an element u of F and an element v of G, decide whether there a homomorphism from F to G taking u to v.

In this case the term "homomorphism problem" is more appropriate. The term "substitution problem" has also been used. In this paper, we will stick with the first formulation.

Let us fix a countably infinite set $X = \{a_1, a_2, a_3, \dots\}$ and take F to be the free group on X. We shall sometimes write a for a_1, b for a_2, and c for a_3.

The endomorphism problem is known to be decidable. Since an endomorphism of F may be preceded and followed by automorphisms, the existence of an endomorphism taking u to v depends only on the automorphism classes of u and v in F. In particular, u and v may be replaced by conjugates of themselves and thus we may assume that u and v are cyclically reduced as group words over X.

The difficulty of solving the endomorphism problem depends heavily on the element u. Thus it is common practice to fix u and talk about the endomorphism problem for u. If $u = a$, then for any v there is a endomorphism of F taking u to v. If $u = a^k$

for some integer $k > 1$, then the time to solve the endomorphism problem for u, that is, to determine whether v is a k-th power in F, is linear in the length $|v|$ of v. If u is the commutator $[b, a] = b^{-1}a^{-1}ba$, then the time to solve the endomorphism problem for u, or equivalently to determine whether v is a commutator in F, is polynomial in $|v|$ with the degree of the polynomial at most 4. However, there is no polynomial time algorithm known for deciding whether an element v of a free group is the product of two cubes. That is, all known solutions of the endomorphism problem for $u = a^3b^3$ require time not polynomial in $|v|$.

This paper surveys what is known about the complexity of the endomorphim problem for various elements u. It formalizes procedures suggested by Edmunds that try to demonstrate polynomial-time complexity for a particular element u and suggests some modest improvements in these procedures. The procedures have been implemented and have been used successfully in some cases.

2. Notation

The free group F is a quotient monoid of the free monoid M on $X^{\pm} = X \cup X^{-1}$, where X^{-1} is a set of formal inverses for the elements of X. As was done above, we normally blur the distinction between *words*, that is, elements of M, and elements of F. However, here we will have to be careful. Elements of X will be called *generators* and elements of $X^{\pm}$ will be called *letters*. A word w is *freely reduced* if it contains no subword of the form xx^{-1} with x in $X^{\pm}$ and w is *cyclically reduced* if all cyclic permutations of w are freely reduced.

There are two different definitions of F. In one, we take F to be the set of freely reduced words and define the product uv to be the freely reduced word freely equivalent to the product of u and v in M. In the other definition, elements of F are equivalence classes of words under free equivalence. If $[u]$ denotes the free equivalence class containing the word u, then the product $[u][v]$ of classes is $[uv]$, where uv is the product of u and v in M. Each approach has its advantages and disadvantages. My preference is for the second approach and this is the one that will be used in this paper.

The set $X^{\pm}$ has a natural involution $^{-1}$, which we extend to M. If u is the word $u_1u_2 \dots u_r$, where the u_i are in $X^{\pm}$, then u^{-1} is the word $u_r^{-1} \dots u_1^{-1}$. The empty word will be denoted by ε.

A generator x *occurs* in a word u if x or x^{-1} is one of the terms of u. Each such term is an *occurrence* of x in u. A *quadratic word* is a word u such that for each generator x the number of occurrences of x in u is 0 or 2.

We shall often need to consider monoid endomorphisms ψ of M. All such endomorphisms will be assumed to be compatible with the inverse map on M. That is, $\psi(u^{-1}) = \psi(u)^{-1}$ for all elements u of M. Such an endomorphism is determined by the images $\psi(x)$ for x in X and these images may be arbitrarily chosen in M. The monoid automorphisms of M correspond to the permutations of $X^{\pm}$ that are compatible with the inverse map.

For any subset A of $X^{\pm}$, we will let τ_A be the endomorphism of M that maps to ε all generators x such that x or x^{-1} is in A and fixes all other elements of X.

Let w_1 and w_2 be words. We shall say that w_2 is an *image* of w_1 if there is an endomorphism σ of M such that $w_2 = \sigma(w_1)$ and w_2 is a *proper image* of w_1 if σ can be chosen so that for all generators x occurring in w_1 the word $\sigma(x)$ is nonempty. We shall also say that w_2 is a *cyclic image* of w_1 if some cyclic permutation of w_2 is an image of w_1. The notion of *proper cyclic image* is defined analogously.

It is an easy exercise to show that if w_1 and w_2 are words such that each is a proper image of the other, then the two words lie in the same orbit of $\mathrm{Aut}(M)$. It is only slightly more difficult to show that if w_1 and w_2 are each proper cyclic images of the other, then w_2 lies in the orbit of $\mathrm{Aut}(M)$ containing some cyclic permutation of w_1.

A cyclically reduced word u is *redundant* if there are distinct letters x and y such that $y \neq x^{-1}$, both x and y occur in u, and all occurrences of x, x^{-1}, y, and y^{-1} in u are, up to a cyclic permutation of u, in subwords xy or $y^{-1}x^{-1}$. A more formal, but perhaps less intuitive, statement of this condition is that $w = \sigma(\tau_y(u))$ is a cyclic permutation of u, where σ is the endomorphism of M that maps x to xy and fixes all letters different from x and x^{-1}. For example, $u = cabcac^{-1}b^{-1}ab$ is redundant, as can be seen by taking $x = b$ and $y = c$. Here $w = abcac^{-1}b^{-1}abc$. If u is redundant and v is any word, then there is an endomorphism of F taking $[u]$ to $[v]$ if and only if there is an endomorphism taking $[\tau_y(u)]$ to $[v]$. A word that is not redundant is *irredundant*. As we try to solve the endomorphism problem for pairs of words u and v, it is enough to consider the case in which u is irredundant.

Let us fix a linear order on $X^{\pm}$, say

$$a_1 < a_1^{-1} < a_2 < a_2^{-1} < a_3 < \cdots.$$

We will compare words using the corresponding length-plus-lexicographic order on words, which is a well-order on M. Given a word u, let $\overline{u}$ denote the first word in this order in the orbit of $\mathrm{Aut}(M)$ containing u and let $\widehat{u}$ be the first word in the set of words obtained from u by applying automorphisms of M and cyclic permutations. Algorithms for computing $\overline{u}$ and $\widehat{u}$ will be described in Section 4. In trying to decide whether there is an endomorphism of F taking $[u]$ to $[v]$, it suffices to assume that $u = \widehat{u}$ and $v = \widehat{v}$, since there are endomorphisms of F taking $[u]$ to $[\widehat{u}]$ and $[\widehat{v}]$ to $[v]$.

3. Background

The solution of the endomorphism problem for $u = a^k$ has been known for a long time.

Proposition 3.1. *Let k be a positive integer and let v be a cyclically reduced word. Then $[v]$ is a k-th power in F if and only if v is a k-th power in M, or equivalently, if and only if v is an image of a^k.*

Given a word v, we can cyclically reduce it in time linear in $|v|$ and then check whether the result is a k-power in M also in linear time.

Statements about running times, such as the ones in the preceding paragraph, need to be taken with a grain of salt. The generators occurring in our words come from an infinite set. Some care must be taken with the input format in order to make precise statements about complexity. We shall tacitly assume that the generators occurring in our words are always among the first million or so elements of X and thus can be represented by fixed precision signed integers.

After Proposition 3.1, the next step in the effort to solve the endomorphism problem was the fascinating result of Wicks in **[mjW62]**.

Proposition 3.2. *Let v be a cyclically reduced word. Then $[v]$ is a commutator in F if and only if there are words s, t, and u such that some cyclic permutation of v is equal to $stus^{-1}t^{-1}u^{-1}$, or equivalently, if and only if v is a cyclic image of $abca^{-1}b^{-1}c^{-1}$.*

In **[peS69]** Schupp provided an algorithm for solving the endomorphism problem for u when the number of generators occurring in u is at most 2. This algorithm is not polynomial in $|v|$ and does not appear practical.

The same paper of Schupp contains a very useful general observation. Let x be an element of X. For any word w, let $|w|_x$ denote the number of terms in w equal to x minus the number of terms equal to x^{-1}. Of course $|w|_x = 0$ for all but finitely many x in X.

Proposition 3.3. *Let u be a word such that the gcd of the numbers $|u|_x$ as x ranges over X is 1. Then every element of F is the image of $[u]$ under some endomorphism.*

In **[ccE75]**, Edmunds defined for any word u a set of words B_u such that the following conditions are satisfied:

1. If v is a freely reduced word, then $[v]$ is the image of $[u]$ under an endomorphism of F if and only if v is a proper image of some element of B_u.

2. If u is a quadratic word, then B_u is finite.

3. There is a procedure for listing the elements of B_u and this procedure terminates if B_u is finite.

In **[ccE79]**, Edmunds took a somewhat different approach. For a word u, he defined a set of words D_u such that the following hold:

1. If v is a cyclically reduced word, then $[v]$ is the image of $[u]$ under an endomorphism of F if and only if v is a proper cyclic image of some element of D_u.

2. If u is a quadratic word, then D_u is finite.

3. There is a procedure for listing the elements of D_u and this procedure terminates if D_u is finite.

Readers should note that in **[ccE75]** Edmunds uses the term *cancellation-free* or *c-free* image for what we have defined to be a proper image. In **[ccE79]**, a c-free image is a proper cyclic image. Because of this slight inconsistency, the term "c-free" will not be used here.

In **[mC81]**, Culler used a topological approach to study the endomorphism problem for quadratic words. He concentrated on the two families of quadratic words

$$s(n) = a_1^2 a_2^2 \dots a_n^2$$

and

$$c(n) = [a_1, a_2][a_3, a_4] \dots [a_{2n-1}, a_{2n}].$$

Quite a bit of work has gone into studying the structure and number of the elements of maximal length in $D_{c(n)}$. The papers **[aaV95]**, **[aaV97]**, and **[BV99]** culminate with an explicit formula for the number of elements of maximal length in $D_{c(n)}$.

4. Some basic algorithms on words

This section collects some useful algorithms for working with words that will be part of efforts to find efficient solutions of the endomorphism problem. The first and simplest of these algorithms is the procedure for cyclically reducing a word. This process is sufficiently familiar that a formal description is unnecessary. It runs in time linear in the length of its argument.

Every orbit of Aut(M) on M has a first element. Given a word v, we can find the first element $\overline{v}$ in the orbit of Aut(M) containing v with the following procedure.

Procedure REP(v)
    # v is the word $v_1 v_2 \dots v_r$.
    $Y := \emptyset$;
    Let f be the unique map from $Y^{\pm}$ to $X^{\pm}$;
    For i from 1 to r do
        If v_i is in $Y^{\pm}$ then $z_i := f(v_i)$;
        else
            $Y := Y \cup \{v_i\}$;
            $j := |Y|$;
            Extend f to $Y^{\pm}$ by $f(v_i) := a_j$ and $f(v_i^{-1}) := a_j^{-1}$;
            $z_i := a_j$;
        fi;
    od;   # For i
    Return $z_1 z_2 \dots z_r$;
End.

For example, if $v = b^{-1}aba^{-1}c^{-1}bc$, then REP($v$) returns $aba^{-1}b^{-1}cac^{-1}$. The running time of REP is $O(|v|)$.

For any word v, we defined $\widehat{v}$ be the first element among the words REP(v_1), where v_1 ranges over the cyclic permutations of v. Using the obvious approach of looking at each cyclic permutation of v in turn, we can construct an algorithm to compute $\widehat{v}$ in time $O(|v|^2)$.

In some of the procedures described in the next section we shall have to test a word v for being redundant. In the context where this test is needed, the word v will satisfy the condition $v = \overline{v}$. Thus the set of generators occurring in v is of the form $\{a_1, a_2, \ldots, a_m\}$ for some integer $m \leq |v|$. This makes it easy to construct a test for redundancy that runs in time $O(|v|)$. One has to record for each generator x that has been encountered so far whether x or x^{-1} has been seen only next to a single letter y and if so, what y is.

Given two words u and v, we shall need to decide whether v is a proper image of u. This is not so easy. For a fixed u the procedure used by the author in implementations is polynomial in $|v|$ but the degree of the polynomial depend on u. Let $u = u_1u_2\ldots u_r$, where the u_i are letters. We are looking for an endomorphism σ of M such that $v = \sigma(u)$ and $\sigma(u_i)$ is nonempty for all i. We start by considering the possibilities for the image of u_1, which must be a nonempty prefix of v. Once we have a choice for $\sigma(u_1)$, we consider the first index i such that u_i is not equal to u_1 or u_1^{-1}. Then $p = \sigma(u_1 \ldots u_{i-1})$ is defined. If p is not a prefix of v, then we try another choice for $\sigma(u_1)$. If $v = pq$ for some word q, then $\sigma(u_i)$ must be a nonempty prefix of q. Now we consider the first index j such that $\sigma(u_j)$ is not specified. If $p = \sigma(u_1 \ldots u_{j-1})$ is not a prefix of v, then we try another image of u_i, or if there are no more possibilities, for $\sigma(u_1)$. If $v = pq$, then $\sigma(u_j)$ is a nonempty prefix of q. Continuing in this way, we obtain a backtrack search for σ. As described, the procedure runs in time $O(|v|^{m+1})$, where m is the number of generators occurring in u. This can be improved. For example, if the image under σ is known for all but one of the generators occurring in u, then the image of that generator is determined. This observation reduces the time to $O(|v|^m)$. Other considerations can be used to reduce the constant implicit in the O-notation. Determining whether or not v is a proper cyclic image of u take time $O(|v|^{m+1})$, since v can have up to $|v|$ distinct cyclic permutations.

5. The approach of Edmunds

This section attempts to consolidate and extend slightly the work described in **[ccE75]** and **[ccE79]** and to formalize the algorithmic ideas implicit there. As remarked previously, the notation in these two papers is not totally consistent. For this reason, readers should be careful in comparing the exposition here with that in either of Edmunds' papers.

In **[ccE75]**, Edmunds gives two definitions of the set B_u, the first nonconstructive, the second constructive. Let T be the set of freely reduced words v in M such that $v = \overline{v}$ and for a word u let E_u be the set of free reductions of images of u. Then B_u is defined to be the set of elements v of $T \cap E_u$ such that v is not a proper image of any other element of $T \cap E_u$.

As a function on M, free reduction is not an endomorphism. However, to provide a constructive definition of B_u, Edmunds introduces certain endomorphisms of M that mimic aspects of free reduction. Suppose that v and w are nonempty, freely reduced words such that vw is not freely reduced. Thus some suffix of v is the inverse of some prefix of w. There are several cases to distinguish.

First, if $v = w$, then there are nonempty words p and q such that $v = pqp^{-1}$. If v and w are different, there are four possibilities:

1. There are nonempty words p, q, and r such that $v = pr$ and $w = r^{-1}q$.

2. There is a nonempty word p such that $v = pw^{-1}$.

3. There is a nonempty word p such that $w = v^{-1}p$.

4. $w = v^{-1}$.

These observations are the motivation for the following definitions. Let v be a word and let x and y be letters such that $x \neq y^{-1}$ and xy is a subword of v. (Later we shall permit xy to be a subword of some cyclic permutation of v.) Let z be a generator that does not occur in v. We define a set $R = R(x, y, v)$ of endomorphisms of M. All elements of R fix all letters not in $\{x, x^{-1}, y, y^{-1}\}$. If $x = y$, then R consists of the single endomorphism that maps x to zxz^{-1}. If $x \neq y$, then R contains four endomorphisms, $\rho_1, \rho_2, \rho_3, \rho_4$, which act as follows:

$$\begin{aligned} \rho_1(x) &= xz, & \rho_1(y) &= z^{-1}y, \\ \rho_2(x) &= xy^{-1}, & \rho_2(y) &= y, \\ \rho_3(x) &= x, & \rho_3(y) &= x^{-1}y, \\ \rho_4(x) &= x, & \rho_4(y) &= x^{-1}. \end{aligned}$$

Given a word v, let I_v be the set of all words of the form $\overline{w}$, where w is the free reduction of $\tau_A(v)$ and A ranges over all subsets of the generators occurring in v. Let J_v be the union of the sets I_w, where w is $\rho(v)$ for some ρ in one of the sets $R(x, y, v)$, where xy is a subword of v.

The following procedure is essentially the one proposed by Edmunds for computing the set B_u.

Procedure EDMUNDS_B(u)
    # u is a word.
    $Q := I_u$;
    Let B be the set of words v in Q such that v is not a proper image

of some element of $Q - \{v\}$;
$Q := B$;
While $Q \neq \emptyset$ do
Choose an element v of Q of minimal length and delete v from Q;
For w in J_v do
If w is a proper image of some element of B then continue fi;
Delete from B and Q all words that are proper images of w;
Add w to both B and Q;
od;
od;
Return B;
End.

The procedure EDMUNDS_B may not terminate, but B_u is the set of words that are added to and never deleted from the set B during the operation of this procedure. If B_u is finite, then EDMUNDS_B does terminate and returns B_u.

If $u = a^3$, then $B_u = \{\varepsilon, a^3, ab^3a^{-1}\}$. More generally, if $u = a^k$ for some integer $k > 1$, then $B_u = \{\varepsilon, a^k, ab^ka^{-1}\}$.

For even moderately complicated quadratic words u, the set B_u can be rather large. For example, if $u = a^2b^2$, then B_u has 43 elements, which begin with

$$\varepsilon,\ a^2,\ a^2b^2,\ abab^{-1},\ aba^{-1}b,\ abba,\ abba^{-1},$$

and end with

$$abcdbec^{-1}eda^{-1},\ abcdb^{-1}edce^{-1}a^{-1},\ abcdcebd^{-1}ea^{-1},\ abcddc^{-1}ebea^{-1}.$$

The sets B_u may contain words that are not cyclically reduced. In **[ccE79]**, Edmunds took a different approach, one that for quadratic words leads to smaller sets of test elements. For a word u, let K_u be the set of irredundant words v obtainable as follows: Choose a subset A of the generators occurring in u, let w be the cyclic reduction of $\tau_A(u)$, and set $v = \widehat{w}$. Set L_u equal to the union of the sets K_v, where $v = \rho(u)$ for some ρ in $R(x, y, u)$ and xy is a subword of some cyclic permutation of u.

The set D_u is defined to be the smallest set D of words such that D contains K_u and for all v in D the set L_v is contained in D.

Here are the major differences in the definitions of B_u and D_u.

1. Elements v of B_u satisfy $v = \overline{v}$, while $v = \widehat{v}$ for elements of D_u.

2. Elements of D_u are cyclically reduced. Elements of B_u may not be cyclically reduced.

3. Elements of D_u need only be irredundant. This is weaker than the corresponding condition for B_u, which says that no element of B_u is a proper image of any other element.

Edmunds has a procedure for listing the elements of D_u analogous to EDMUNDS_B.

```
Procedure EDMUNDS_D(u)
    # u is a word.
    D := K_u;
    Q := D;
    While Q ≠ ∅ do
        Choose an element v of Q of minimal length and delete v
            from Q;
        For w in L_v do
            If w is not in D then add w to both D and Q fi;
        od;
    od;
    Return D;
End.
```

Note that once words are added to D they are never removed. The set D_u is the set of words added to D at some time during the operation of EDMUNDS_D. If D_u is finite, then EDMUNDS_D terminates and returns D_u.

For the quadratic word $u = a^2b^2$ the set D_u is much smaller than B_u. There are six words in D_u, namely

$$\varepsilon,\ a^2,\ a^2b^2,\ abab^{-1},\ a^2bc^2b^{-1},\ abacb^{-1}c.$$

Finiteness of B_u or D_u implies that the endomorphism problem for u is decidable in polynomial time.

Proposition 5.1. *If u is a fixed word for which B_u or D_u is finite and v is a variable word, then the free group endomorphism problem for u and v is decidable in time polynomial in $|v|$.*

Proof. We shall only consider the case that B_u is finite. The case in which D_u is finite is completely analogous.

Since u is fixed, we may assume that B_u is known. We first freely reduce v. Then we have to decide whether v is a proper image of some element w of B_u. The discussion in the previous section showed that for a fixed w this could be done in time polynomial in $|v|$. Since $|B_u|$ is a fixed constant, it follows that the time to perform the entire procedure is $O(|v|^m)$, where m is the maximum number of generators occurring in an element of B_u. □

If D_u is finite, then we can determine whether $[v]$ is an endomorphic image of $[u]$ in time $O(|v|^{n+1})$, where n is the maximum number of generators occurring in an element of D_u. For $u = aba^{-1}b^{-1}$, the value of n is 3 and one can determine whether $[v]$ is a commutator in time $O(|v|^4)$.

Although D_u works well for quadratic words, for other words it can be too big. For example, if $u = a^3b^2$, then B_u has just two elements, ε and a. However, D_u is very large and is almost certainly infinite. The reason for this difference is that every element of F is an endomorphic image of $[u]$. Once EDMUNDS_B finds that a is in B, it removes all other nonempty words and refuses to put any more words into B. The procedure EDMUNDS_D rather quickly finds that a is in D but is unable to recognize that the computation can be stopped.

There is a rather obvious way to define a set C_u that has the best properties of both B_u and D_u. Consider the following procedure.

```
Procedure EDMUNDS_C(u)
    # u is a word.
    Q := K_u;
    Let C be the set of words v in Q such that v is not a proper cyclic
        image of some element of Q − {v};
    Q := C;
    While Q ≠ ∅ do
        Choose an element v of Q of minimal length and delete v from Q;
        For w in L_v do
            If w is a proper cyclic image of some element of C then
                continue fi;
            Delete from C and Q all words that are proper cyclic
                images of w;
            Add w to both C and Q;
        od;
    od;
    Return C;
End.
```

As before, C_u is the set of elements added to C during the operation of EDMUNDS_C and never removed. If C_u is finite, then EDMUNDS_C terminates and returns C_u.

Proposition 5.2. *Let u and v be words with v cyclically reduced. Then $[v]$ is an endomorphic image of $[u]$ if and only if v is a proper cyclic image of some element of C_u.*

The proof of Proposition 5.2 is a minor variation of the proofs of the corresponding results about B_u and D_u. The set C_u is a subset of both B_u and D_u. The endomorphim problem for $[u]$ and $[v]$ is solvable in time polynomial in $|v|$ if C_u is finite.

If $u = a^2b^2$, then $C_u = D_u$. If $u = a^3b^2$, then $C_u = B_u$.

6. Experimental results

To try to get some insight into the complexity of the endomorphism problem for short words u, an effort was made to determine the sets C_u for all elements u of M with length at most 6. By Proposition 3.3, if u is a word for which some generator has exactly one occurrence, then $C_u = \{\varepsilon, a\}$ and the endomorphism problem for $[u]$ is solvable in constant time. If $|u| \leq 6$ and u is not covered by this observation, then either at most two generators occur in u or $|u| = 6$ and u is quadratic. Since the endomorphism problem is solvable in polynomial time for quadratic words, it was decided to concentrate on words involving at most two generators.

A prototype implementation of the method of Whitehead was used to determine representatives for the orbits of $\mathrm{Aut}(F)$ containing elements of length at most 6 involving at most two generators. To each representative, a C implementation of EDMUNDS_C was applied. The results are summarized on Table 1.

Class Rep u	Nonempty elements of C_u
a	a
aa	aa
aaa	aaa
$aaaa$	$aaaa$
$aabb$	aa, $aabb$, $abab^{-1}$ $aabccb^{-1}$, $abacb^{-1}c$
$aba^{-1}b^{-1}$	$aba^{-1}b^{-1}$, $abca^{-1}b^{-1}c^{-1}$
$aaaaa$	$aaaaa$
$aaabb$	a
$aabab^{-1}$	aaa, $aabab^{-1}$, $aababb$, $ababcbac^{-1}$, $aabcb^{-1}abccb^{-1}$
$aaba^{-1}b^{-1}$	a
$aaaaaa$	$aaaaaa$
$aaaabb$	???
$aaabab^{-1}$	$aaaa$, $aaabab^{-1}$, $aabababb$, $abababcbac^{-1}$, $aabcb^{-1}abcb^{-1}abccb^{-1}$
$aaaba^{-1}b^{-1}$	???
$aaabbb$	???
$aabaab^{-1}$	$aaaa$, $aabaab^{-1}$, $aababbab$, $ababcbabac^{-1}$, $aabcb^{-1}abccb^{-1}abcb^{-1}$
$aabab^{-1}b^{-1}$	a
$aaba^{-1}a^{-1}b^{-1}$	???
$aaba^{-1}bb$	a
$aaba^{-1}b^{-1}b^{-1}$	a

Table 1: Experimental results

In only four of the 20 cases did the program not terminate with a finite set. These four cases are noted with question marks in the table. In all other cases considered, the endomorphism problem for the class representative is solvable in polynomial time.

An examination of the words being constructed in the inconclusive runs suggested that termination was not likely. The case of a^3b^3 was particularly interesting. There were clearly patterns in the words being produced. In agreement with a result in **[ccE79]** the number of generators occurring in a word added to the set C never exceeded 3. The number of words in C of a given length n seemed to be polynomial in n. If these observations could be confirmed and if it could be shown that the lengths increase more or less steadily, then polynomial time complexity in the case of a^3b^3 could also be asserted.

References

[BV99] R. Bacher, A. Vdovina, Counting 1-vertex triangulations of oriented surfaces, in: Proceedings International Conference on Formal Power Series and Algebraic Combinatorics (Barcelona 1999).

[mC81] M. Culler, Using surfaces to solve equations in free groups, Topology 20 (1981), 133–145.

[ccE75] C. C. Edmunds, On the endomorphism problem for free groups, Comm. Algebra 3 (1975), 1–20.

[ccE79] C. C. Edmunds, On the endomorphism problem for free groups II, Proc. London Math. Soc. 38 (1979), 153–168.

[gsM82] G. S. Makanin, Equations in a free group (Russian), Izv. Akad. Nauk SSSR Ser. Mat. 46 (1982), 1199–1273, 1344.

[peS69] P. E. Schupp, On the substitution problem for free groups, Proc. Amer. Math. Soc. 23 (1969), 421–423.

[aaV95] A. Vdovina, Constructing of orientable Wicks forms and estimation of their number, Comm. Algebra 23 (1995), 3205–3222.

[aaV97] A. Vdovina, Products of commutators in free products, Internat. J. Algebra Comput. 7 (1997), 471–485.

[mjW62] M. J. Wicks, Commutators in free products, J. London Math. Soc. 37 (1962), 433–444.

Department of Mathematics
Rutgers University
New Brunswick, NJ 08903, U.S.A.
sims@math.rutgers.edu

Talks

L. Babai
A polynomial-time theory of black-box groups

A. Betten
Intersection numbers of designs

A. Borovik
Involutions in probabilistic recognition of finite simple groups

S. Bratus
Constructive recognition of a black box group isomorphic to $PSL(n, q)$, $SL(n, q)$, or a perfect central extension of $PSL(n, q)$, $n > 3$, $q > 4$

P. Brooksbank
Constructive recognition of the classical groups in their natural representation

M. Conder
Fast recognition of classical groups over large fields

G. Cooperman
Parallel GAP

B. Eick
Infinite polycyclic groups in GAP 4

J. Fulman
A generating function approach to the enumeration of cyclic and separable matrices in finite classical groups

G. Havas
Experiments in coset enumeration

G. Hiss
Low degree representations of quasi-simple groups

D. Holt
Computing automorphism groups of finite groups

A. Hulpke
Computing maximal subgroups by the homomorphism principle

C. Leedham-Green
Matrix group recognition and beyond

S. Linton
Why coset enumeration is not about enumerating cosets

E. Luks
Better than brute force

K. Lux
Treating the exceptional cases of the MeatAxe

F. Lübeck
On p-singular elements in Chevalley groups in characteristic p

D. Maslen
Computational harmonic analysis for wreath products of symmetric groups

A. Miasnikov
Genetic algorithms and groups

G. Michler
A new computer construction of the irreducible 112-dimensional 2-modular representation of Janko's group J_4

T. Miyazaki
Deterministic polynomial-time algorithms for certain matrix group problems

S. Murray
Generating random elements in matrix groups

A. Niemeyer
Recognising normalisers of symplectic-type r-subgroups of $GL(r, q)$

P. Neumann
Simple suggestions about matrix groups

E. O'Brien
Computing automorphism groups of p-groups

H. Pahlings
Orbits on vector spaces and the $k(GV)$–problem

I. Pak
What do we know about product replacement algorithm?

C. Praeger
Cyclic matrices and the Meataxe

L. Pyber
Asymptotic results for linear groups

A. Roy
Symmetry breaking predicates for search problems

R. Schwingel
The stabiliser of a set of subspaces

C. Sims
Computing Wicks forms

L. Soicher
GRAPE 4.0

***Software demonstrations*:** A. Betten, G. Cooperman, A. Miasnikov, L. Soicher, W. Unger

List of participants

László Babai, University of Chicago, Chicago, IL 60637, U.S.A.
laci@cs.uchicago.edu

Robert Beals, IDA Center for Communications Research, Princeton, NJ 08540, U.S.A
beals@ccr-p.ida.org

Anton Betten, Universität Bayreuth, D-95440 Bayreuth, Germany
Anton.Betten@uni-bayreuth.de

Alexandre Borovik, UMIST, Manchester M60 1QD, England
borovik@lanczos.ma.umist.ac.uk

Maiia Botvinnik, University of Oregon, Eugene, OR 97403, U.S.A.
mbotvinn@math.uoregon.edu

Sergey Bratus, Northeastern University, Boston, MA 02115, U.S.A.
sbratus@ccs.neu.edu

Peter Brooksbank, University of Oregon, Eugene, OR 97403, U.S.A.
brooks@geometry.uoregon.edu

Andrea Caranti, Università di Trento, 38100 Trento, Italy
caranti@science.unitn.it

Marston Conder, University of Auckland, Auckland, New Zealand
conder@math.auckland.ac.nz

Gene Cooperman, Northeastern University, Boston, MA 02115, U.S.A.
gene@ccs.neu.edu

Bettina Eick, Gesamthochschule Kassel, D-34109 Kassel, Germany
eick@mathematik.uni-kassel.de

Larry Finkelstein, Northeastern University, Boston, MA 02115, U.S.A.
laf@ccs.neu.edu

Jason Fulman, Dartmouth College, Hanover, NH 03755, U.S.A.
fulman@coos.dartmouth.edu

Volker Gebhardt, Universität Essen, D-45326 Essen, Germany
gebhardt@exp-math.uni-essen.de

Viktor Grinberg, Northeastern University, Boston, MA 02115, U.S.A.
victor@ccs.neu.edu

Larry Grove, University of Arizona, Tucson, AZ 85721, U.S.A.
grove@math.arizona.edu

Mitch Harris, University of Illinois, Chicago, IL 60680, U.S.A.
maharri@cs.uiuc.edu

George Havas, University of Queensland, Brisbane QLD 4072, Australia
havas@it.uq.edu.au

Gerhard Hiss, RWTH Aachen, D-52056 Aachen, Germany
hiss@math.rwth-aachen.de

Derek Holt, University of Warwick, Coventry CV4 7AL, England
dfh@maths.warwick.ac.uk

Alexander Hulpke, University of St. Andrews, St. Andrews KY16 9SS, Scotland
current address: The Ohio State University, Columbus, OH 43210, U.S.A.
ahulpke@math.ohio-state.edu

William Kantor, University of Oregon, Eugene, OR 97403, U.S.A.
kantor@bright.uoregon.edu

Walter Kim, University of Chicago Chicago, IL 60637, U.S.A.
wkim@midway.uchicago.edu

Inna Korchagina, The Ohio State University, Columbus, OH 43210, U.S.A.
inna@math.ohio-state.edu

Charles Leedham-Green, Queen Mary and Westfield College, London E1 4NS, England
C.R.Leedham-Green@qmw.ac.uk

Mark Lewis, Kent State University, Kent, OH 44242, U.S.A.
lewis@mcs.kent.edu

Paul Li, University of Chicago, Chicago, IL 60637, U.S.A.
paulli@math.uchicago.edu

Steve Linton, University of St. Andrews, St. Andrews KY16 9SS, Scotland
sal@dcs.st-and.ac.uk

Frank Lübeck, RWTH Aachen, D-52056 Aachen, Germany
luebeck@math.rwth-aachen.de

Eugene Luks, University of Oregon, Eugene, OR 97403, U.S.A.
luks@cs.uoregon.edu

Klaus Lux, University of Arizona, Tucson, AZ 85721, U.S.A.
klux@math.arizona.edu

Kay Magaard, Wayne State University Detroit, MI 48202, U.S.A.
kaym@math.wayne.edu

David Maslen, Dartmouth College, Hanover, NH 03755, U.S.A.
maslen@cs.dartmouth.edu

Alexei Miasnikov, City College, CUNY New York, NY 10031, U.S.A.
alexei@rio.sci.ccny.cuny.edu

Gerhard Michler, Universität Essen, D-45326 Essen, Germany
archiv@exp-math.uni-essen.de

Takunari Miyazaki, University of Oregon, Eugene, OR 97403, U.S.A.
miyazaki@cs.uoregon.edu

Scott Murray, University of Chicago, Chicago, IL 60637, U.S.A.
murray@math.uchicago.edu

Joachim Neubüser, RWTH Aachen, D-52056 Aachen, Germany
jneubues@math.rwth-aachen.de

Peter Neumann, The Queen's College, Oxford OX1 4AW, England
peter.neumann@queens.ox.ac.uk

Mike Newman, Australian National University, Canberra ACT 2601, Australia
newman@maths.anu.edu.au

Alice Niemeyer, University of Western Australia, Nedlands WA 6907, Australia
alice@maths.uwa.edu.au

Eamonn O'Brien, University of Auckland, Auckland, New Zealand
obrien@math.auckland.ac.nz

Gretchen Ostheimer, Tufts University, Medford, MA 02155, U.S.A.
current address: Hofstra University, Hempstead, NY 11550, U.S.A.
gostheim@emerald.tufts.edu

Herbert Pahlings, RWTH Aachen, D-52056 Aachen, Germany
hpahling@math.rwth-aachen.de

Igor Pak, Yale University, New Haven, CT 06520, U.S.A.
paki@pascal.math.yale.edu

Cheryl Praeger, University of Western Australia, Nedlands WA 6907, Australia
praeger@maths.uwa.edu.au

László Pyber, Hungarian Academy of Sciences, 1364 Budapest, Hungary
pyber@math-inst.hu

Amitabha Roy, University of Oregon, Eugene, OR 97403, U.S.A.
aroy@cs.uoregon.edu

Sambuddha Roy, University of Chicago, Chicago, IL 60637, U.S.A.
samroy@cs.uchicago.edu

Ruth Schwingel, Queen Mary and Westfield College, London E1 4NS, England
R.Schwingel@qmw.ac.uk

Ákos Seress, The Ohio State University, Columbus, OH 43210, U.S.A.
akos@math.ohio-state.edu

Charles Sims, Rutgers University, New Brunswick, NJ 08903, U.S.A.
sims@math.rutgers.edu

Michael Slattery, Marquette University, Milwaukee, WI 53233, U.S.A.
mikes@sylow.mscs.mu.edu

Stephen Smith, University of Illinois, Chicago, IL 60680, U.S.A.
smiths@math.uic.edu

Leonard Soicher, Queen Mary and Westfield College, London E1 4NS, England
L.H.Soicher@qmw.ac.uk

Ronald Solomon, The Ohio State University, Columbus, OH 43210, U.S.A.
solomon@math.ohio-state.edu

William Unger, University of Sydney, Sydney NSW 2006, Australia
billu@maths.usyd.edu.au

Charles Wright, University of Oregon, Eugene, OR 97403, U.S.A.
wright@math.uoregon.edu

Yechezkel Zalcstein, National Science Foundation, Washington, DC 20550, U.S.A.
zzalcste@nsf.gov

Bei Fragen zur Produktsicherheit wenden Sie sich bitte an:
If you have any questions regarding product safety,
please contact:

Walter de Gruyter GmbH
Genthiner Straße 13
10785 Berlin
productsafety@degruyterbrill.com